Small Ring Compounds in Organic Synthesis III

Synthesis and Synthetic Applications of 1-Donor Substituted Cyclopropanes with Ethynyl, Vinyl and Carbonyl Groups

J. R. Y. Salaün

Donor-Acceptor-Substituted Cyclopropanes: Versatile Building Blocks in Organic Synthesis

H.-U. Reissig

Functionalised Cyclopropanes as Synthetic Intermediates

M. S. Baird

AKADEMIE-VERLAG BERLIN

Topics in Current Chemistry · Vol. 144

Small Ring Compounds in Organic Synthesis III

Small Ring Compounds in Organic Synthesis III

With 1 Figure and 11 Tables

AKADEMIE-VERLAG BERLIN
1988

Die Originalausgabe erscheint im Springer-Verlag
Berlin · Heidelberg · New York · London · Paris · Tokyo
in der Schriftenreihe „Topics in Current Chemistry“,
Volume 144

Vertriebsrechte für die sozialistischen Länder:
Akademie-Verlag Berlin

Vertriebsrechte für alle Staaten mit Ausnahme der sozialistischen Länder:
Springer-Verlag Berlin · Heidelberg · New York · London · Paris · Tokyo

ISBN 3-540-18368-X Springer-Verlag Berlin · Heidelberg · New York
ISBN 0-387-18368-X Springer-Verlag New York · Heidelberg · Berlin

ISBN 3-05-500503-1 Akademie-Verlag Berlin

Erschienen im Akademie-Verlag Berlin, DDR-1086 Berlin, Leipziger Straße 3—4
Lizenznummer: 202 · 100/506/88
Printed in the German Democratic Republic
Gesamtherstellung: VEB Druckerei „Thomas Müntzer“, 5820 Bad Langensalza
LSV 1275
Bestellnummer: 763 834 3 (3072/144)
15600

Editorial Board

Prof. emer. Dr. Dr. h.c. mult. Georg Wittig
1897–1987

Georg Wittig 16. 6. 1897–26. 8. 1987

Georg Wittig, Professor Emeritus died August 26th, 1987 in Heidelberg a few weeks after his ninetieth birthday. With him the members of the international community of organic chemists have lost one of their greatest representatives of this century.

Georg Wittig was born in Berlin. He studied chemistry in Marburg/Lahn where he obtained his Ph.D. and became Privatdozent in 1926. A few years later he moved to the Technische Hochschule Braunschweig. In 1937 he was appointed Professor Extraordinarius at Freiburg/Breisgau and in 1944 Professor Ordinarius and Director of the Chemical Institute at the University of Tübingen. In 1956 he succeeded Karl Freudenberg at the University of Heidelberg where he formally retired in 1967.

Wittig is best known for his work on phosphorus ylides, which condense with carbonyl compounds to form alkenes. The WITTIG REACTION has obtained great value far beyond its usefulness in academic research. By means of this reaction vitamins, hormones, pharmaceuticals and many other desirable classes of organic compounds are nowadays synthesized on a large industrial scale. The most familiar examples are Vitamin A and β-carotene. Here we find an impressive example of fundamental research finally leading to important and wide ranging applications.

The WITTIG REACTION was discovered in connection with the successful synthesis of pentacovalent derivatives of the elements of the fifth main group.

Other outstanding contributions of Wittig are the observation of the halogen-metal-exchange reaction and the discovery of dehydrobenzene as early as in 1939 and 1942, respectively. This work stimulated fundamental work in the field of carbanion chemistry and led for example to the concept of the *ate* complexes, which is now generally accepted. In extension of the dehydrobenzene studies the class of small cycloalkynes was found: many of these intermediates with a strained carbon-carbon triple bond can be easily generated and are now versatile building blocks in organic syntheses. In a rather late stage of his academic career Wittig developed the directed aldol condensation, an extremely elegant method for the carbon-carbon-bond formation between carbonyl compounds in a manner which was not possible previously. It is interesting to note that some of his last publications were devoted to the formation of diradials, a topic in which he was already engaged at the very beginning of his research.

Georg Wittig is also well known for his engagement regarding scientific literature. For a long period he was a member of the editorial boards of many national and international journals. From 1963 for instance he was one of the editors of Topics in Current Chemistry.

His own impressive activities are documented in more than 300 scientific papers published between 1924 and 1980.

In 1979 Georg Wittig together with Herbert C. Brown was awarded the Nobel prize for Chemistry. The Royal Swedish Academy honored Brown and Wittig "for their development of boron and phosphorus compounds, respectively, into important reagents in organic synthesis". Apart from the Nobel prize Georg Wittig received a great number of awards, honorary doctorates and other appointments. The honorary doctorate he received from Sorbonne, Paris in 1957 as the first German after World War II was an acknowledgement of his merits in reestablishing the reputation of German science after the war.

During his long career as scientist and academic teacher, more than 300 graduate students and post-doctoral co-workers were associated with Wittig. Many of them now hold important positions in industry and at universities throughout the world. Those of us who were privileged to accompany Georg Wittig for a few years on his long way through the fascinating field of organic chemistry were deeply impressed by his sincere and honest personality. He set a high standard of expectation on his students' abilities, qualifications, enthusiasm and decency as well as being always open for personal problems and most helpful in solving them.

On November 17th 1987 the Chemische Gesellschaft zu Heidelberg held a colloquium in memoriam of Georg Wittig attended by respresentatives of the University and the City of Heidelberg, members of the faculty of chemistry, colleagues and former students in reverence to his esteemed personality and his great contributions to science.

Werner Tochtermann

Table of Contents

Table of Contents of Volume 133

Table of Contents of Volume 135

Synthesis and Synthetic Applications of 1-Donor Substituted Cyclopropanes with Ethynyl, Vinyl and Carbonyl Groups

Jacques R. Y. Salaün

Laboratoire des Carbocycles, UA 478 du CNRS, Université de Paris-Sud, 91405 Orsay Cedex, France

Table of Contents

Substituted on the same carbon by an electron donating group and an adjacent multiple bond, cyclopropanes provide building blocks of unprecedented synthetic potential. They are readily available along different routes involving: cyclopropanone hemiacetals, oxaspiropentanes, alkylidenecyclopropanes, 1-heterosubstituted lithiocyclopropanes, α-enone silyl enol ethers, 1,3-dichloroacetone and 1-hydroxycyclopropylcarbonyl derivatives as main sources.

First of all, they undergo acid-induced $C_3 \rightarrow C_4$ ring expansion to four-membered rings, in particular to the 2-vinylcyclobutanone system, which is an efficient precursor of C_5, C_6 and C_8 homologous rings by subsequent acid- and base-induced, thermal, or photolytic ring enlargements. On the other hand, they undergo thermal $C_3 \rightarrow C_5$ ring expansion providing cyclopentanone enol ethers or derivatives with high chemo-, regio- and stereoselectivity.

The usefulness of these ring enlargements has been illustrated by the total synthesis of some natural products with four- (grandisol), five- (α-cuparenone, prostanoid, jasmanoid, methylenomycin B, spirovetivane, aphidicolin, dicranenone), six- (β-selinene, compactin) and eight-membered rings (poitediol). Regioselective α or α′ monomethylation of α-enones is based on the base-induced ring opening of trimethylsiloxycyclopropanes whereas, stereoselective alkylation can be obtained by the base-induced ring opening of cyclobutanones (grandisol, deoxypodocarpate). Finally, these attractive building blocks provide a convenient key to enter the field of the homoenolate chemistry.

1 Introduction

Although cyclopropane derivatives have been known for more than 100 years [1], is was not until about 1960 that the exploration of the cyclopropane chemistry really began. In fact the utility of such building blocks for organic synthesis has been recognized only in the recent years.

The chemical reactivity of a cyclopropane ring, closely resembles that of an olefinic double bond: both groups interact with neighbouring π-electron systems and p-electron centers, add acids, halogens and ozone, undergo catalytic hydrogenation and cycloaddition, form metal complexes, etc. [2]. More specifically, the three-membered ring can undergo ring openings induced by solvolysis, bases, electrophiles, nucleophiles, metals as well as thermally and photolytically [3]. With an adjacent electron deficient center, the cyclopropane ring readily ring enlarges to four-membered ring derivatives [3]. Substituted by a vinylic moiety it undergoes $C_3 \rightarrow C_5$ ring enlargement to five-membered ring derivatives, a process which provides an efficient three-carbon annelation method [3,4]. The cyclopropane ring undergoes interconversions with related cyclobutane and open-chain derivatives [5], but it is also produced in high yields in a ring contraction of four-membered cyclic systems bearing both electron donating and leaving groups, in a vicinal arrangement [3].

Like that of most other functional groups, the reactivity of the cyclopropane moiety can be strongly influenced by the substituents on the ring. Thus, for instance, specific reactivity originates from combinations of a three-membered ring with an adjacent multiple bond. Furthermore, cyclopropanes with multiply bonded groups and an electron donating substituent on the same carbon exhibit unexpected and unprecedented synthetic utility. This review summarizes the preparation of such building blocks and exemplifies their use in organic synthesis.

2 Preparation of 1-Donor Substituted Ethynylcyclopropanes

2.1 From Cyclopropane Hemiacetal

Cyclopropanone chemistry has received considerable attention in recent years [6]; nevertheless, the chemistry of this unusually reactive class of ketones has previously found limited use in synthesis mainly because of the difficulties encountered in the preparation and handling of such strained systems. However, the discovery of derivatives capable of delivering the parent three-membered ring ketone or equivalent species *in situ*, has brought the development of cyclopropanone chemistry a decisive step forward. Among them, the cyclopropanone hemiacetal *3*, which is now readily available [7], became a substrate of choice in a number of useful chemical transformations including the formation of cyclopropanols, methylenecyclopropanes, cyclobutanones, β-lactames, γ-butyrolactones, cyclopentanones, pyrroline derivatives, etc. [8]. Previously prepared by the tedious and quite hazardous addition of diazomethane to ketene in the presence of one equivalent of methanol [9] or ethanol [10], the cyclopropanone hemiacetal can now be obtained readily in high yield by simple methanolysis of 1-ethoxy-1-(trimethylsiloxy)cyclopropane *2*, the product of the acyloin-type cyclization of commercial ethyl 3-chloropropanoate *1* by sodium in refluxing ether in the presence of chlorotrimethylsilane, Eq. (1) [7].

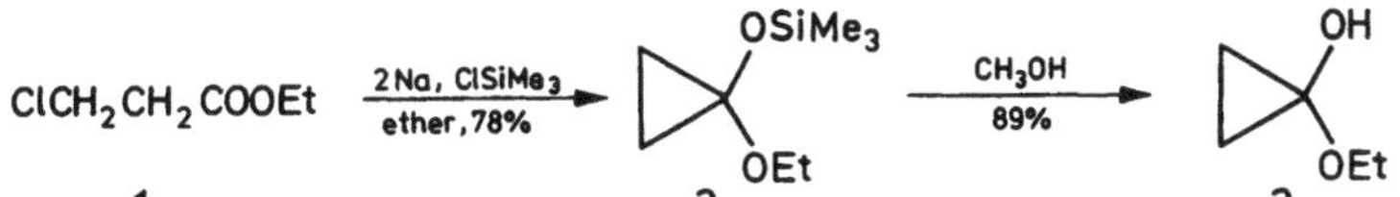

1 *2* *3* (1)

Another access to 1-alkoxy-1-siloxycyclopropanes *5*, precursors of substituted cyclopropanone hemiacetals *6*, was developed with the addition of carbenes, generated from alkylidene iodides and diethylzinc, to the trimethylsilyl enol ethers of carboxylic esters *4*, Eq. (2) [11].

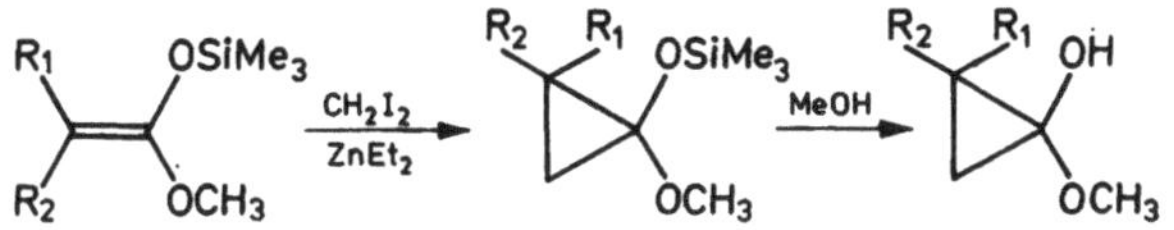

4 *5* *6* (2)

Similarly, the addition of the Simmons-Smith reagent (CH_2I_2 + Cu-Zn couple) to 1-ethoxyvinylacetate (or benzoate) provided 1-ethoxycyclopropylacetate, which upon reaction with methanol or ethanol yielded *3* or the corresponding methyl hemiacetal [12].

The chemical properties of cyclopropanone hemiacetals have been reviewed [8]. Besides oxidative cleavage with low oxidation potential metals (Cu^{II}, Fe^{III}, Ce^{II}, ...) or with molecular oxygen and peroxides, these hemiacetals underwent acid and base induced ring openings. Furthermore, they were subject to nucleophilic attack by various reagents providing an efficient pathway to 1-substituted cyclopropanols. Therefore, the ethyl hemiacetal of cyclopropanone *3* now constitutes a convenient and storable source of the parent ketone *7*, on the basis of the equilibrium shown in Eq. (3) [10,12].

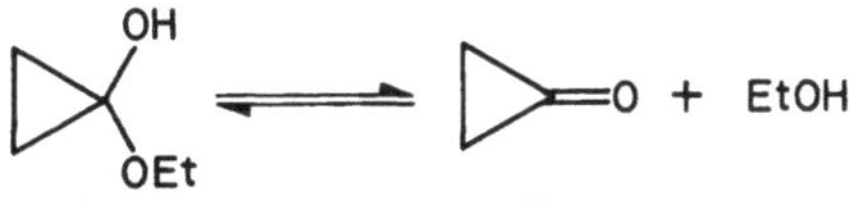

3 *7* (3)

However, owing to this formal equilibrium, two equivalents of the acetylenic Grignard reagents *8*, were required to obtain the propargylic cyclopropanols *9*, Eq. (4) [10,13].

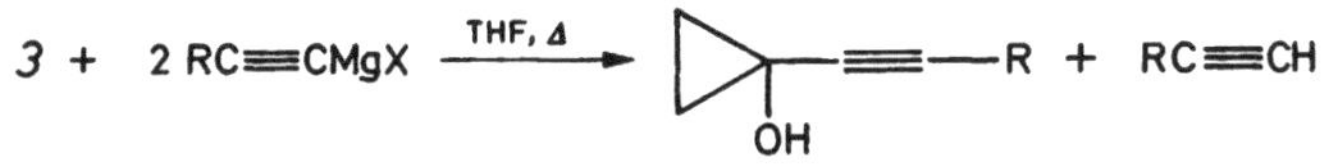

8 21-86% *9*

R = H(10), OEt(14)
CH_3, C_6H_5, $pCH_3C_6H_4$, $pCH_3OC_6H_4$, 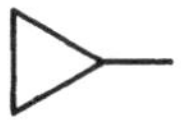
$CH_2{=}CH$(15) (4)

On the other hand, the cyclopropanone hemiacetal *3* did not react with nucleophilic lithium reagents such as lithium cyanide [16], ethynyllithium [16] or aryllithium [17].

This difficulty was overcome by first treating *3* with an equimolar amount of methylmagnesium iodide, to convert it into a species, most likely the magnesium derivative *10*, which did react with various organolithium reagents to give the expected 1-substituted cyclopropanols *11*, Eq. (5) [16,17].

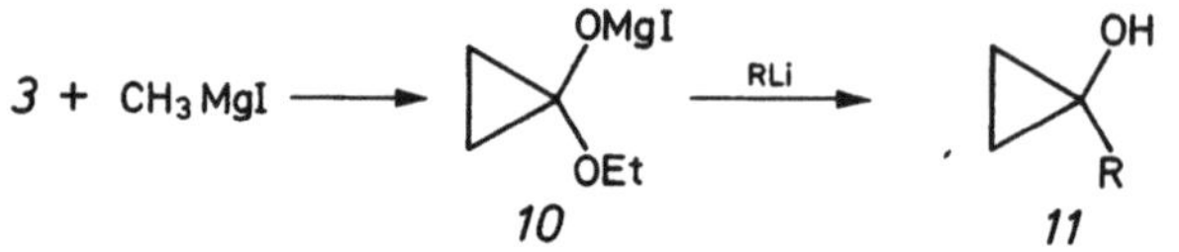

10 11 (5)

For instance, the trimethylsilylbutadiynyl lithium derivative *12*, resulting from the selective monodesilylation of bis-silylbutadiyne [18], when treated with the magnesium salt *10* provided good yields of the 1-(trimethylsilylbutadiynyl) cyclopropanol *13*, which was successfully used as a prostaglandin precursor, (*vide infra*, Sect. 5.5.2.1) Eq. (6) [15].

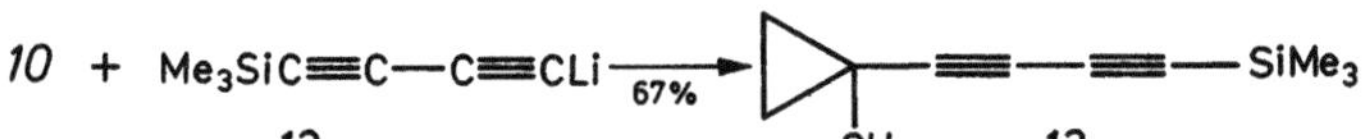

12 13 (6)

2.2 From Tetrachlorocyclopropene

Addition of dichlorocarbene to trichloroethylene gave pentachlorocyclopropane which was smoothly dehydrochlorinated by aqueous potassium hydroxide into tetrachlorocyclopropene *14* [19]. On heating at 150–180 °C, *14* provided an efficient source for tetrachlorovinylcarbene *15*, which could be trapped intermolecularly by olefins to give the 1-chloro-1-(trichlorovinyl)cyclopropanes *16*, Eq. (7) [20].

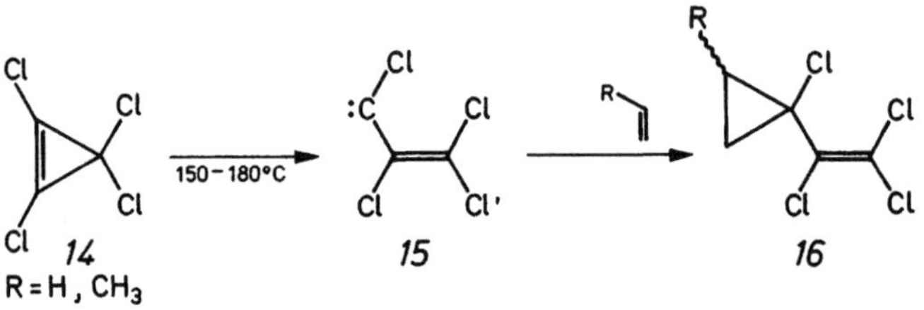

14 15 16 (7)

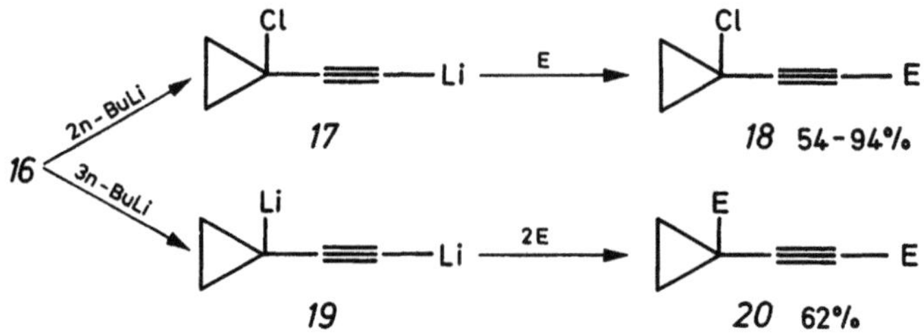

17 18 19 20 (8)

Upon reaction with two moles of *n*-butyllithium the adducts *16* gave the 1-chlorocyclopropylethynyllithium reagents *17* which could be trapped by a wide variety of electrophiles (CO_2, $ClCO_2Me$, Me_3SiCl, H_2CO, R_1COR_2, NCS, CH_3SSCH_3 ...) to give the corresponding cyclopropylacetylene derivatives *18*. Further metalation with three equivalents of *n*-BuLi gave the dianion *19* which, on electrophilic substitution, led to difunctional cyclopropylacetylenes *20* with any desirable combination of donor or acceptor functional groups, Eq. (8) [21].

2.3 From 1,1,3,3- and 1,2,3,3-Tetrachlor propene and 1,1,2,3,3-Pentachloropropane

Alternatively, 1,1,3,3- and 1,2,3,3-tetrachloropropene *21* and *22* or 1,1,2,3,3 pentachloropropane *23* underwent dehydrochlorination with potassium t-butoxide to give, probably through 1,3,3-trichloropropyne *24*, the dichlorovinylidenecarbene *25* which was trapped by olefins to lead to the dichloroethenylidenecyclopropanes *26*. Then, the highly reactive allene *26* added electrophiles as well as nucleophiles such as methoxides to give, for instance, 1-methoxy(2-chloroethynyl)cyclopropane *27*, Eq. (9) [22].

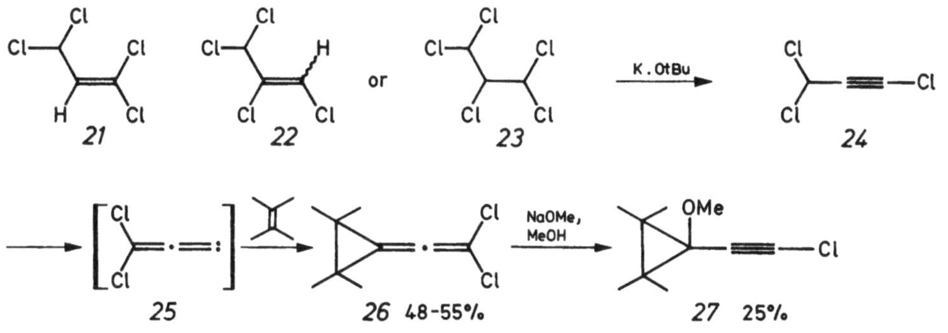

(9)

2.4 From Ethynylcarbenes

The 3,3-dimethyl-5-alkynyl-3H-pyrazoles *28*, obtained by addition of 2-diazopropane to diacetylene, were irradiated to give the alkynylvinylcarbenes *29* and *30* the relative reactivity of which depended mainly on the substituent R on the triple bond. These carbenic species were trapped by cyclopentadiene or furan to give the ethynylcyclopropane derivatives *31* and *32*, respectively, Eq. (10) [23].

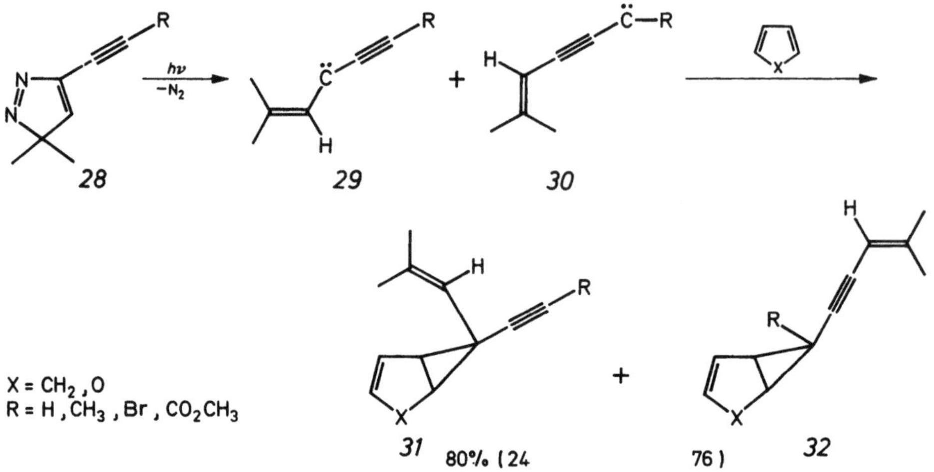

28 29 30 31 32

(10)

The pyrolysis at 140–150 °C of the lithium salts of diethynyl ketone tosylhydrazones *33* led to the formation of the triplet diethynylcarbenes *34* which were trapped by olefins to give, in a nonstereospecific reaction, the 1,1-dialkynylcyclopropane derivatives *35a, b*, Eq. (11) [24].

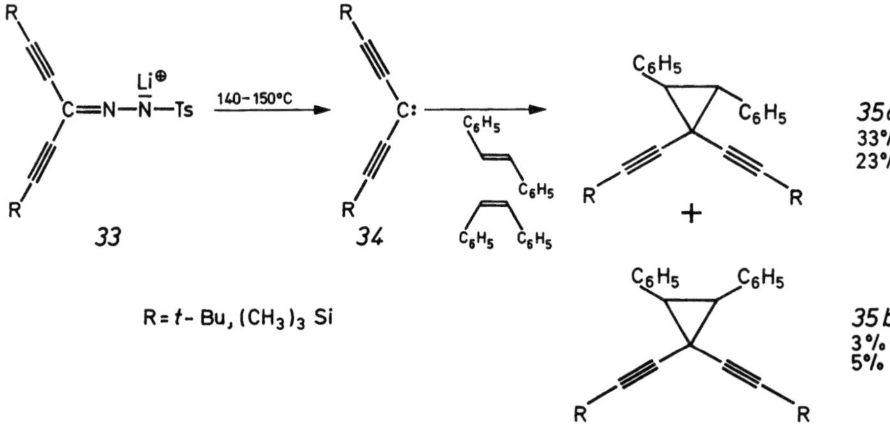

33 34 35a 35b

(11)

2.5 From Ethynyl Vinyl Oxiranes

Upon heating to 300–350 °C the ethynyl vinyl oxiranes *36*, obtained by condensation of vinylsulfonium ylides with acetylenic carbonyl compounds, underwent a Cope-transposition into the oxacycloheptadiene *37*, followed by a Claisen-type reaction leading to a cis, trans mixture of 2-ethynylcyclopropanecarboxaldehydes *38*, Eq. (12) [25].

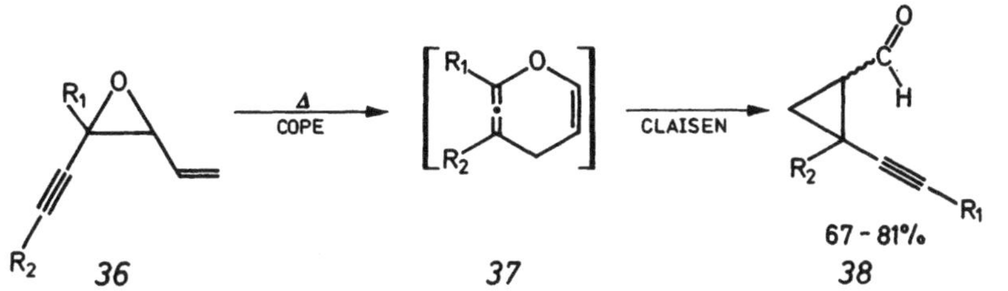

(12)

3 Reactivity of 1-Donor Substituted Ethynylcyclopropanes

3.1 $C_3 \rightarrow C_4$ Ring Expansions

Contrary to the 1-vinylcyclopropanol derivatives which easily underwent an acid catalyzed or thermally induced $C_3 \rightarrow C_4$ ring expansion as discussed in Sect. 5.1.1, the acetylenic cyclopropanols *9* were surprisingly unreactive towards acids. Thus, on heating for two hours to 55 °C in acidic (0.75N HCl) 50% aqueous dioxane containing a catalytic amount of mercuric chloride or upon treatment with m-chloroperbenzoic acid (MCPBA) for 12 hours at room temperature, the alcohol *9* (R = H) was recovered unchanged. However, treatment of *9* with an alcohol free solution of t-butylhypochlorite (t-BuOCl) in chloroform resulted in an exothermic reaction which yielded 2-chloromethylenecyclobutanone *39* as the only isolable product, Eq. (13) [10].

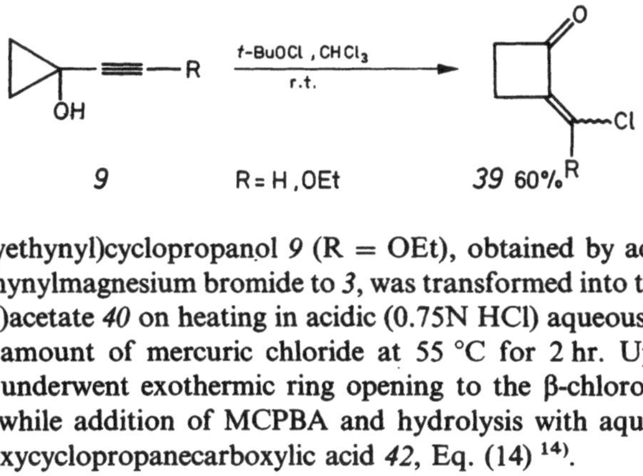

(13)

1-(Ethoxyethynyl)cyclopropanol *9* (R = OEt), obtained by addition of an excess of ethoxyethynylmagnesium bromide to *3*, was transformed into the ethyl (1-hydroxycyclopropyl)acetate *40* on heating in acidic (0.75N HCl) aqueous dioxane containing a catalytic amount of mercuric chloride at 55 °C for 2 hr. Upon treatment with t-BuOCl it underwent exothermic ring opening to the β-chloroethyl ethoxyethynyl ketone *41*, while addition of MCPBA and hydrolysis with aqueous acetone led to the 1-hydroxycyclopropanecarboxylic acid *42*, Eq. (14) [14].

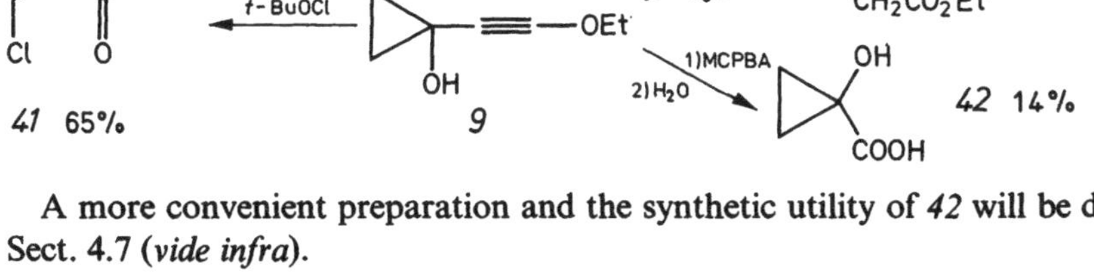

(14)

A more convenient preparation and the synthetic utility of *42* will be discussed in Sect. 4.7 (*vide infra*).

On the other hand, 1-(phenylethynyl)cyclopropanol *9* (R = Ph) underwent a $C_3 \rightarrow C_4$ ring expansion and subsequent decarboxylation when treated with MCPBA to yield the 2-phenylcyclobutanone *47*, likely via the intermediate 2-(1-hydroxy-cyclopropyl)-2-phenyl ketene *44*, formed by migration of the cyclopropyl group in the vinyl cation *43*. The ketene *44* thus resulting could be attacked by a second equivalent of MCPBA and ring expanded to the β-ketoacid *46* which would easily decarboxylate to yield *47*, Eq. (15) [14].

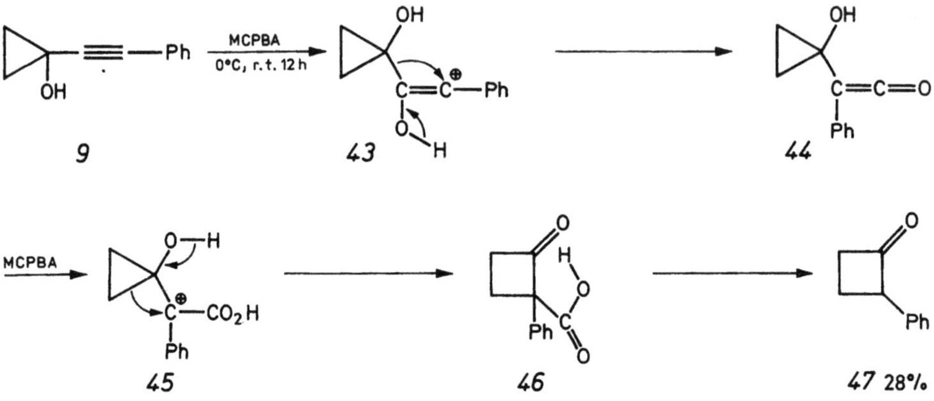

(15)

3.2 $C_3 \rightarrow C_5$, C_6 and C_7 Ring Expansions

A mixture of cis- and trans-2-methylethynylcyclopropanes *48* and *49*, prepared by the catalyzed (CuCl) addition of diazomethane to 3-penten-1-yne, upon heating to 530 °C underwent rearrangement to 3- and 4-methylenecyclopentenes *51* and *52*, likely via the intermediate allylallene *50*, and to 1,3-cyclohexadiene *54* and benzene *55*, likely via the intermediate 1,3,5-hexatriene *53*, Eq. (16) [26].

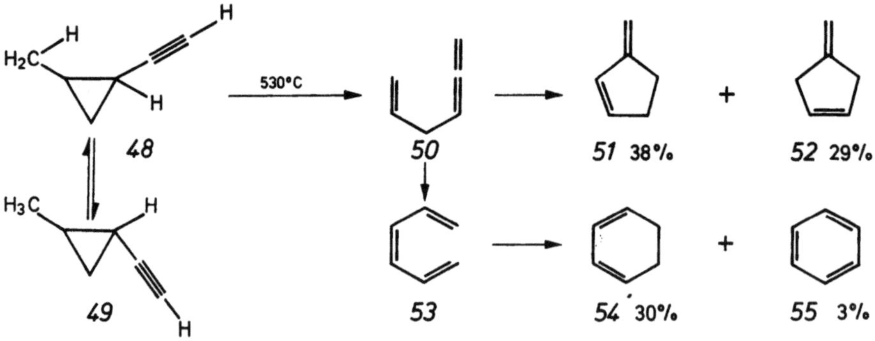

(16)

The thermal isomerization of 1-ethynyl-2-vinylcyclopropane *57* has been also reported. Thus, the dimer *59* was formed in the pyrolysis of the tosylhydrazone salt of syn- and anti-tricyclo[4.1.0.0^{2,4}]heptan-6-one *56* via the cycloheptatriene *58*, the product of a Cope rearrangement of cis- *57* Eq. (17) [27].

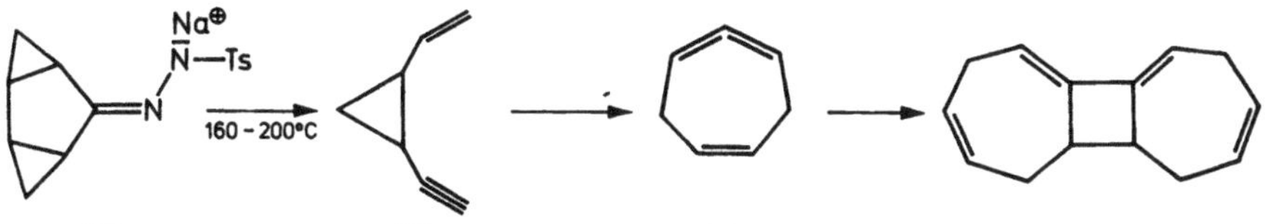

56 *57* *58* *59* (17)

It is noteworthy that, contrary to 1-trimethylsiloxyvinylcyclopropanes which readily undergo thermal $C_3 \rightarrow C_5$ ring expansion (*vide infra*, Sect. 5.5), the O-silylated 1-ethynylcyclopropanols *9* were recovered unchanged, on heating to 600 °C [28].

3.3 Solvolysis

As the 1-ethynylcyclopropanols *9* were remarkably stable to acids (*vide supra*, Sect. 3.1), it appeared interesting to investigate the solvolytic behaviour of the corresponding tosylates [13]. As a matter of fact, it was known that simple cyclopropyl derivatives usually underwent concerted ionization and disrotatory ring opening to allyl cations [29]. Such a ring opening however, can be prohibited by steric constraints [30] or conjugative interactions with donor substituents [86]; thus for instance, 1-cyclopropylcyclopropyl chloride [32] or tosylate [31] to a certain extent yielded unrearranged solvolysis products, i.e., 1-cyclopropylcyclopropanols. Thus it appeared worthwhile to determine the extent to which an adjacent ethynyl group would be able to delocalize the positive charge of a cyclopropyl cation and prevent the ring opening.

The only products of solvolysis of the tosylate *60* from the 1-ethynylcyclopropanol *9*, with R = CH_3 was the allyl derivatives *61* (R = CH_3) from the ring opening of the cyclopropane ring, while unrearranged cyclopropanols (or derivatives) *62* were obtained in high yields when R = cyclopropyl or aryl, Eq. (18) [13].

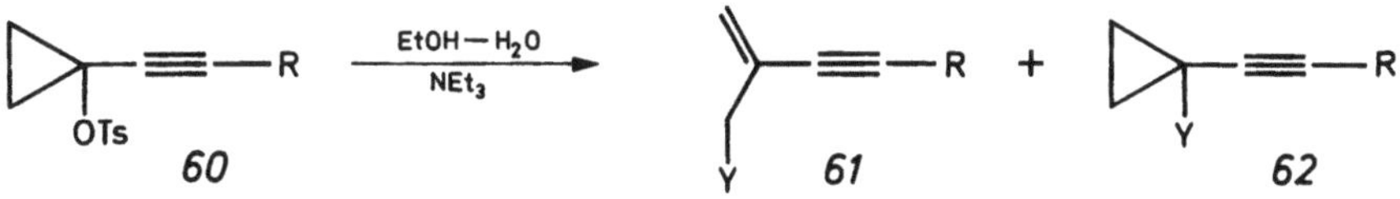

R	Y	61	62
R = CH_3	Y = OH , OEt	90%	—
C_6H_5		15	85
$pCH_3C_6H_4$		5	95
$pCH_3OC_6H_4$		0	100
		6	90

(18)

The solvolysis product distribution and the kinetic data (solvent effect [33]: m = 0.583 — 0.505, substituent effect [34]: $\varrho = -2.98$) were clearly consistent with a SN_i' ionization process involving the anchimeric assistance of the triple bond (k_Δ) [35] leading to the resonance stabilized cation *63* with its positive charge delocalized through the adjacent triple bond into the aryl (or cyclopropyl) group [13].

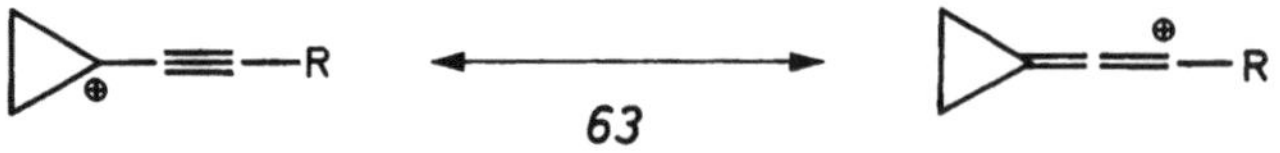

63

It was concluded that the stabilization of the positive charge of *63*, by delocalization over the three carbons of the mesomeric propargyl allenyl system, entailed a powerful electron releasing substituent at the allenyl end [13]. However, although in theory electron releasing substituents might render 1-substituted cyclopropyl cations more stable than their 2-substituted allyl counterparts [36], the expected ring closure of a 2-substituted allyl cation such as *e.g.* *64* to the cyclopropyl cation *65* has not been observed experimentally, Eq. (19) [37].

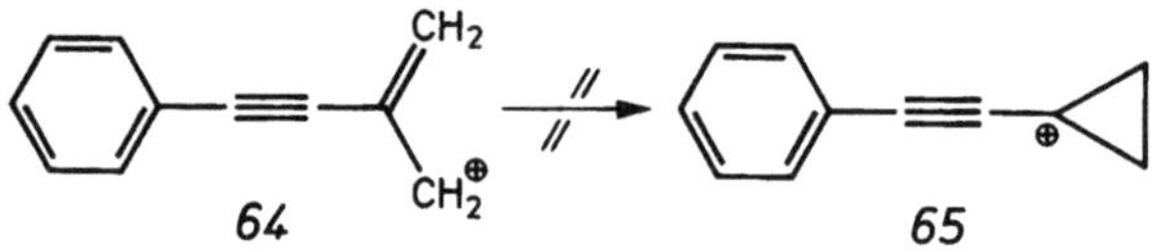

64 65

(19)

3.4 [2 + 4] Cycloadditions

Several of these cyclopropylacetylenes constitute convenient substrates for catalyzed or uncatalyzed cycloaddition reactions. For instance, the methyl 3-[1-chlorocyclopropyl]propiolate *66* is a reasonably reactive dienophile which underwent thermal reaction with various 1,3-dienes to yield [2 + 4] cycloadducts, such as *67*, Eq. (20) [20].

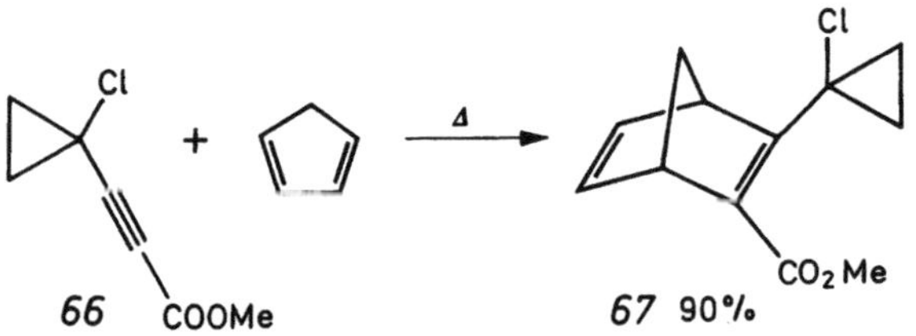

66 67 90%

(20)

In fact from the synthetic point of view, the 1-ethynylcyclopropanols *9* have been mainly used as precursors of 1-donor substituted vinylcyclopropanes, the preparation and utility of which are discussed in the following sections.

4 Preparation of 1-Donor Substituted Vinylcyclopropanes and Cyclopropylcarbonyl Compounds

Because of their exceptional reactivity and reactivity pattern, 1-donor substituted vinylcyclopropanes and cyclopropylcarbonyl compounds, appear to be very attractive building blocks, endowed with unexpected synthetic potential, which has not yet been fully explored. Various routes to these challenging compounds have successfully been tested recently.

4.1 From Cyclopropanone Hemiacetal

As first reported by Wasserman, the addition of two equivalents of vinylmagnesium bromide in refluxing THF to the cyclopropanone hemiacetal *3* led to the 1-vinylcyclopropanols *68* in 64% yields Eq. (21) [10].

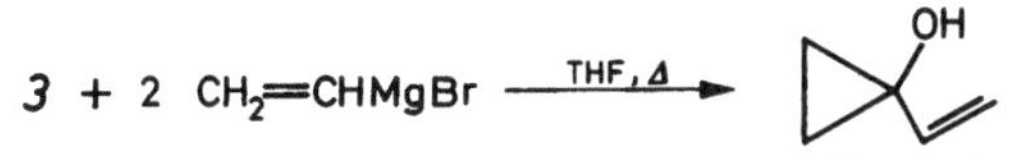

68 64%

(21)

Since the hemiacetal *3*, can now readily be prepared from cheap commercially available starting materials (*vide supra*, Eq. (1)) [7], this reaction constitutes a convenient source of 1-vinylcyclopropanols. Otherwise, the 1-ethynyl-cyclopropanols *9*, also easily available from *3* or from its magnesium salt *10* (*vide supra*, Eq. (4) and (6) underwent either lithium aluminium hydride reduction in refluxing THF to lead exclusively to the E-1-vinylcyclopropanols *69* or reduction with dicyclopentadienyltitanium hydride in ether at 0 °C, prepared from isobutylmagnesium halides and a catalytic amount of dicyclopentadienyl titanium dichloride ($(\eta^5\text{-}C_5H_5)_2TiCl_2$), to yield exclusively the Z isomer *70*, Eq. (22) [15,39].

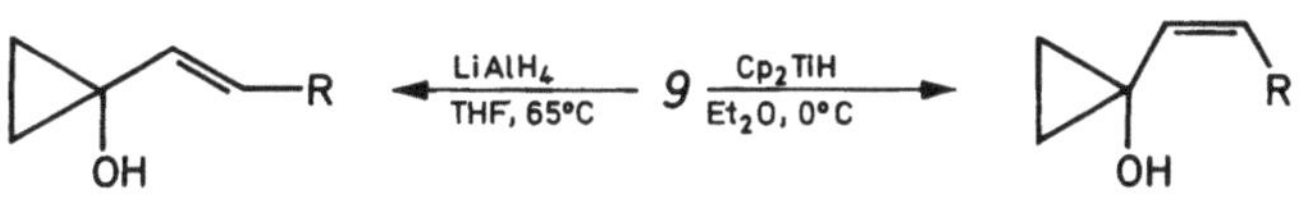

69 93% *70* 100%

(22)

4.2 From 1,3-Dichloroacetone

Following a known synthesis of 1-alkylcyclopropanols [40], addition of vinylic Grignard reagents *72* to commercially available 1,3-dichloroacetone *71* led to the magnesium salt of a 1,3-dichloro alcohols *73* which ring closed to 1-vinylcyclopropanols *74* induced by the highly reactive low valent iron formed *in situ*, by the simultaneous addition of etheral solutions of ethylmagnesium bromide and anhydrous ferric chloride, Eq. (23) [41].

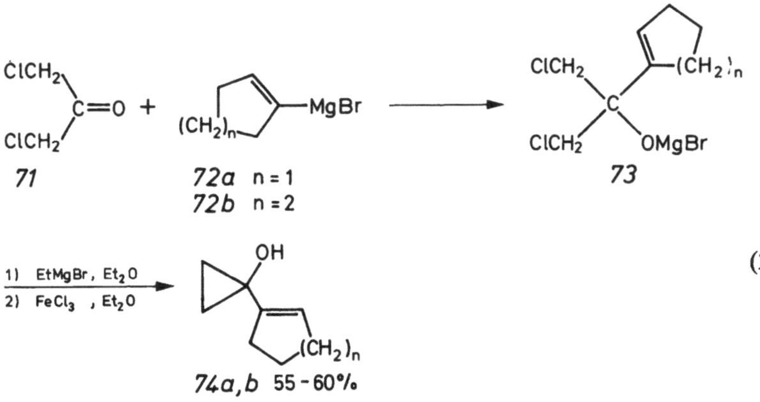

71 *72a* n = 1 *72b* n = 2 *73*

74a,b 55–60%

(23)

4.3 From Simmons-Smith Cyclopropanation of α-Enone Enol Ethers

The selective cyclopropanation of the α-enone silyl enol ether *75*, by methylene iodide and the zinc-silver couple [42], is remarkable. Only the double bond bearing the trimethylsiloxy group reacted to yield the 1-trimethylsiloxy vinylcyclopropane *76* when not more than 1.1 equivalent of the Simmons-Smith reagent was used, but the bis-cyclopropanation product *77* was obtained in good yield with an excess (3 equivalents) of the cyclopropanating reagent, Eq. (24) [42].

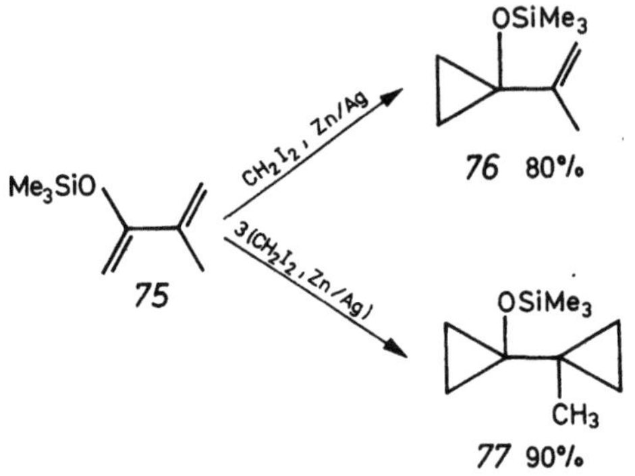

(24)

Subsequent treatment with methanol led to 1-vinyl- and 1-cyclopropylcyclopropanols respectively and with IM methanolic sodium hydroxide to the corresponding ketones by ring opening of the silyl cyclopropyl ether moieties, in excellent yields [42]. Specific α or α'-monomethylation of conjugated cycloalkenones was then possible through such siloxyvinylcyclopropanes intermediates. Thus, from testosterone *78*

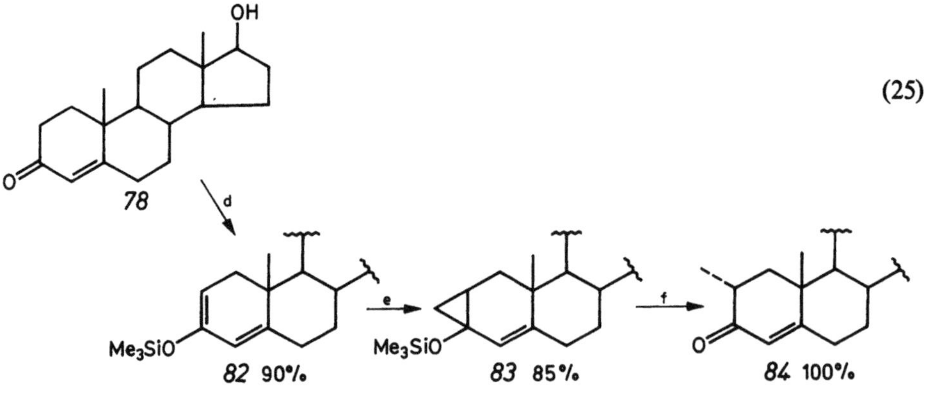

(25)

the silyl enol ethers *79* and *82* have been obtained, practically pure upon treatment with triethylamine and trimethylsilyl chloride (TMSCl), or with lithium diisopropylamide and TMSCl, respectively. Thence, chemoselective cyclopropanation with CH_2I_2 and the Zn/Ag couple followed by alkaline hydrolysis of the trimethylsiloxyvinylcyclopropane intermediates *80* and *83* led either to 4-methyl *81* or to 2-α-methyl testosterone *84* in excellent yields, Eq. (25)[42].

Reagents: a) NEt_3, $ClSiMe_3$, DMF; b) Zn/Ag, CH_2I_2, Et_2O, reflux, pyridine, 52%; c) NaOH, EtOH, reflux 10 hr, 88%; d) LDA, $ClSiMe_3$, THF-Et_2O, 90%; e) Zn/Ag, CH_2I_2, Et_2O, reflux 18 hr, 85%; f) NaOH, EtOH, reflux 48 hr, 100%.

Simmons-Smith regioselective cyclopropanation of α-enone alkyl enol ethers also provided 1-alkoxyvinylcyclopropanes in high yields[43].

4.4 From Dye-Sensitized Photooxygenation of Alkylidenecyclopropanes

The dye-sensitized (eosin) photooxygenation at —50 °C of alkylidenecyclopropanes *85*, easily available by the Wittig reaction of 3-bromopropyltriphenylphosphonium ylide[44] and suitable ketones, gave the hydroperoxides *86*, which were reduced *in situ* by an equivalent of triphenylphosphine[45] to 1-alkenylcyclopropanols *87* in good yields. When PPh_3 was replaced by pyridine (5%) *86* rearranged exclusively

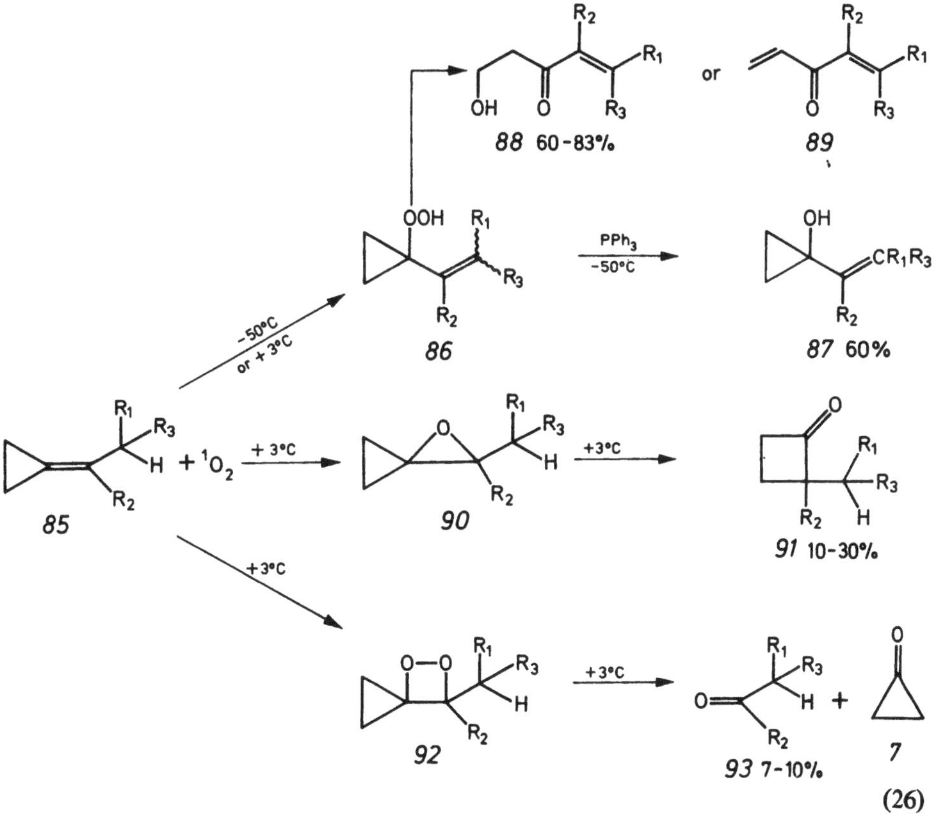

(26)

to β'-hydroxy α-enones *88*, whereas in the absence of pyridine the α,α'-dienones *89*, dehydration products of *88*, were formed competitively. On the other hand, the photo-oxygenation of the olefins *85* at 3 °C led to three types of products: the hydroxyenones *88* arising from the thermal rearrangement of the unstable hydroxyperoxides *86*, the cyclobutanones *91*, likely through the well known ring expansion of the oxaspiropentanes *90* [41,46] formed by addition of singlet oxygen to the olefins *85* [47] and the ketones *93* from the thermal fragmentation of the intermediate dioxetanes *92*, Eq. (26) [48].

4.5 From Oxaspiropentanes

Oxaspiropentanes have been synthesized by the epoxidation of methylenecyclopropanes with peracetic [49], peroxybenzimidic [50], with p-nitroperbenzoic [46] and m-chloroperbenzoic acid [51]. The parent oxaspiropentane *95*, a convenient precursor of cyclobutanone [46], was obtained from the peracid oxidation of a methylene chloride solution of methylenecyclopropane *94*, Eq. (27) [46,51].

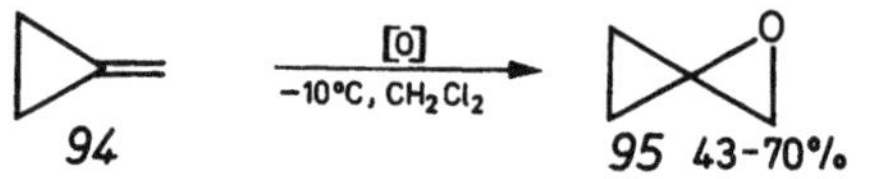

(27)

Besides by these epoxidations, oxaspiropentanes have been prepared through the nucleophilic addition of 1-lithio-1-bromocyclopropanes to ketones at low temperature. Thus for example, the dibromocyclopropane *96* prepared by addition of dibromocarbene to cyclohexene [52] underwent metalation with butyllithium to give the lithiobromocyclopropane *97* which was converted into the oxaspiropentane *98* upon simple addition to cyclohexanone, Eq. (28) [53,54].

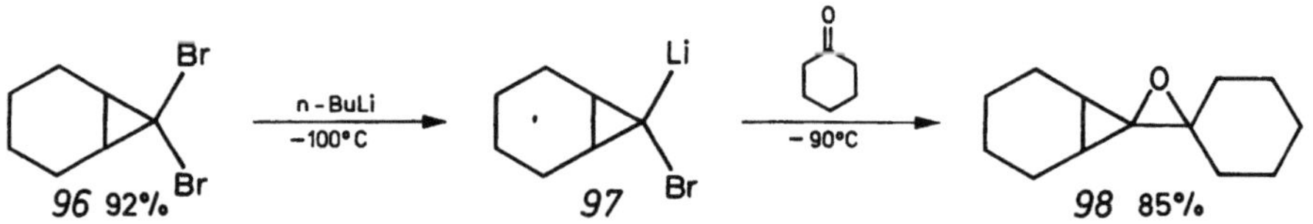

(28)

The intermediacy of such oxaspiropentanes has been proposed in the addition of diazomethane to ketones [50] and in the reaction of dimethyloxosulfonium methylide with α-haloketones [55]. In contrast to phosphorous ylides, sulfur ylides usually condense with carbonyl compounds to yield epoxides, thus reaction of the N,N-dimethylaminophenyloxosulfonium cyclopropylide *99* with cyclohexanone produced the dispiroepoxide *100* which rearranged to the spiro [3.5] nonan-1-one *101* upon isolation by gas chromatography, Eq. (29) [56].

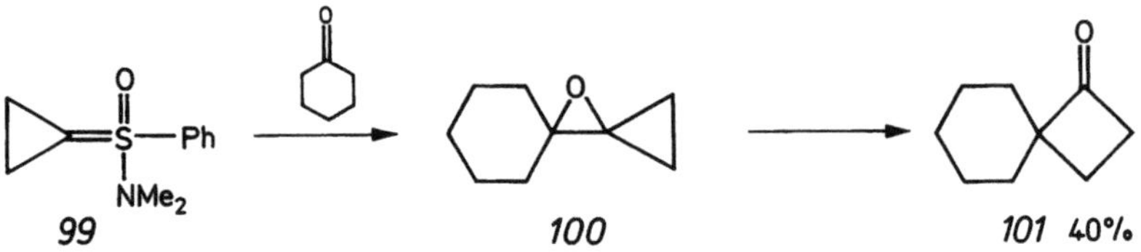

(29)

Oxaspiropentanes have been obtained from the cyclopropylide *103*, prepared by treatment of cyclopropyldiphenylsulfonium tetrafluoroborate *102* either with sodium methylsulfinyl carbanion in dimethoxyethane at —45 °C or with potassium hydroxide in dimethylsulfoxide at 25 °C. While the reaction of the ylide *103* with α,β-unsaturated carbonyl compounds has resulted in selective cyclopropylidene transfer to the α,β-carbon-carbon double bond leading to spiropentanes, condensation of *103* with non-conjugated aldehydes and ketones led to oxaspiropentanes such as *104*, which have been isolated in 59–100% yields, Eq. (30) [57].

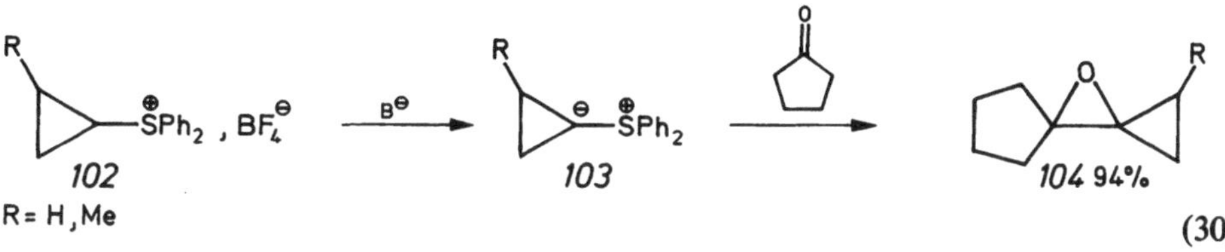

(30)

The direction of the base induced ring opening of oxaspiropentanes proved to be highly depending on the nature of the base and solvent. Thus the epoxide *100* opened either mainly to 1-(1-cyclopropenyl) cyclohexanol *106* on reaction with lithium di-isopropylamide in ether or mainly to the expected 1-(1-cyclohexenyl) cyclopropanol *105* on reaction with lithium diethylamide in pentane, Eq. (31) [57].

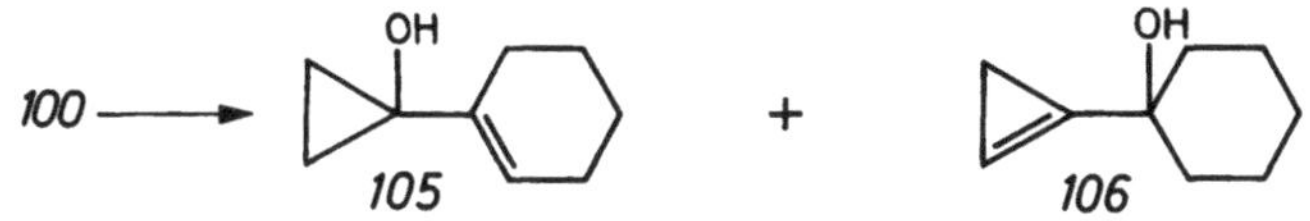

	105	106
t-BuNHLi, Ether	1	1.23
iPr2NLi, Ether	1	7.60
Et2NLi, Ether	1.5	1
Et2NLi, Pentane	9	1

(31)

These results have been rationalized on the basis of the stereochemical features of the reaction [57]. The regioselectivity can also be influenced by kinetic or thermodynamic control of the reaction. Thus, exposure of the oxaspiropentane *107* to a sterically hindered base (LDA) for a relatively short time in a hydrocarbon solvent resulted in the abstraction of a proton of the methyl group by kinetic preference, leading almost exclusively to the 1-vinylcyclopropanol *108*, in contrast, use of a sterically less demanding base for a longer time in hexane provided exclusively the thermodynamically more stable 1-vinylcyclopropanol *109* as a 98/2 trans/cis mixture, Eq. (32) [58].

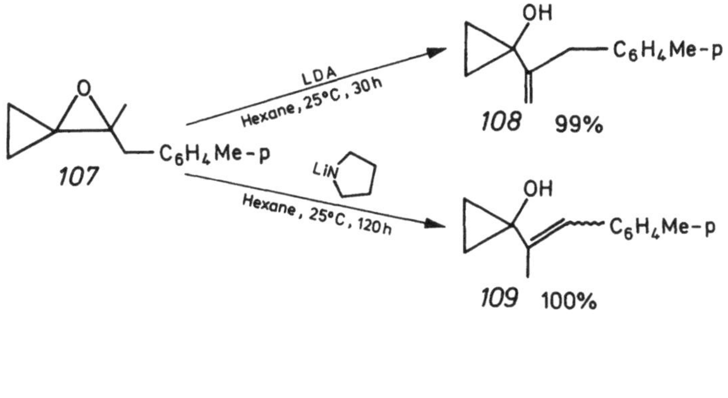

(32)

A severe limitation of this method, however, is the failure of the ylide *103* to yield oxaspiropentanes (*vide supra*) from α,β-unsaturated ketones and the poor yields of vinylcyclopropanes obtained from its reactions with hindered ketones or with conformationally rigid six-membered rings. Moreover, attempts to extend the oxaspiropentane ring opening to compounds containing an adjacent tertiary center have failed; thus, oxaspiropentane *110* did not lead to *111*, Eq. (33) [57].

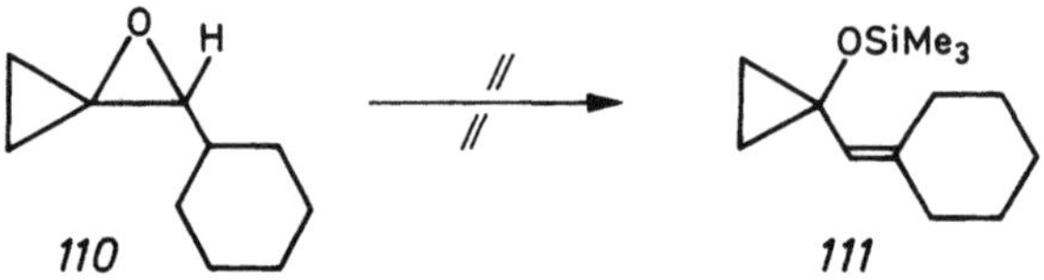

(33)

It has also been shown that oxaspiropentanes undergo smooth ring opening with sodium phenylselenide at room temperature to give β-hydroxyselenides, which, upon oxidation with m-chlorobenzoic acid at —78 °C and ring enlargement at —30 °C in the presence of pyridine led directly to cyclobutanones. However, in the case of the oxaspiropentane *112* prepared from an aldehyde, the intermediate *113* mainly eliminated selenoxide to give the 1-vinylcyclopropanol *114*, which was not converted to the cyclobutanone *115* under these conditions, Eq. (34) [59].

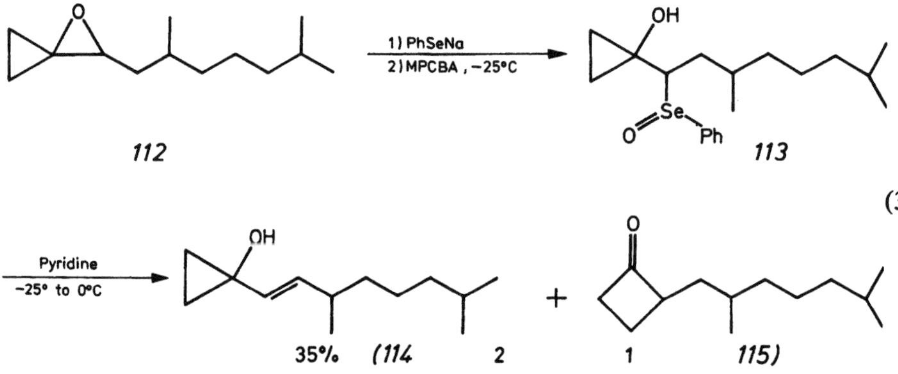

(34)

4.6 From 1-Heterosubstituted Lithiocyclopropanes

4.6.1 1-Arylthiocyclopropyllithium

Treatment of cyclopropyl phenyl sulfide [60] with *n*-butyllithium in tetrahydrofuran at 0 °C for 2 hr led to 1-phenylthiocyclopropyllithium *116* [61], which added to ketones

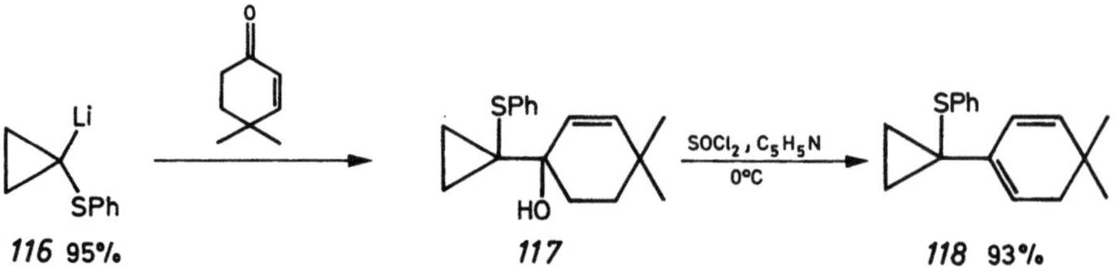

(35)

at 0 °C to yield hydroxysulfide *117*[62]; this, upon treatment with thionyl chloride in pyridine at 0 °C, was dehydrated without complications arising from ring expansion (*vide infra*) to give the phenylthiovinylcyclopropane *118*, Eq. (35)[62].

Simple saturated ketones, however, were recovered unchanged even when a twofold excess of the lithiocyclopropyl phenyl sulfide *116a* was employed apparently due to problems of enolization, but hindered ketones or α-enones underwent complete carbonyl condensation[61].

Reaction of the lithiocyclopropyl aryl sulfides *116a–c* with dimethylformamide gave the 1-arylthiocyclopropanecarboxaldehydes *119a–c*, which were treated at —78 °C with the enolate of cyclohexanone, for instance, to form the aldol products *120a–c* in 83–86% yield. Subsequent dehydration of *120a–c*, which could be complicated by retroaldol or ring enlargement reactions, required the use of a mixture of phosphorus oxychloride and hexamethylphosphorus triamide to give the expected vinylcyclopropanes *121a–c* in 71–76% yields, Eq. (36)[63a].

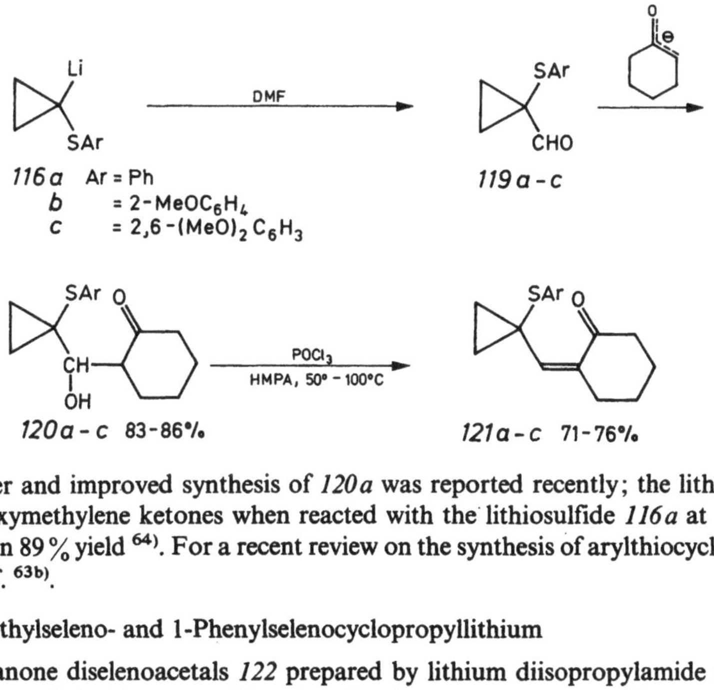

(36)

A simpler and improved synthesis of *120a* was reported recently; the lithium salt of α-hydroxymethylene ketones when reacted with the lithiosulfide *116a* at —78 °C gave *120a* in 89% yield[64]. For a recent review on the synthesis of arylthiocyclopropanes see Ref. [63b].

4.6.2 1-Methylseleno- and 1-Phenylselenocyclopropyllithium

Cyclopropanone diselenoacetals *122* prepared by lithium diisopropylamide induced ring closure of 3-chloro 1,1-di(methylseleno)- or 3-chloro-1,1-di(phenylseleno)-propane, have been transformed into the corresponding 1-selenocyclopropyllithium derivatives *123* upon treatment with *n*-butyllithium in THF at —78 °C. These intermediates have been trapped at —78 °C with aldehydes and ketones to produce the corresponding β-hydroxyselenides *124* in good yields, Eq. (37)[66].

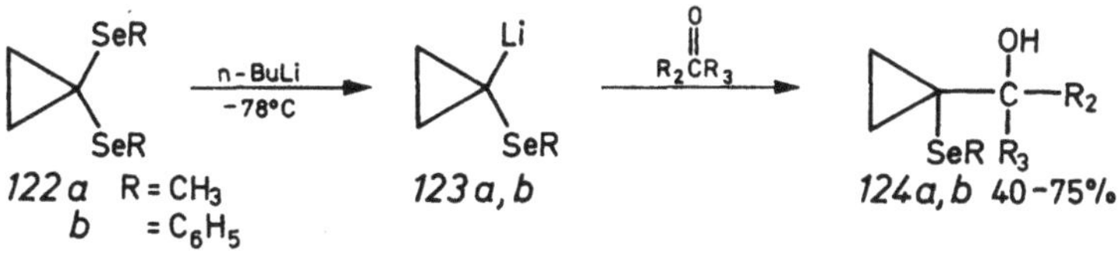

(37)

Dehydration of *124* to the expected 1-seleno-1-vinylcyclopropanes was succesful only with tertiary alcohols of this type; it required the use of thionyl chloride in the presence of triethylamine, pyridine or hexamethylphosphorus triamide followed by reaction with potassium t-butoxide in DMSO, or the use of the Burgess reagent $[CH_3O_2CN^-SO_2\overset{+}{N}Et_3]$ [67] in toluene at 110 °C. Thus, dehydratation of *e.g.* the selenohydroxide *125* with this reagent led to a mixture of the 1-methylselenovinylcyclopropanes *126* and *127*, Eq. (38) [68].

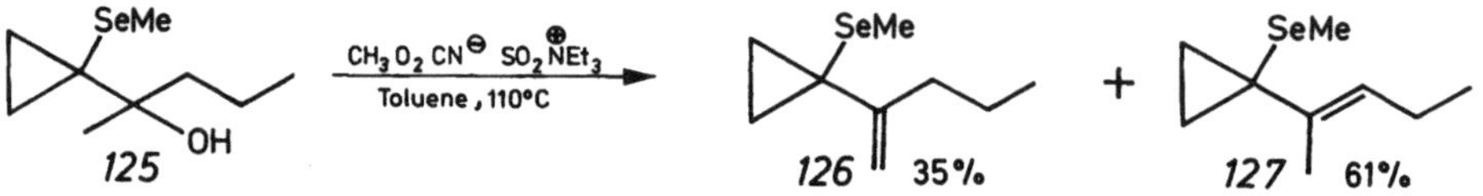

(38)

This lack of generality and regioselectivity has been overcome, however, by using the 1-selenocyclopropyl aldehydes *128a, b* prepared in high yield either upon reaction of *123* with dimethylformamide or by reduction of the phenylselenocyclopropyl *129* [69] with diisobutylaluminum hydride [70]. Subsequent olefination of aldehydes *128* with the suitable phosphorus ylides produced the desired 1-seleno-1-vinylcyclopropanes *130* in high yield with preferred (Z)-stereochemistry, Eq. (39) [70a].

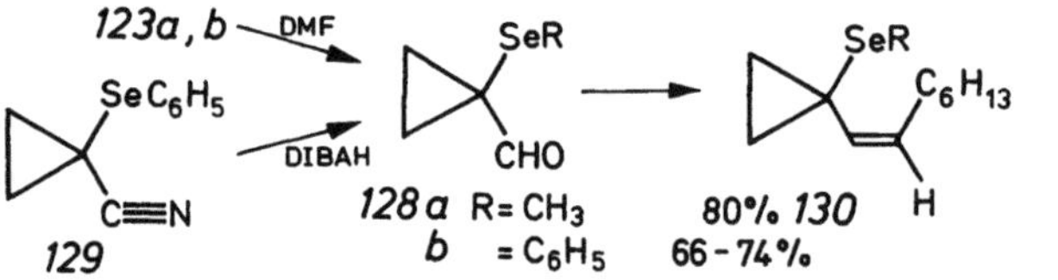

(39)

On the other hand, reaction of the α-selenoalkyllithium *131* with the aldehydes *128* at —78 °C (method A), or reaction of the α-selenoaldehyde *132* with the α-lithiocyclopropylselenide *123* (method B) led to the hydroxy β,β′-diselenides *133* in high yields. Subsequently, regioselective elimination of the hydroxy and seleno groups to give the desired 1-seleno-1-vinylcyclopropanes occurred when reacting *133* with PI_3 (or P_2I_4) and triethylamine at 0 °C with predominant formation of the (E)-isomer *134* (55 and 98% by method A and B, respectively), Eq. (40) [70a]. For a recent review on the synthesis of 1-metallo-1-selenocyclopropanes see Ref. [70b].

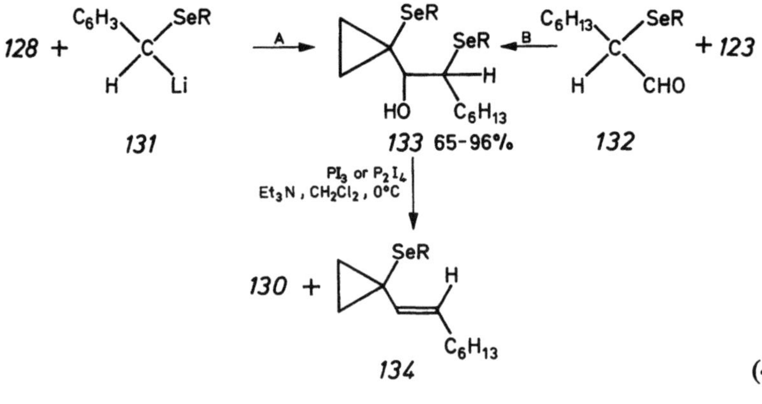

(40)

4.6.3 1-Alkoxycyclopropyllithium

Since an alkoxy group stabilizes an adjacent positive charge better than an arylthio group, an improved reactivity was expected from compounds of type *116* with an alkoxy group. To prove this, 1-methoxy-1-phenylthiocyclopropane *135* was prepared by successive treatment of 1-phenylthiocyclopropyllithium *116a* with iodine and sodium carbonate in refluxing methanol. Reductive lithiation with two equivalents of lithium 1-(dimethylamino)naphthalenide (LDMAN) in THF at —78 °C led to 1-methoxycyclopropyllithium *136*, Eq. (41) [71].

(41)

The O,S-cyclopropanone acetal *135* was also obtained by reacting cyclopropyl phenyl sulfide with trichloroisocyanuric acid (chloreal), silver nitrate and cadmium carbonate in methanol, or by reacting cyclopropanone bis-phenylthioacetal with mercuric chloride and mercuric oxide in methanol at 100 °C [72,73]. The reductive desulfurization of *135* with LDMAN, however, required rather careful control of the experimental conditions. Therefore, other more practical approaches to *136* have recently been investigated. Thus, reaction of tri-*n*-butylstannyllithium (or magnesium chloride) with the magnesium salt *10* [16,17] of cyclopropanone hemiacetal *3* [8] (*vide supra*, Sect. 2.1) afforded after protection of the hydroxyl group, low yields of the (1-methoxymethoxycyclopropyl)tri-n-butylstannane *137*; this could smoothly be transmetalated with *n*-BuLi in THF at —78 °C to provide the l-alkoxycyclopropyllithium reagent *138*, Eq. (42) [74].

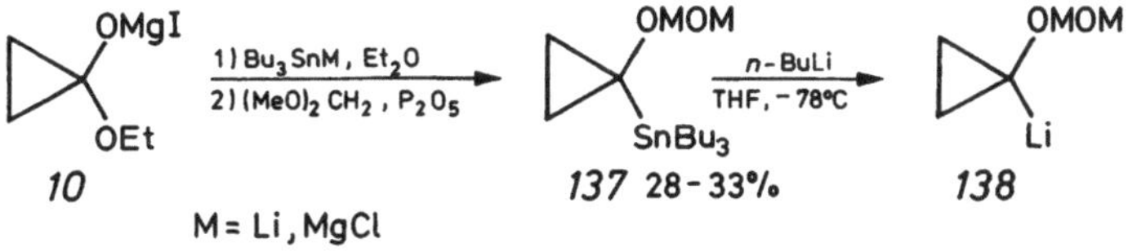

(42)

The most convenient preparative scale precursor of 1-alkoxycyclopropyllithium reagents was found to be 1-bromo-1-ethoxycyclopropane *139*, prepared in good yields by reaction of 1-ethoxy-1-trimethylsiloxycyclopropane *2* [7] with phosphorus tribromide in the absence of pyridine. Addition of *139* to *t*-butyllithium (2 equivalents) in ether at —78 °C resulted in immediate and exothermic halogen metal exchange to form the expected 1-ethoxycyclopropyllithium *140*, Eq. (43) [74].

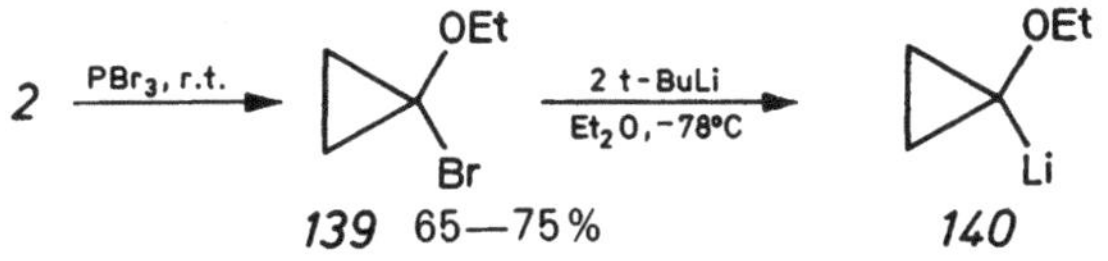

(43)

The 1-alkoxycyclopropyllithiums *136* and *140* have been added to a variety of conjugated aldehydes and ketones to produce cyclopropyl carbinols such as *141*, which, when treated directly in acidic media (*e.g.*, 10% HBF_4 in wet tetrahydrofuran) underwent ring expansion to the challenging 2-vinylcyclobutanone *142* (*vide infra*, Sect. 5, Eq. (44)) [43,64,73,74].

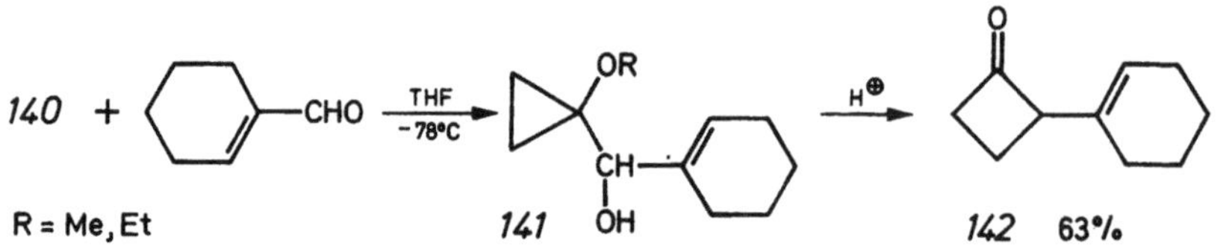

(44)

4.6.4 1-Trimethylsilylcyclopropyllithium

Trialkylsilyl substituents behave in a twofold manner, showing the properties of both electron donor and acceptor groups. It is well established that such substituents strongly favour carbenium ion development at the β-carbon (the β-effect), but exert a weak electron-attracting effect at the α-carbon atom [75]. Trimethylsilylcyclopropane *143* [76], contrary to cyclopropyl phenyl sulfide (*vide supra*, Sect. 4.6.1) could not be deprotonated at its cyclopropyl position under a variety of conditions, including prolonged exposure to sec-butyllithium and tetramethylethylenediamine in THF. On the other hand, 1-trimethylsilylcyclopropyl phenyl sulfide *145*, readily available either from the addition of chlorotrimethylsilane to 1-phenylthiocyclopropyllithium *116a* [61,77], or by sequential treatment of 1,3-di(phenylthio)propane with two equivalents of *n*-Buli [78] and Me_3SiCl, underwent reductive lithiation either with lithium naphthalenide in THF at −78 °C [79] or with LDMAN at −45 °C [77] to give 1-trimethylsilylcyclopropyllithium *144*, Eq. (45).

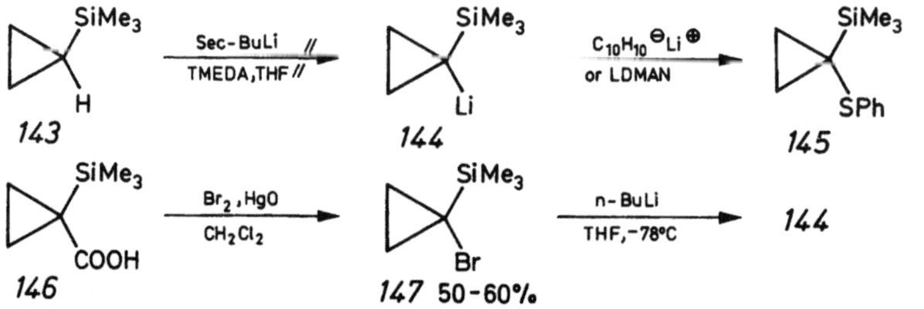

(45)

(46)

Similarly, *144* has been obtained from the reaction of 1-trimethylsilylcyclopropyl methyl selenide with n-BuLi [80]. The α-bromosilane *147* underwent lithiation with *n*-BuLi in THF at −78 °C to provide *144* with superior efficiency to any other method, Eq. (46)) [81]. *147* was prepared in large quantities by the Hunsdiecker degradation of the 1-trimethylsilylcyclopropanecarboxylic acid *146*, obtained by successively reacting the commercially available cyclopropanecarboxylic acid with *n*-BuLi (2 equivalents) and $ClSiMe_3$ [82]. Uneventfully, *144* added to carbonyl compounds, except for cyclopentanone where enolate anion formation competed; the 1-trimethylsilylcyclopropylcarbinols *148* underwent acid-induced dehydration to the expected 1-trimethylsilylvinylcyclopropanes *149* [79,81] while base induced elimination (KH, diglyme, 90 °C) led to cyclopropylidenecycloalkanes *150* [77], Eq. (47).

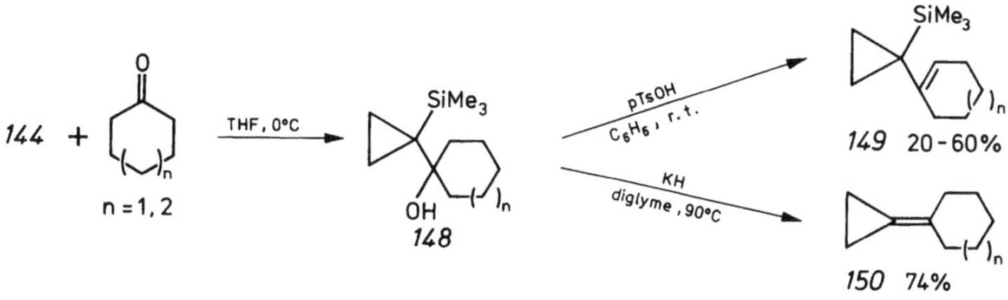

(47)

The addition of α-lithiovinyltrimethylsilane *151* [83], generated from α-bromovinyltrimethylsilane [84] with *t*-BuLi (1.5 equivalents) at −78 °C in ether, to ketones and aldehydes was also investigated. The allylic alcohols *152* thus obtained underwent smooth cyclopropanation when the modified Simmons-Smith procedure utilizing EtZnI [85] was applied. The cyclopropylcarbinols *153* were directly dehydrated without rearrangement upon exposure to catalytic amounts of p-TsOH in benzene at 20 °C to give *149* in yields of 52–75%, Eq. (48) [79,81].

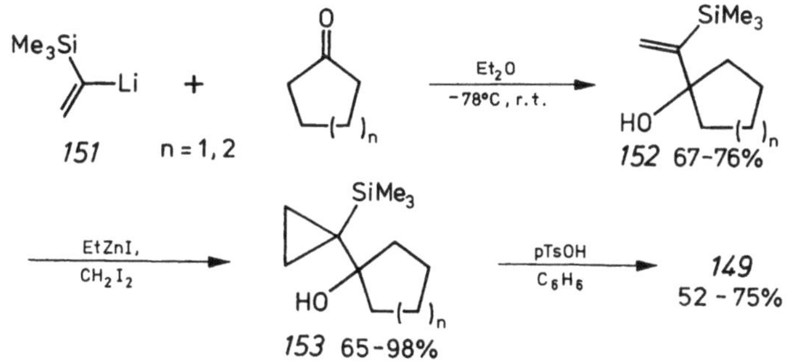

(48)

A number of functionalized 1-trimethylsilylcyclopropanes have become readily accessible along this and other routes [86]. Among them, the 1-trimethylsilylcyclopropane carboxaldehyde *156* was obtained from the spiroalkylation of trimethylsilylacetonitrile *154* upon successive treatment with 1 equivalent of lithium diisopropylamide (LDA), 1.5 equivalent of 1,2-dibromoethane and finally with a second equivalent of LDA. Subsequent diisobutylaluminum hydride (DIBAH) reduction of *155* followed by hydrolysis of the resulting imine with dilute sulfuric acid gave the aldehyde *156* in high yields, Eq. (49) [86].

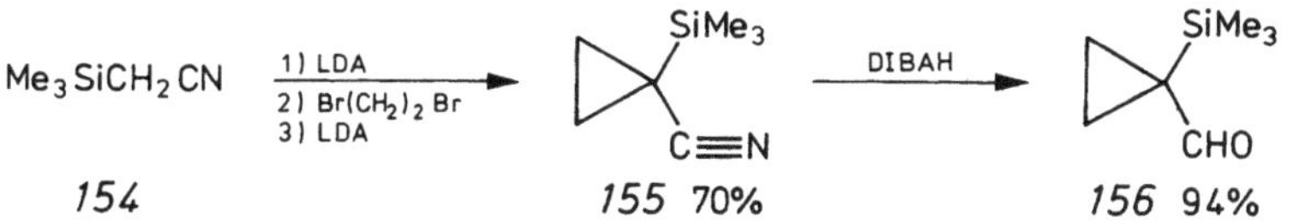

(49)

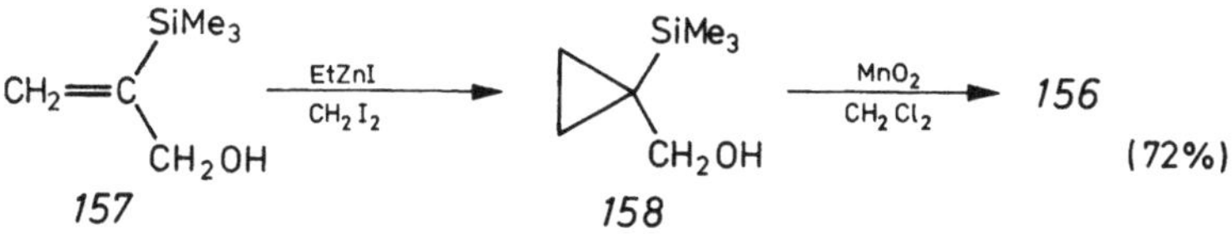

(50)

The bifunctional cyclopropane *156* was also prepared by modified Simmons-Smith cyclopropanation [85] of 2-trimethylsilyl-2-propen-1-ol *157* [84] followed by oxidation of the cyclopropylcarbinol *158* with activated manganese dioxide [88], in 72% overall yield, Eq. (50) [86,89]. Coupling of the aldehyde *156* with 2,6-dimethylcyclohexenone *159* [90] induced by the low valent titanium reagent from $TiCl_3$ and zinc-copper couple (or lithium metal) provided the silylated cyclopropyldiene *160*, in 50–60% yield, Eq. (51) [89,91].

156 + 

159 *160* 50–60% (51)

The aldehyde *156* has also been condensed with α-lithioselenides to lead to β-hydroxyselenides which underwent regioselective elimination (*vide supra*, Sect. 4.6.2, Eq. (40) [70]) upon treatment with methanesulfoxyl chloride and triethylamine providing complementary methods for gaining access to 1-silylated vinylcyclopropanes [91].

4.7 From 1-Hydroxycyclopropylcarbonyl Compounds

The high synthetic potential of 1-hydroxycyclopropanecarboxylic acid *42* with its two functionalities on the same carbon of the three-membered ring has only recently been recognized [92].

As a matter of fact, *42* is readily accessible, starting with the acyloin condensation of succinic esters in the presence of trimethylsilylchloride to provide 1,2-disiloxycyclobutene *161* in high yields [93]. Bromination of *161* in pentane at low temperature, led to the 1,2-cyclobutanedione *162*, which underwent acid or base induced ring contraction to 1-hydroxycyclopropanecarboxylic acid *42* [94]. More conveniently, *42* was prepared in a one-pot reaction by first adding bromine in CH_2Cl_2 at −10 °C and then ice-water to *161*; the hydroxyacid *42* was obtained by continuous extraction in 94% yield, Eq. (52) [95,96].

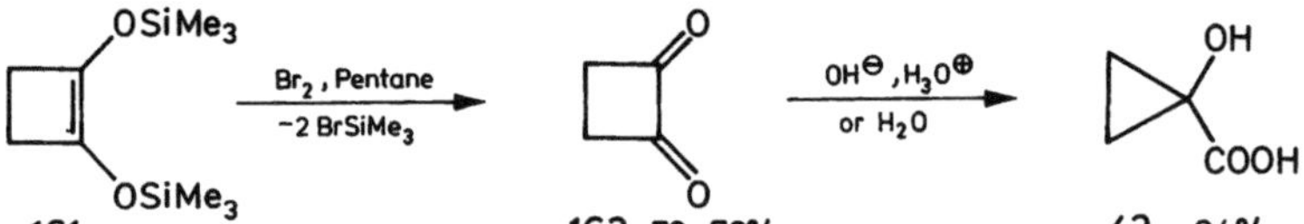

161 *162* 70–76% *42* - 94% (52)

Other conceivable routes to *42*, for instance the oxidation of the lithium salt of α-lithiocyclopropanecarboxylic acid [97] with molecular oxygen [98], have failed [39].

Cyclopropanols in general, can well serve as homoenolate anion precursors, i.e., the β-anion of ethyl propionate [99], however, to avoid the easy base or acid induced ring opening the hydroxyl function of *42* must be protected when necessary. On simple addition of one equivalent of 3,4-dihydro-2H-pyran to a CH_2Cl_2 solution of the α-hydroxy acid *42*, the tetrahydropyranyl ether *163* was obtained exclusively, in the absence of an acid catalyst. Then addition of two equivalents of methyllithium to the acid *163* gave the methyl ketone *164*, which underwent Wittig olefination *e.g.* with p-methylbenzylidenetriphenylphosphorane, to produce an (E/Z) mixture of the vinylcyclopropane *165*; deprotection was achieved by simple action of ethanol in the presence of 10% of pyridinium p-toluenesulfonate (PPTS)[101] and gave the 1-vinylcyclopropanol *166*, Eq. (53)[92].

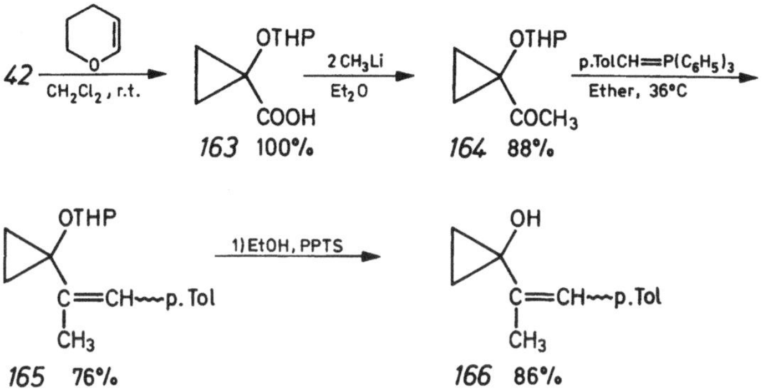

(53)

Like 1-acylcyclopropanols in general, the 1-hydroxycyclopropanecarboxaldehyde *167*, the counterpart of l-arylthio- *119a–c*, 1-phenyl(methyl)seleno- *128a–b* and 1-trimethylsilylcyclopropanecarboxaldehyde *156* (*vide supra*) could not be isolated, as it readily expanded to the corresponding 2-hydroxycyclobutanone *168*[41]. The labelling pattern in deuterium oxide-NaOD has suggested an equilibrium between the two isomeric acyloins *167* and *168* favouring the four-membered ring. Eq. (54)[102].

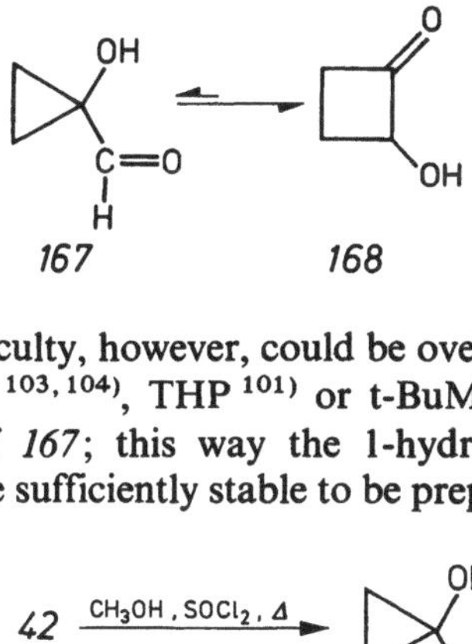

(54)

This difficulty, however, could be overcome by first protecting the hydroxyl group as a MEM[103,104], THP[101] or t-BuMe$_2$Si ether[105] before forming the carbonyl function of *167*; this way the 1-hydroxycyclopropanecarboxaldehyde derivatives *171a–c* were sufficiently stable to be prepared and handled. To this end, esterification

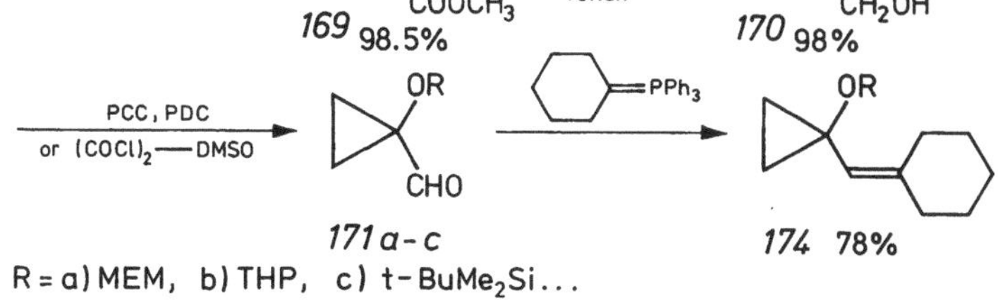

(55)

of *42* with methanol and a catalytic amount of thionyl chloride [106] gave *169*, which was protected either with β-methoxyethoxymethyl chloride [103, 104], dihydropyran [101] or t-butyldimethylsilyl chloride [105] and then reduced with lithium aluminium hydride to give the cyclopropylcarbinols *170*. Finally, oxidation of *170* either with pyridinium chlorochromate (PCC) [107], pyridinium dichromate (PDC) [108] or with oxalyl chloride activated dimethylsulfoxide [109] produced the expected aldehydes *171a–c* in 88, 68 and 98% yields, respectively, Eq. (55) [92, 96, 110, 111].

Another route to the aldehyde *171d* (R = EtO(CH_3)CH-) involved the cyclopropanone cyanohydrin *173a*, which was prepared either from the labile cyclopropanone *7* and hydrocyanic acid [112] or by the addition of lithium cyanide to the magnesium salt of the cyclopropanone hemiacetal *10* [16]. Reaction of *173a* with excess ethyl vinyl ether at room temperature gave the acetal *173b*, in quantitative yield [113]; *173b* had previously been obtained in 62% yield by the ring closure of 2-chloropropionaldehyde cyanohydrin ethoxyethyl ether *172* with sodium bis-trimethylsilylamide [14]. Subsequent reduction of *173b* with sodium dihydro-bis(2-methoxyethoxy)aluminate [115] led to the aldehyde *171d* Eq. (56) [113].

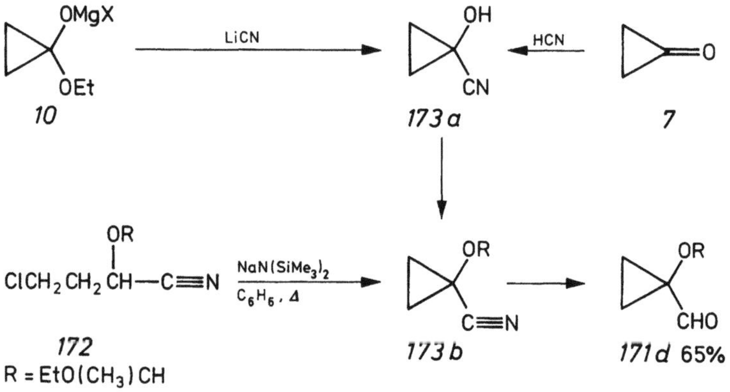

(56)

The protected 1-hydroxycyclopropanecarboxaldehyde *171* is an efficient and convenient precursor to 1-donor substituted vinylcyclopropane derivatives. Thus, for instance simple Wittig reaction of the aldehyde *171b* with cyclohexylidenephosphorane prepared from the corresponding phosphonium iodide and potassium *t*-butoxide in THF gave the vinylcyclopropane *174* in high yields Eq. (55), a precursor of spirovetivanes (*vide infra*), which could not be obtained by the ring opening of the oxaspiropentane *110* (*vide supra*, Sect. 4.5, Eq. (33)) [111].

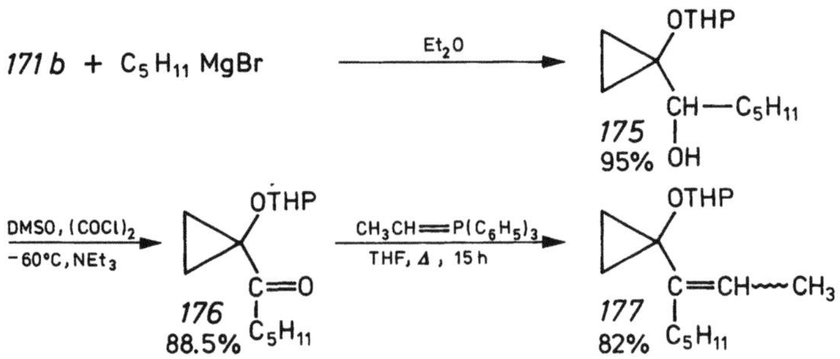

(57)

Another approach to such vinylcyclopropanes involved the addition of Grignard reagents, *e.g.* *n*-amylmagnesium bromide to *171b* leading to the cyclopropylcarbinol *175*. Oxidation at —60 °C in methylene chloride with DMSO-$(COCl)_2$ [109] led to the ketone *176*, which was treated with ethylidenetriphenylphosphorane for instance to produce in 69% overall yield from *171b* a (Z, E) mixture of the disubstituted vinylcyclopropanes *177*, precursors of dihydrojasmone (*vide infra*, Sect. 5.5).

4.8 From Optically Active 1-Hydroxyvinylcyclopropanes

The cyclopropane moiety is present in a large number of natural products [116] and pharmaceutically interesting compounds [117], most of which are optically active. For instance, such units have been found in the side chain of sterols from marine sources [118] and have been synthesized along a stereocontrolled route from (+)α-pinene involving the stereospecific ring contraction of a cyclobutanol p-toluenesulfonate [119]. Moreover, optically active vinylcyclopropane units have been found in Hormosirene (or Dictyopterene B) *178*, a specific sex attractant of several brown algae of the Australian shelf, which displays an intense ocean smell and in Dictyopterene A *179*, a minor constituent of the pheromone bouquets [120].

(−) *178* (1R,2R)

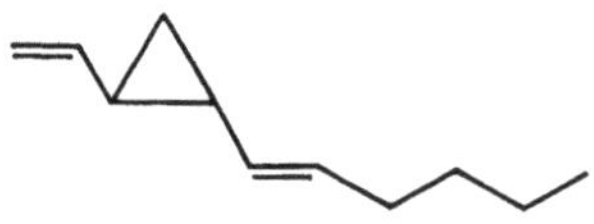
(+) *179* (1R,2R)

The synthesis of optically active *178* and *179* have been performed starting with a resolution of the *trans* 2-vinylcyclopropanecarboxylic acid [121]. Highly stereoselective S_CN' reactions of chiral homoallylic esters derived from (+)camphor [122], and of optically active allylic [123] benzoates involving chirality transfer have also been reported for the enantioselective preparation of *178* and *179*, respectively. Optically active 1-hydroxycyclopropylcarbonyl compounds and 1-hydroxyvinylcyclopropane derivatives were available from chiral succinic esters, following the synthetic scheme displayed in equations 52–55 (*vide supra*, Sect. 4.7). Thus, (+)-(R)dimethyl methylsuccinate *180* (97% e.e.) readily obtained by enzymatic resolution of the racemic ester [124] upon acyloin condensation in the presence of $ClSiMe_3$ led to the (+)-(R)-3-methyl-1,2-bis(trimethylsiloxy) cyclobutene *181* ($[\alpha]_D$ = 19.05°, c = 2.34 (CCl_4)). Then, bromination and subsequent hydrolysis of *181* [95] quantitatively gave, after continuous extraction with ether, (—)-(1S,2R)-1-hydroxy-2-methylcyclopropanecarboxylic acid *182a* ($[\alpha]_D$ = —57°, c = 1.38 (CCl_4)) which was esterified with methanol and thionyl chloride to give *182b* ($[\alpha]_D$ = —32.76°, c = 1.4 (CCl_4)). Determination of the enantiomeric ratios by ^{1}H NMR in the presence of $Eu(hfc)_3$ [126] has shown that the ester *182b* had >97% e.e. [125]. Therefore, practically no racemisation *i.e.*, no enolisation occurred during the acyloin condensation of *180* which involved radical anions as intermediates [93]. Protection of the hydroxy group of *182b* ($tBuMe_2SiCl$, imidazole, DMF) [105], reduction with DIBAH in toluene at —70 °C followed by oxidation (DMSO-$(COCl)_2$) [109] (*vide supra*, Sect. 4.8, Eq. (55)) gave the aldehyde *183* ($[\alpha]_D$ = —45°, c = 2.1 ($CHCl_3$)) in 94% overall yield. Subsequent addition of triethylphosphoacetatecarbanion in THF led in 88% yield to the optically active (—)-(1S, 2R) 1-siloxyvinylcyclopropane *184* a ($[\alpha]_D = -2.45°$, c = 2 (CCl_4)) and after removal of the protective group (CH_3OH, $ClSiMe_3$), to the trans-1-vinylcyclopropanol *184b*, whose ^1H NMR analysis in the presence of $Eu(hfc)_3$ [126] showed an enantiomeric excess of 87%, Eq. (58) [125].

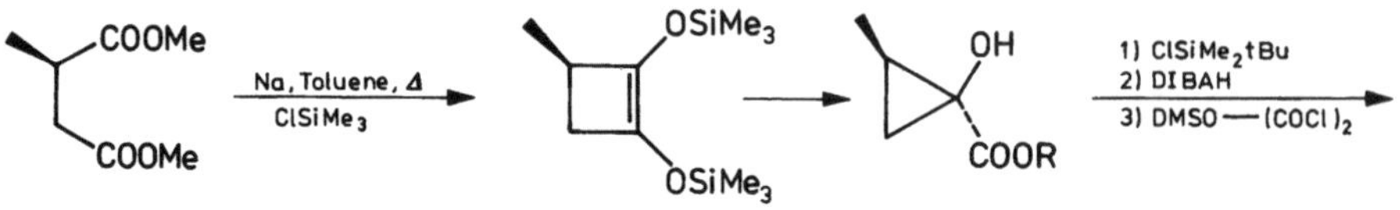

(+)-(R)-*180* (97%ee) (+)-(R)-*181* 89% (−)-(1S, 2R)-*182a* R = H
b = CH_3 (97%ee)

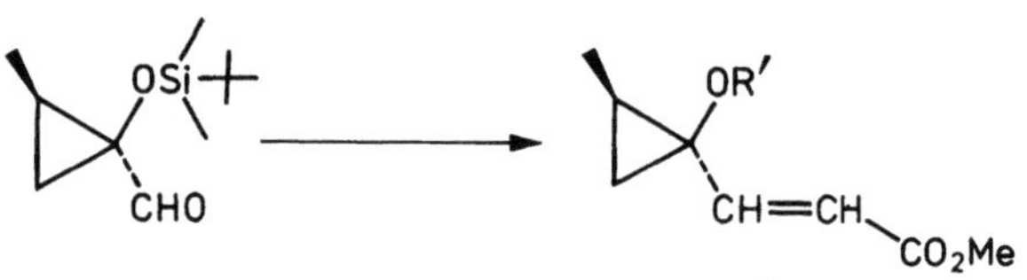

(58)

(−)-(1S, 2R)-*183* (−)-(1S, 2R)-*184a* R′ = t-BuMe₂Si
b = H (>87%ee)

4.9 Miscellaneous Methods

Lithium (phenylthiocyclopropylcuprate) *185*, prepared by the reaction of cuprous thiophenolate with cyclopropyllithium in THF at −78 °C, was added to β-iodo-enones. For example, treatment with 3-iodo-2-cyclohexen-1-one *186* provided the corresponding β-cyclopropyl α,β-unsaturated ketones *187* with high efficiency, Eq. (59) [127].

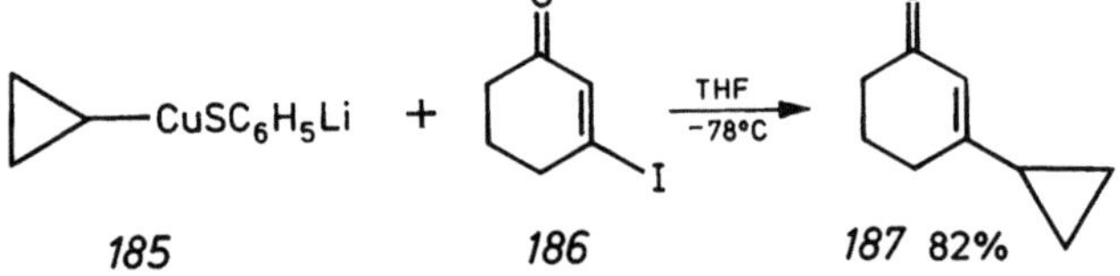

(59)

185 *186* *187* 82%

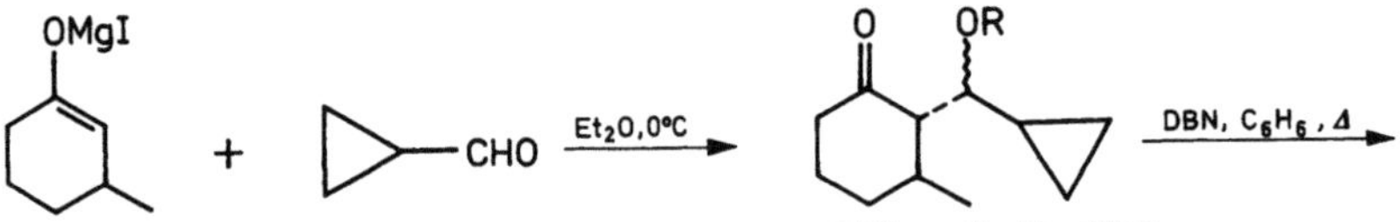

188 *189* *190a* R = H 97%
b = Ac 98%

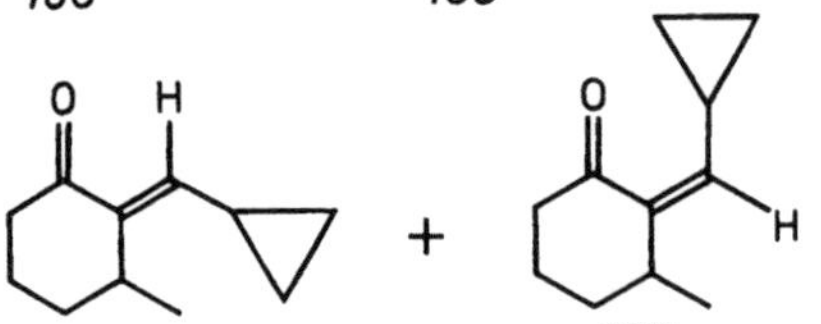

(60)

191 (13 : 1) *192*
78%

Otherwise, cuprous iodide-catalyzed addition of methylmagnesium iodide to 2-cyclohexen-1-one in ether at 0 °C, followed by trapping of the resultant enolate anion *188* with cyclopropanecarboxaldehyde *189*, afforded the two diastereomers of the cyclopropylcarbinol *190a*. Further transformation into the corresponding acetates *190b* (acetic anhydride, pyridine), followed by treatment with 1,5-diazabicyclo[4.3.0]-non-5-ene (DBN) in refluxing benzene, provided in 78 % yield, a mixture of the desired β-cyclopropyl enones *191* and *192*, in a ratio of 13:1, Eq. (60) [127].

In analogy to the preparation of the cyclopropanone hemiacetal *3* (*vide supra*, Eq. (1) [7]), reductive cyclization of the piperidide of 3-chloropropionic acid *193* with sodium in ether in the presence of $ClSiMe_3$, gave the 1-piperidino-1-trimethylsilyloxy-cyclopropane *194a* which was converted to the cyclopropanol *194b* upon treatment with methanolic tetrabutylammonium fluoride. Both *194a* and *194b* can be used for the ready generation of cyclopropane derivatives; thus the silyl ether *194a* could be reacted directly with the vinylic Grignard reagents *195* to provide the vinyl cyclopropane derivative *196*, (Eq. (61) [128]).

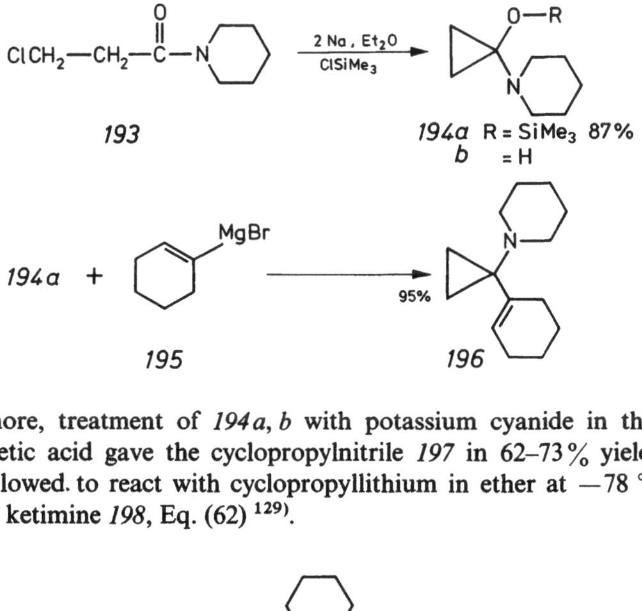

(61)

Furthermore, treatment of *194a, b* with potassium cyanide in the presence of aqueous acetic acid gave the cyclopropylnitrile *197* in 62–73 % yield. This nitrile was then allowed to react with cyclopropyllithium in ether at —78 °C to give the cyclopropyl ketimine *198*, Eq. (62) [129].

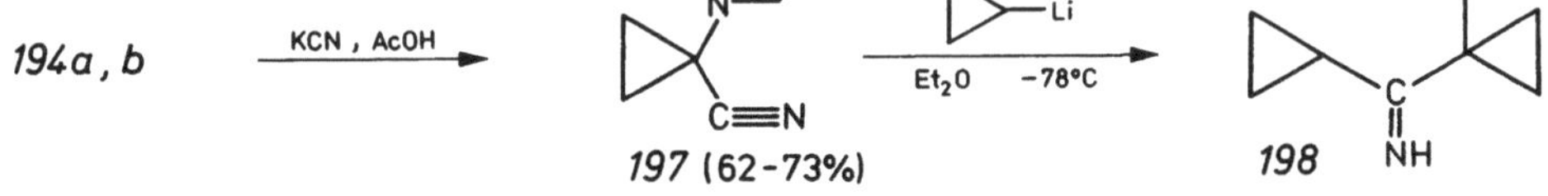

(62)

While the nucleophilic addition of 1-lithio-1-bromocyclopropanes to ketones gave oxaspiropentanes, precursors of 1-donor substited vinylcyclopropane derivatives (*vide supra*, Sect. 4.5, Eq. (28)), addition of n-BuLi at low temperature to 1,1-dibromocyclopropane *199* (prepared in 75 % yield from the addition of dibromocarbene to isobutene) led to the carbenoïd *200* which was added to N-methylformanilide to give the 1-bromocyclopropanecarboxaldehyde *201*. Wittig olefination with triphenylmethylenephosphorane produced the 1-bromovinylcyclopropane *202*, and upon treatment either with *n*- or *t*-butyllithium, the 1-vinylcyclopropyllithium *203*, which was able to undergo electrophilic substitution. For instance, addition of carbon dioxide and diazomethane, gave the cyclopropanecarboxylic acid *204a* and ester *204b*, successively, Eq. (63) [130].

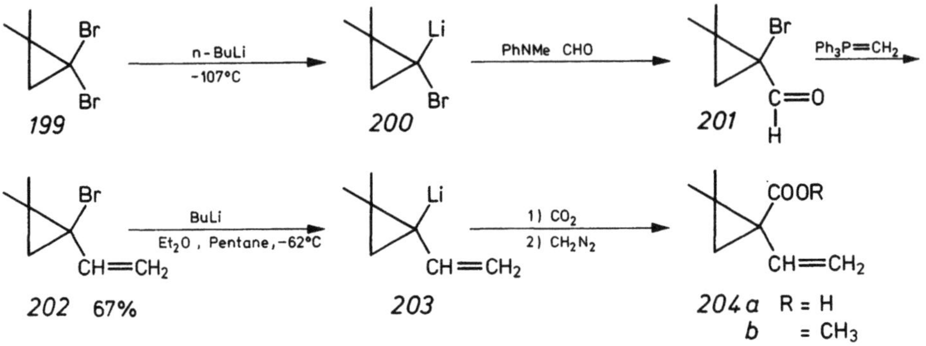

(63)

5 Reactivity and Synthetic Applications of 1-Donor Substituted Vinylcyclopropanes and Cyclopropylcarbonyl Compounds

5.1 Involving $C_3 \rightarrow C_4$ Ring Expansions

The ring enlargement of properly activated cyclopropane moities is now one of the major methods of forming cyclobutane derivatives. Thus, cyclopropyl groups adjacent to an electron deficient center X (carbon or heteroatom) underwent $C_3 \rightarrow C_4$ ring expansion to four-membered ring derivatives, involving cation, radical or carbene intermediates. Some of these reactions, however, afforded mixtures of cyclobutyl, cyclopropylmethyl and 3-butenyl compounds, which severely limited their synthetic applicability [131]. On the other hand, cyclopropanes *205* with an electron donor substituent Y on the same carbon underwent specific and thereby synthetically valuable $C_3 \rightarrow C_4$ ring expansion to cyclobutanones *206* or related four-membered ring compounds, Eq. (64) [43,132].

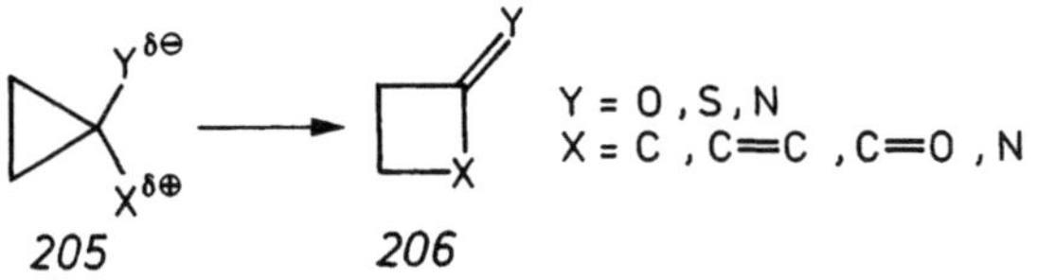

(64)

Indeed, cyclobutanones constitute challenging synthetic intermediates for a whole host of applications including cyclopentanone, cyclohexanone and cyclooctanone formation, olefin and diene synthesis, geminal alkylation, reductive acylation, among others, ... (*vide infra*).

5.1.1 From 1-Vinylcyclopropanol and Vinylogous Derivatives

First of all, it has been shown that the 1-vinylcyclopropanols *207* undergo ring expansion to cyclobutanones *209* with a large variety of electrophilic reagents. Thus, with hydrobromic and perbenzoic acid and with t-butylhypochlorite, 2-alkyl-, -2-hydroxymethyl- and 2-chloroalkylcyclobutanones *209*, respectively, were obtained, through 1-hydroxycyclopropylcarbinyl cation intermediates *208* Eq. (65) [10,133].

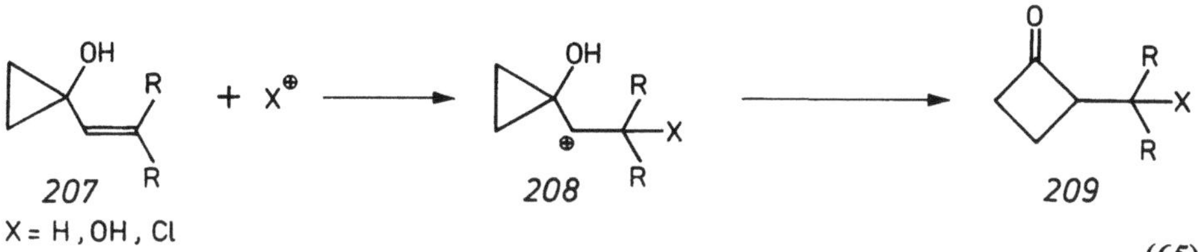

(65)

With paraformaldehyde and dibenzylamine hydrochloride in refluxing ethanol 1-vinylcyclopropanol *68* underwent a Mannich-type reaction to yield the corresponding 2-(2-dibenzylaminoethyl)cyclobutanone *210*, Eq. (66) [10].

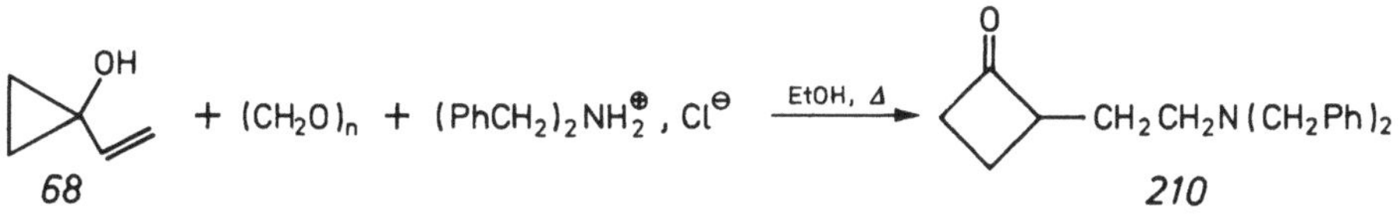

(66)

When 2-methyl-1-vinylcyclopropanol *211* was treated with concentrated sulfuric acid at 0 °C, the rearrangement took place to the extend of 40% within 5 min to give a mixture of 2,4-trans- *212*, 2,4-cis- *213*, 2,3-trans- *214* and 2,3-cis-dimethylcyclobutanone *215* Eq. (67) [133].

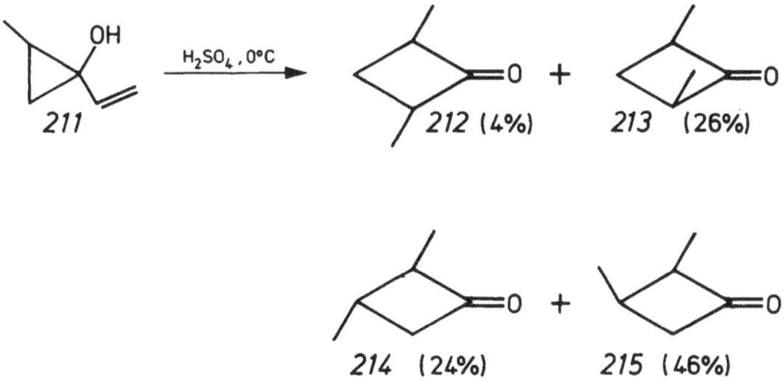

(67)

In a non polar solvent, the acid-catalyzed ring expansion of *211* took a different course. Thus, when the cyclopropanol *211* was treated with dry HBr in methylene chloride at 0 °C for 5 min, only the 2,3-dimethylcyclobutanones *214* and *215* were formed in 83% yield with a trans/cis ratio (*214*/*215*) of 3:1. It has been verified that under all the acidic conditions employed, the 2,3 and 2,4-isomers did not interconvert;

starting from either 2,3-isomer, cis-trans isomerization occurred in the HBr/CH_2Cl_2 medium and resulted in the formation of product mixtures with the same trans/cis ratio [133]. The observation that 2,3-dimethyl-substituted products apparently were favored in the acid-catalyzed rearrangement was consistent with a preferred migration of the more highly substituted carbon atom, also observed in the peracid oxidation of methylenecyclopropanes [134]. The vinyl carbinol *211* was recovered unchanged after heating at 110 °C for 17 hr in pyrolysis tubes which had been previously washed with concentrated ammonium hydroxide solution [133]. On the other hand, on heating at 100 °C either in the liquid phase (sealed tube, not washed with base) or in the gas phase (gas chromatography), the vinyl cyclopropanols *74a, b* were converted, in nearly quantitative yield, into the corresponding spirocyclobutanones *216a, b* (Eq. (68)) [41].

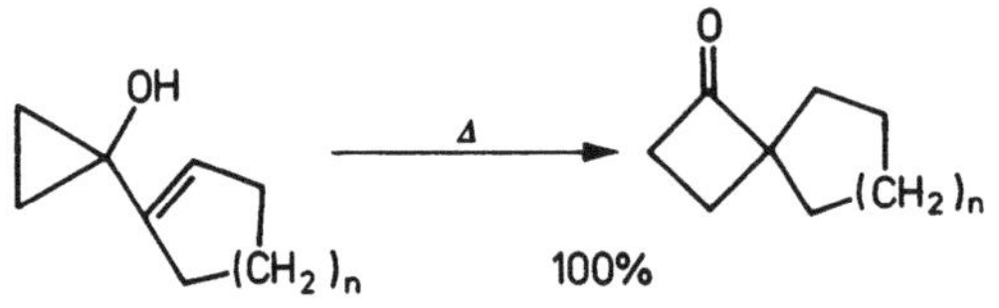

74a, n = 1
b, n = 2

216a, b

(68)

The stereochemistry and mechanism of such a thermally induced ring expansion have been investigated by rearranging a deuterium labelled cyclopropanol. As a matter of fact, a thermally allowed $[2_a + 2_a + 2_s]$ concerted process [135] would imply an *anti-addition* to the double bond of *74a* (hydroxyl H—C^5 and C^3—C^4 bonding) (Fig. 1). On the other hand, an intramolecular addition of the hydroxyl proton to the π-orbital of the double bond at C^5, would result in the formation of a positive charge at C^4. As it has been shown that the delocalization of the positive charge of the cyclopropylcarbinyl cation would prevent the free rotation around the C^1—C^4 axis [136], such a process would imply a *syn-addition* (Fig. 2). Finally an intermolecular transfer of a proton on the double bond at C^5, analogous to the electrophilic addition (*vide supra*, Eq. (65)) would probably lead to a mixture of products of *syn-* and *anti-addition* (Fig. 3).

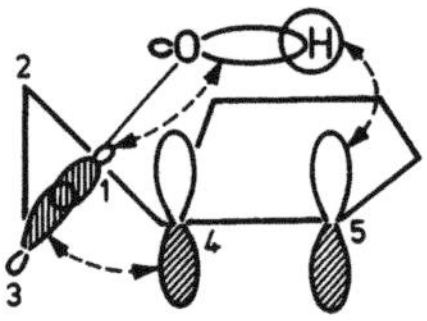

Fig. 1

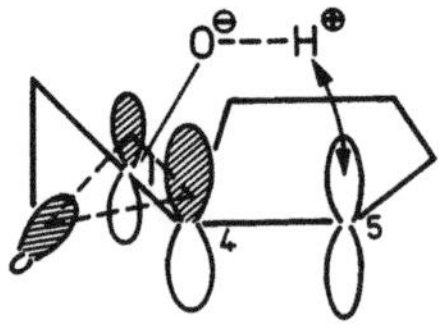

Fig. 2

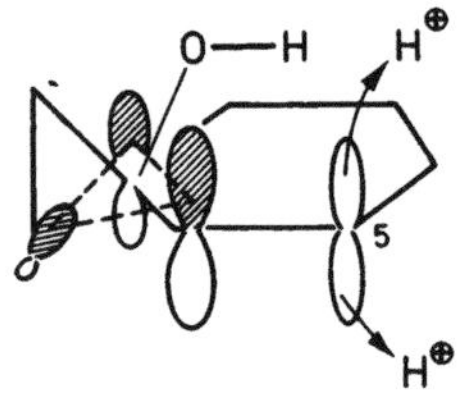

Fig. 3

The thermal rearrangement in a sealed tube at 100 °C of the $[D_5]$-labelled cyclopropanol *217* (prepared from perdeutero-α,α'dichloroacetone, (*vide supra*, Sect. 4.2, Eq. (23)) led to the expected 2,2,3,3,5-pentadeuteriospiro[3.4] octan-1-one *218*, Eq. (69) [41]

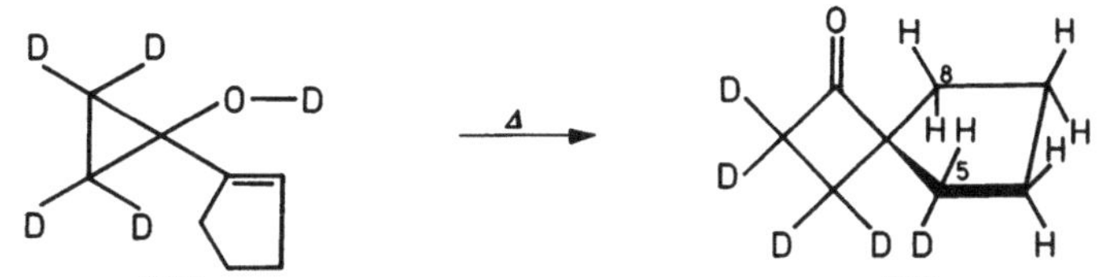

217 218 (69)

The *anti*-orientation of the deuterium at C^5 on the spirocyclobutanone *218* was proven unambiguously by examination of the ^1H NMR spectra of this ketone coordinated with tris-(dipivalomethano)europium; effectively, two protons at C^5 and C^8 were markedly shifted downfield ($\Delta\delta = 3$ ppm) implying their syn orientation with respect to the carbonyl group. Therefore, the thermal ring enlargement *217* → *218* seemed to involve an intramolecular stereospecific cis-addition on the double bond (Fig. 2) which, consequently ruled out a concerted process [41].

Addition of triethylphosphonoacetate carbanion in THF to the 1-tetrahydropyranyloxycyclopropanecarboxaldehyde *171b* gave the trans-vinylcyclopropanecarboxylate *219* in 88% yield; reduction of the ester with DIBAH to minimize conju-

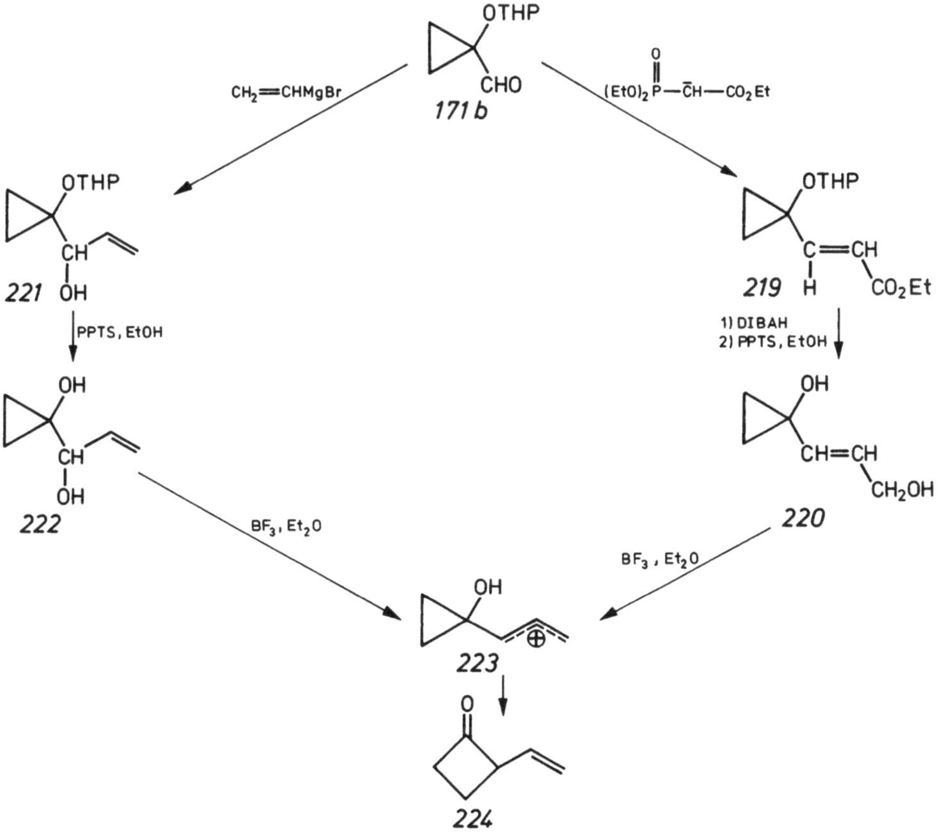

Scheme 1. Synthesis of 2-vinylcyclobutanone [96]

gate reduction, deprotection of the OH group (PPTS, EtOH) [101] led to the allylic diol *220*. On the other hand, addition of vinylmagnesium bromide to aldehyde *171b* provided the allyl alcohol *221* and after removal of the THP group [101] the 1-(1-vinylcarbinol)cyclopropanol *222*. Upon addition of a catalytic amount of BF_3—Et_2O both vinylogous diols *222* and *220* underwent quantitative dehydration and $C_3 \rightarrow C_4$ ring expansion to 2-vinylcyclobutanone *224*, within 15 min at room temperature as monitored by t.l.c., most likely via the intermediate formation of the same cyclopropylcarbinyl cation *223*, (Scheme 1) [96].

Peterson olefination of the aldehyde *171b* with 1-trimethylsilylprop-3-ynyl lithium [137], furnished a (Z, E) mixture of enynes *225* in 85% yield. Desilylation with KF in DMF [138], metalation with n-butyllithium, condensation with hexanal and lithium aluminum hydride reduction gave the conjugated dienol *226* in 72% overall yield from *171b*. Treatment of *226* with 10 mol. % of PPTS [101] in ethanol at 55 °C led directly, within 30 min, as monitored by t.l.c. to (E,E)-2-(1,3-nonadienyl)-cyclobutanone *227* (Eq. (70)) [96]. This $C_3 \rightarrow C_4$ ring enlargement occurred under rather mildly acidic conditions since the pH of a 1 M aqueous solution of PPTS is 3.0 [101].

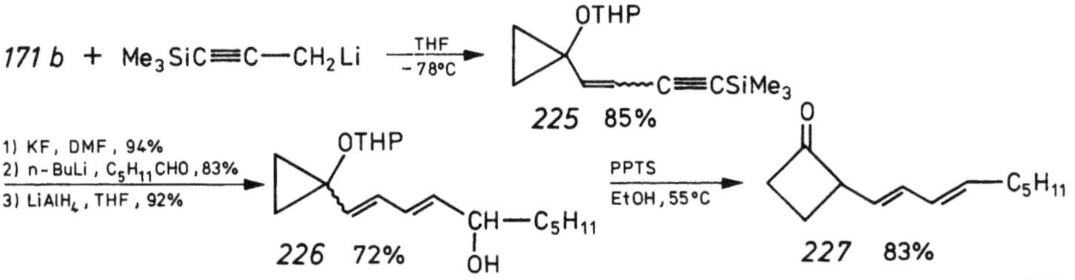

(70)

Furthermore, such a $C_3 \rightarrow C_4$ ring expansion could even be induced by lithium chloride. Thus, the cyclopropylcarbinol *228*, prepared by addition of acetylenic Grignard reagents to the cyclopropanecarboxaldehyde *171a* in 80–90% yield [110], was transformed into the tosylate *229* upon successive treatment with one equivalent of methyllithium in *ether* at 0 °C and with one equivalent of tosyl chloride at —40 °C, lithium chloride being formed as by-product. The formation of tosylate *229* appeared, however, to be strongly dependent upon the nature of the solvent; effectively, the same

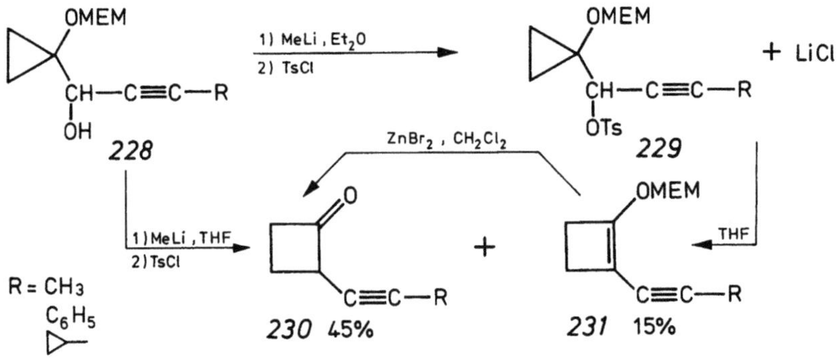

(71)

reaction (*i.e.* addition of CH_3Li and tosyl chloride to *228*) carried out in *tetrahydrofuran*, led directly to the 2-alkynylcyclobutanone *230* (45% yield) and to the 1-methoxyethoxymethyl-2-alkynylcyclobutenyl ether *231* (15%), Eq. (71) [110].

Moreover, removal of ether under vacuum and addition of tetrahydrofuran to the crude mixture of tosylates *229* and LiCl effected the ring expansion to cyclobutanones *230* and cyclobutenyl ethers *231*. Upon treatment with anhydrous zinc bromide in methylene chloride, enol ethers *231* also underwent cleavage to give the expected cyclobutanones *230*. In this way 2-alkynylcyclobutanones *230* with different substituents on the triple bond (i.e., R = CH_3, C_6H_5, cyclopropyl, etc.) were obtained in 55–60% yield [110]. It was obvious from these experiments that LiCl in THF was effective to cleave the MEM ethers *228*, while MEM ethers usually require zinc-bromide [103–104], and induce their ring expansion. This likely involves, after ionization of the tosylates in THF, the intermediacy of the cyclobutyl cyclopropylcarbinyl carbenium ion system *232* [110].

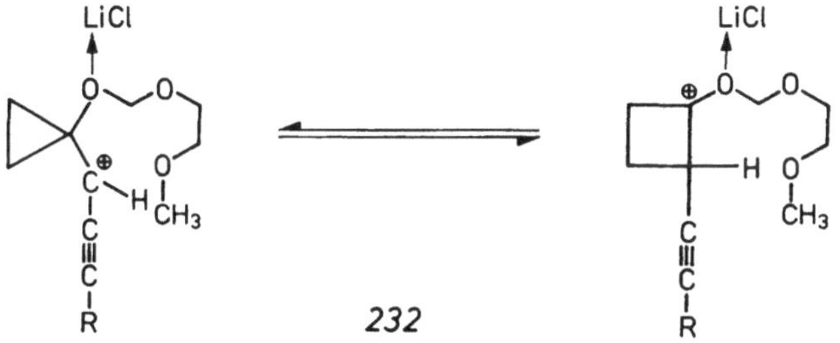

232

5.1.2 From 1-Arylthiocyclopropyl Derivatives

The adducts of 1-arylthiocyclopropyllithium *116* [61] to aldehydes and ketones, upon treatment with p-toluenesulfonic acid in refluxing benzene under anhydrous conditions or with the Burgess reagent [67], underwent ring expansion to 1-phenylthiocyclobutenes, which may be hydrolyzed to cyclobutanones, desulfurized to cyclobutanes or thermolyzed to dienes. Thus, the cyclopropylcarbinol *233*, adduct of *116a* to *t*-butyl methyl ketone, was rearranged to the cyclobutanone enol thioether *234*, Eq. (72) [139].

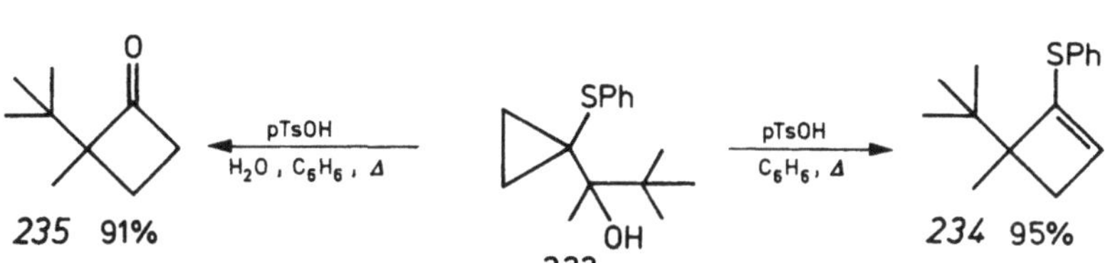

235 91% *233* *234* 95% (72)

Direct rearrangement to cyclobutanones can be achieved under four sets of conditions: a) with 48% aqueous fluoroboric acid in ether, at room temperature; b) with anhydrous stannic chloride in methylene chloride, normally at room temperature; c) with p-toluenesulfonic acid in benzene saturated with water, at reflux and d) with trimethyloxonium tetrafluoroborate in methylene chloride (Meerwein's reagent). For instance, the cyclopropylcarbinol *233* was rearranged to 2-*t*-butyl-2-methylcyclobutanone *235*, upon treatment with one equivalent of pTsOH in water-saturated benzene at reflux for 1.5 hr, Eq. (72) [139]. In a slightly different approach, the alcohol was converted into a better leaving group to allow ring expansion under milder conditions. Thus, the alkoxide generated by the addition of *116a* to cyclohexanone was quenched with O-phenylenephosphorochlorodite to give the phosphite *236*. In contrast to the tosylate *229* (*vida supra*, Sect. 5.1.1, Eq. (71)), simple warming of these adducts in THF did not lead to rearrangement; the addition of a catalytic amount of concentrated HCl to aqueous THF at reflux for 15 hr was necessary to achieve ring enlargement to the spiro[3.5]non-1-one *216b*, Eq. (73) [139]. (See for comparison Eqs. (68) and (71)).

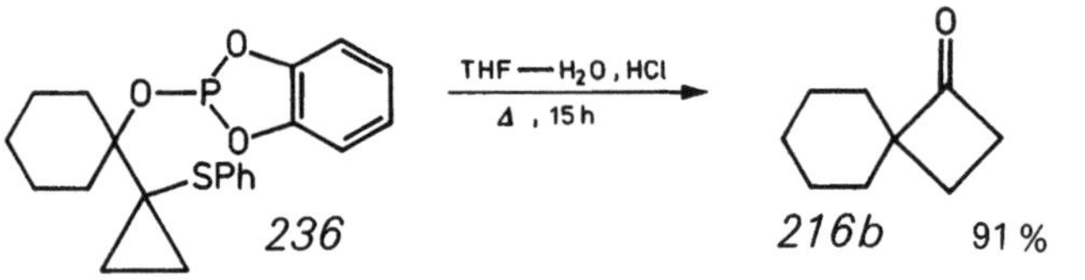

(73)

The utility of this $C_3 \rightarrow C_4$ ring expansion was demonstrated by the total synthesis of (±) grandisol *241* a constituent of the sex pheromone of the bool weevil [140], based on a double cyclobutanone annelation (237 → 240) using *116a* as spiroannelating reagent, followed by a stereospecific haloform cleavage of a cyclobutanone ring, allowing the introduction of the side chains of this monoterpene with the proper cis stereochemistry (Scheme 2) [139]. For a comparable synthesis of *241* also based on a $C_3 \rightarrow C_4$ ring expansion see Ref. [43].

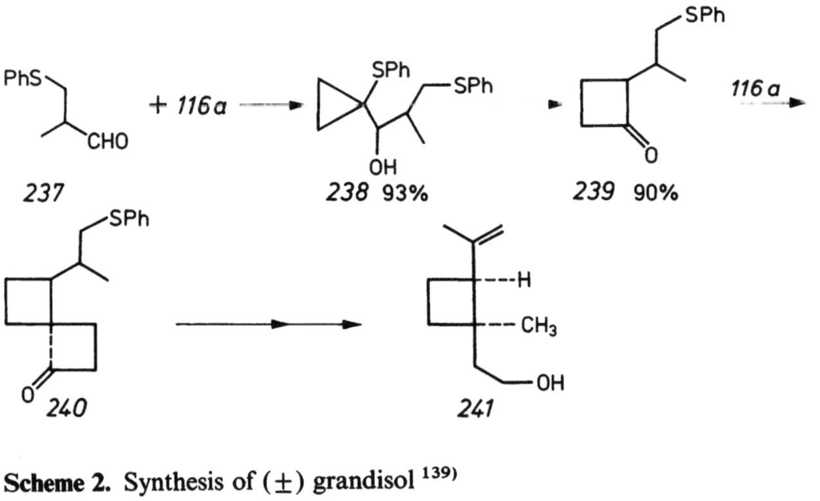

Scheme 2. Synthesis of (±) grandisol [139]

The challenging 2-vinylcyclobutanone system (*vide infra*, Sects. 5.3 and 5.4) was also available from the $C_3 \rightarrow C_4$ ring expansion of 1-arylthiocyclopropyl vinylcarbinols *242*. Thus, upon addition to a mixture of ether and 48% fluoroboric acid, *242* gave rise to the 2-vinylcyclobutanone derivative *243* in most cases, however, in moderate yields, (Eq. (74)) [63].

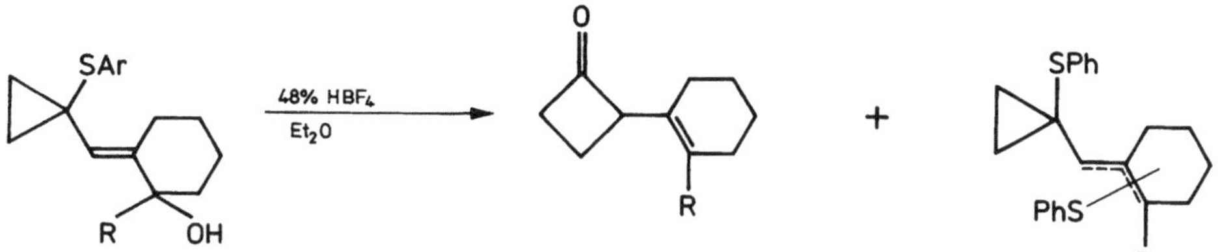

242 *243* 20-60% *244*

a Ar = Ph, R = H, C_4H_9
b Ar = 3,6-$(MeO)_2C_6H_3$

(74)

Since thiophenol was formed in the reaction, this by-product trapped the intermediate cation to give the bis(phenylthio)vinylcyclopropane *244* and so limited the formation of the desired cyclobutanone. To overcome this problem, a substitution pattern providing electronic acceleration for the cyclopropyl bond migration but also a steric bulk to inhibit the nucleophilicity of the thiol was required. For this purpose, 1-(2,6-dimethoxyphenylthio)vinylcyclopropanes such as *242b* were prepared; the yield and cleanliness of the reaction were effectively increased, allowing by this route the isolation of pure cyclobutanones *243* [63].

It has recently been reported that β-(1-phenylthio)cyclopropyl enones *245* were more conveniently prepared from the lithium salt of β-hydroxymethylene ketone and the phenylthiocyclopropyllithium *116a* (*vida supra*, Sect. 4.6.1); upon treatment with Lewis acids (e.g., $AlCl_3$, $SnCl_4$, $TiCl_4$, etc.) in CH_2Cl_2 or better with refluxing 50% aqueous trifluoroacetic acid, they were converted to γ-ketocyclobutanones such as *246*, Eq. (75) [64].

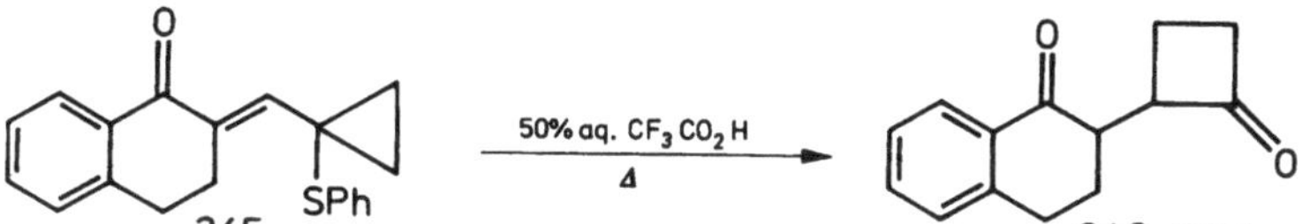

245 *246* 95%

(75)

For a recent review see Ref. [63b].

5.1.3 From 1-Methylseleno- and 1-Phenylselenocyclopropyl Derivatives

It has been shown that 1-methyl (or phenyl) selenocyclopropylcarbinol derivatives, analogous to arylthiocyclopropylcarbinols, undergo either acid induced dehydration into 1-selenovinylcyclopropanes (*vide supra*, Sect. 4.6.2, Eq. (38)) or directly ring expansion, upon treatment with the Burgess reagent ($CH_3O_2CN^-$, $SO_2N^+Et_3$, toluene, 110°, 4 hrs) [67], *e.g.* *247* to the regioisomeric selenocyclobutenes *248* and *249*, Eq. (76) [68].

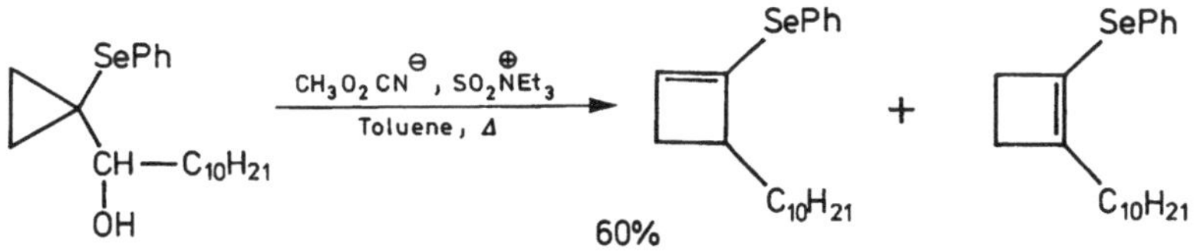

247 *248* (15:85) *249*

(76)

An unusual reactivity was observed for the 1-methylselenocyclopropylcarbinol *250* bearing a p-tolyl group α to the hydroxyl function. It underwent ring enlargement to 2-methyl-2-p-tolylcyclobutanone *251* upon simple treatment with p-TsOH in refluxing aqueous benzene for 12 hr. Substituted by a hydrogen or an alkyl group, the corresponding 1-methylselenocyclopropylcarbinol remains unchanged under these conditions. The cyclobutanone *251* was then treated with the bulky 2-lithio-2-selenopropane to provide in 66% yield the β-hydroxyselenide *252* as a 5/1 mixture of stereoisomers which was rearranged to α-Cuparenone *253* when treated with thallium ethoxide in $CHCl_3$ (57%), with silver tetrafluoroborate on alumina in CH_2Cl_2 (69%), or with methyl fluorosulfonate in ether, (82%), (Scheme 3) [141].

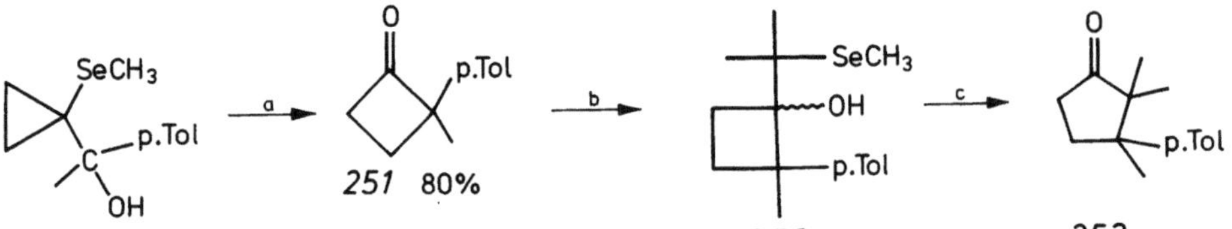

250 252 253

Scheme 3. Synthesis of α-Cuparenone

Reagents: a) pTsOH, C_6H_6—H_2O, 80 °C, 12 hr, 80%; b) $Me_2CSeMeLi$, Et_2O, −78 °C, 66%; c) TlOEt, $HCCl_3$, 20 °C, 21 hr, 57%; or $AgBF_4$, Al_2O_3, CH_2Cl_2, 20 °C, 3 hr, 69%; or CH_3OSO_2F, ether, 20 °C, 1 hr, 82%.

β-Cuparenone was also synthetized following a similar strategy [141]. The synthetic applications of 1-metallo-1-selenocyclopropanes have recently been reviewed [70b].

5.1.4 From 1-Alkoxycyclopropyl Derivatives

The replacement of the phenylthio- or phenylseleno group by an alkoxy group gave, as expected, better results in the formation of cyclobutanone derivatives due to the greater ability of oxygen to stabilize a positive charge and the ready acid-catalyzed hydrolysis of the intermediate enol ether. Thus, the addition products of 1-alkoxy-cyclopropyllithium *140* to conjugated aldehydes or ketones were treated directly with 10% HBF_4 in wet THF (one volume of aqueous 48% HBF_4 mixed with 4 volumes of THF) to yield 2-vinylcyclobutanones in satisfactory yields (*vide supra*, Sect. 4.6.3, Eq. (44)). For other synthetic applications of the ring expansion of alkoxycyclopropanes see Ref. [43b].

5.2 Involving $C_3 \rightarrow C_4 \rightarrow C_5$ Ring Expansions

The rearrangements reported herein concern only the 2-vinylcyclobutanone derivatives which have also been obtained from the cycloaddition of vinylketenes to simple olefins [142,143]. For such a stepwise $C_3 \rightarrow C_4 \rightarrow C_5$ ring enlargement, not involving a 2-vinylcyclobutanone, see for instance Scheme 3 (Sect. 5.1.3).

5.2.1 Acid Induced

As a matter of fact, 2-vinylcyclobutanones were able to undergo further $C_4 \rightarrow C_5$ or $C_4 \rightarrow C_6$ ring expansions depending on the nature of the substituents on the system.

Thus, for instance, in the presence of a 10:1 mixture of methanesulfonic acid and phosphorus pentoxide (Eaton's reagent)[144], the 1-methyl-1-vinylcyclobutanone *254b* underwent predominantly $C_4 \rightarrow C_5$ ring enlargement to the spirocyclopentenone *257b* (51 %) and to a lesser extent $C_4 \rightarrow C_6$ ring expansion to the fused cyclohexenone *261b* (13 %); while the normethyl analogue *254a* yielded exclusively the 1,3-rearrangement product *261a* (65 %) (Eq. (77))[145].

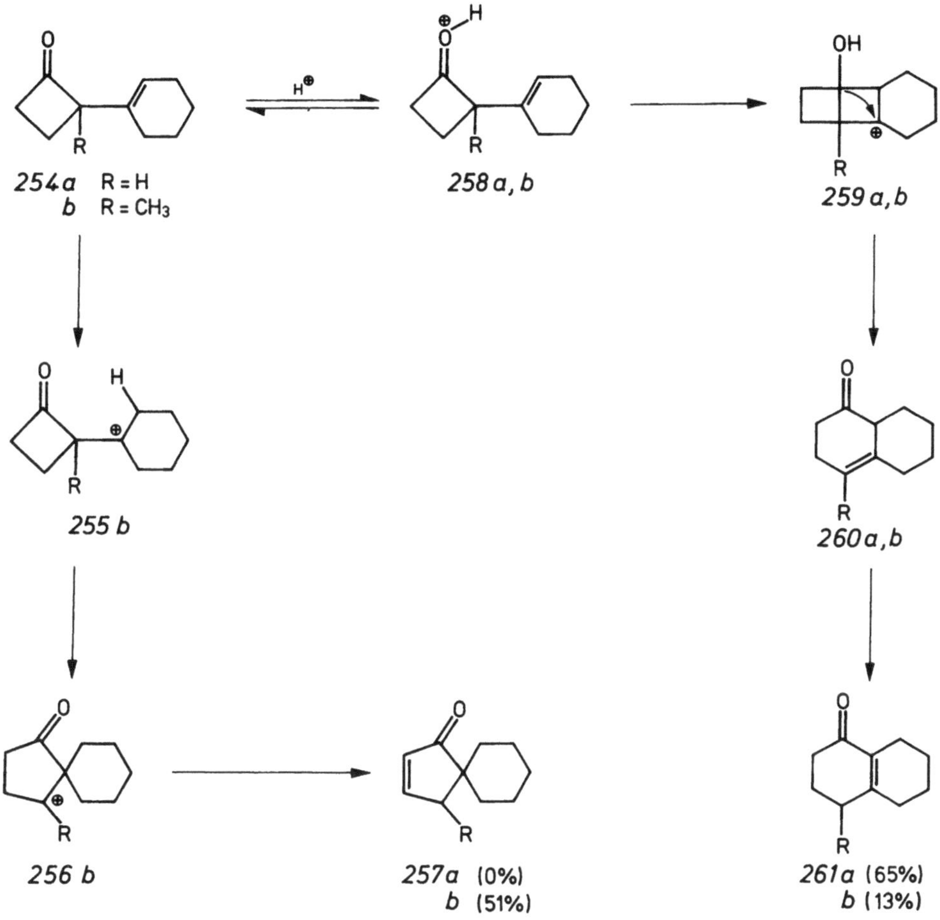

(77)

This result can be rationalized assuming protonation of the double bond of *254b* to lead to the cyclobutylcarbinyl cation *255b*, which subsequently rearranged to the tertiary cyclopentyl cation *256b* and by deprotonation and double bond migration the conjugated spirocyclopentenone *257b* was produced. On the other hand, protonation of the carbonyl group and electrophilic attack of the resulting cation *258a, b* on the olefinic linkage can lead to the strained bicyclic cation *259a, b* which would be expected to spring open to *260a, b*. Finally acid catalyzed migration of the double bond into the conjugated position led to *261a, b* (Eq. (77))[145].

The 2-alkyl-2-vinylcyclobutanones required for this rearrangement were also readily prepared under milder conditions from cyclopropanol derivatives. Thus, the vinylogous alcohols *262a, b* and *263a, b* were obtained from the cyclopropyl methyl ketone *164* (*vide supra*, Sect. 4.7, Eq. (53)) following the sequence used for the preparation of 2-vinylcyclobutanone (*vide supra*, Sect. 5.11, Scheme 1). Upon treatment either with 0.1 equivalent of BF_3-Et_2O in $CHCl_3$ at room temperature for 15 min, or with 0.1 equivalent of the 10:1 mixture $MeSO_3H—P_2O_5$ (Eaton's reagent) [144] in ether at room temperature for 5 min both alcohols were converted quantitatively into the 2-vinylcyclobutanones *264a, b* (R' = H, C_4H_9). Furthermore, treatment of neat *262a, b* or *263a, b* with 10:1 $MeSO_3H—P_2O_5$ [144] (17 equiv.) at room temperature led directly, (as did the cyclobutanones *264a, b* under the same conditions) either to dihydrojasmone *265a* (R' = C_4H_9) in 65–90% yields, or to 2,3-dimethylcyclopentenone *265b* a precursor of methylenomycin B [146], in 55–68% yield, (Scheme 4) [147]. (See scheme 11 for an other synthesis of dihydrojasmone involving a $C_3 \rightarrow C_5$ ring expansion of a 1-hydroxyvinylcyclopropane derivative).

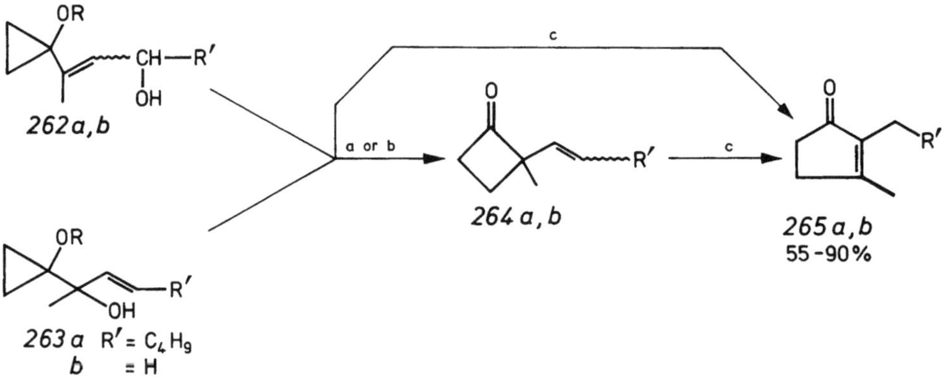

Scheme 4. Syntheses of dihydrojasmone and of a methylenomycin B precursor
Reagents: a) $BF_3—Et_2O$, $CHCl_3$, r.t., 15 min, 100%; b) $MeSO_3H—P_2O_5$, Ether, r.t., 5 mn; c) neat $CH_3SO_3H—P_2O_5$, r.t.

5.2.2 Base Induced

In practice, addition of 1-(lithioethyl)phenyl selenoxide *267* to spiro[3.5]non-5-en-1-one *266* [63,73], followed by neutralisation with acetic acid and heating in refluxing THF cleanly afforded the vinylic cyclobutanol *268* in 72% yield, as a 8:1 mixture of diastereomers. On the other hand, when the reaction mixture was refluxed in THF without prior neutralisation, an unexpected regiospecific $C_4 \rightarrow C_5$ ring expansion occured within minutes, leading to the cyclopentanone *270* and its α-selenenylated derivatives *271*. Conversion of the α-selenenylated ketone *271* into *270* could be accomplished most conveniently by treatment of the crude reaction mixture with aluminum amalgam providing the cyclopentanone *270* in 71% yield from *266*. The mechanism of this ring expansion is apparently a direct pinacol-like rearrangement of the initial adduct *269*, in which the more highly substituted carbon migrates preferentially (Eq. (78)) [148].

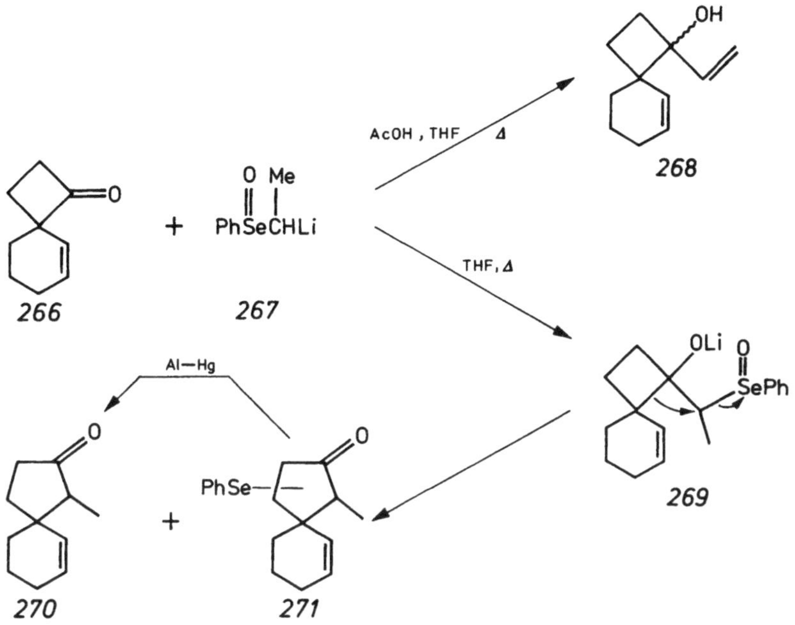

(78)

5.2.3 Photolytic

The spirocyclobutanones *272–274a, b*, incorporating a cyclohexa-, cyclohepta- and cyclooctadiene moiety respectively, have been synthetized following mainly the same methodology, *i.e.* by the acid induced (50% HBF_4) $C_3 \rightarrow C_4$ ring expansion of a 1-methoxyvinylcyclopropylcarbinol, prepared from 1-methoxycyclopropyllithium *136*. Upon photolysis in acetonitrile ((λ > 347 nm) in the presence of Michler's ketone as

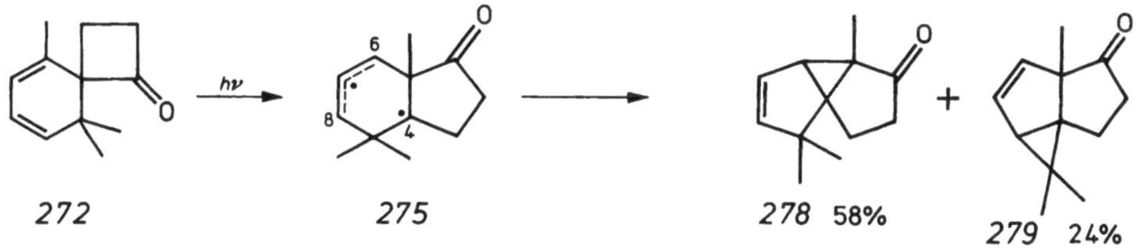

(79)

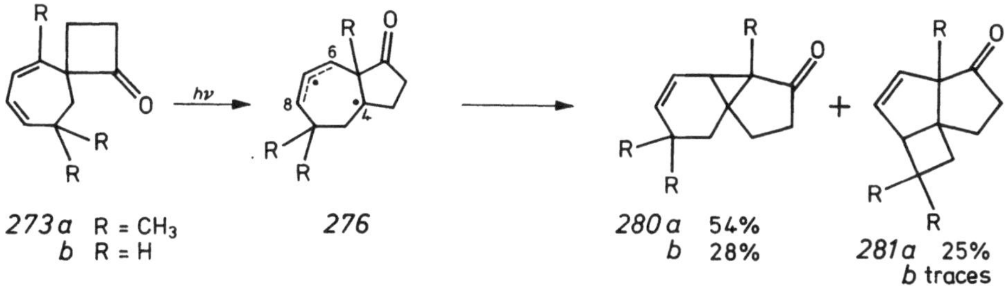

(80)

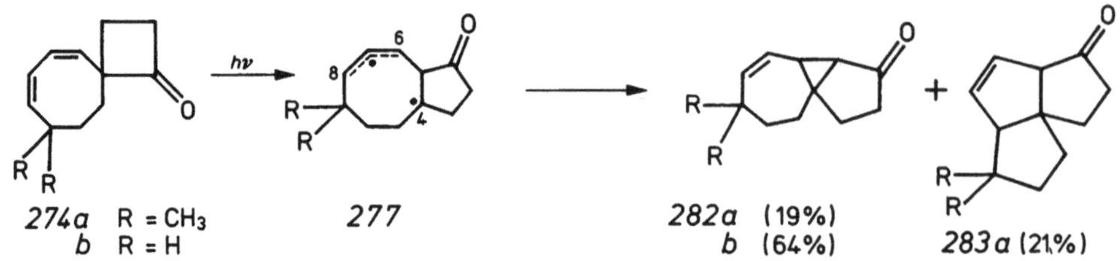

(81)

a triplet sensitizer, these 2-vinylcyclobutanones underwent an oxa-di-π-methane rearrangement [149], which included as primary steps a 1,2-acyl shift involving one or both double bonds of the diene system and leading to the intermediates *275*, *276* and *277* respectively. The tricyclic cyclopentanones *278–283* arose from these intermediates by ring closure either between C^4 and C^6 or C^4 and C^8 (Eqs. (79–81)) [150].

The presence of a gem-dimethyl group as in *272*, *273a* and *274a* dramatically changed the photoproduct distribution, since only these substrates led to the products *279*, *281a* and *283a* resulting from vinylogous ring closure. Substrates *273b* and *274b* without methyl substitution gave only products of rearrangement involving one double bond [150].

5.3 Involving $C_3 \rightarrow C_4 \rightarrow C_6$ Ring Expansions

5.3.1 Acid Induced

The acid-induced $C_3 \rightarrow C_4 \rightarrow C_6$ rearrangement has been summarized above, together with the acid induced $C_3 \rightarrow C_4 \rightarrow C_5$ ring expansions, (*vide supra*, Sect. 5.2.1, Eq. (77)).

5.3.2 Base Induced

The lithium and potassium salts of 2-vinylcyclobutanols underwent $C_4 \rightarrow C_6$ ring expansion at 25–70 °C, providing an efficient method for the synthesis of 3-cyclohexenol derivatives. Thus, reduction of 2-phenyl-2-vinylcyclobutanone *284a* with 2 equivalents of Li-sec Bu_3BH [151] in THF at —78 °C produced a mixture of diastereomeric cyclobutanol salts, which rearranged to 4-phenyl-3-cyclohexenol *286a* upon warming to room temperature. On the other hand, upon reduction of 2-methyl-2-vinylcyclobutanone *284b* with Li- or K-sec Bu_3BH, the system resisted rearrangement even at 70 °C in the presence of hexamethylphosphoric triamide. However, when this mixture was treated with a slight excess of methyllithium, the resulting salts underwent smooth rearrangement to *286b* [152]; alternatively, when *284b* was treated directly

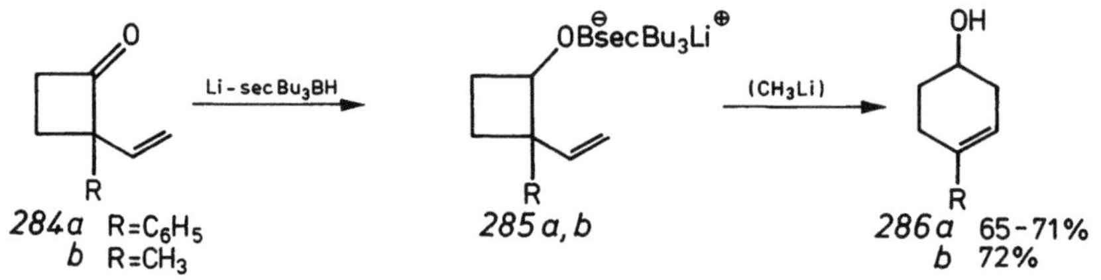

(82)

with 1.15 equiv. each of Li-sec-Bu_3BH and CH_3Li in THF—H_2O—HMPT (9:1:5) at 70 °C for 7 hr, 4-methyl-3-cyclohexenol *286b* was obtained in 72% yield, Eq. (82) [142].

Apparently, the borate complex *285a, b* generated by reduction of *284b* was stable to the rearrangement; methyllithium served to liberate the more reactive free lithium alkoxide. In fact, this rearrangement was effected in good yield at room temperature by employing the potassium vinylcyclobutanol salts, readily generated by reaction of 2-vinylcyclobutanones with 1.15 equivalents each of Li-sec-Bu_3BH and CH_3Li in the presence of excess potassium ethoxide in THF—H_2O—HMPT [142].

The utility of this stepwise $C_3 \rightarrow C_4 \rightarrow C_6$ ring expansion has been demonstrated by the synthesis of the optically active eudesmane sesquiterpene (—)-β-selinene *295* starting from the commercially available (—)-perillaldehyde *287* [153]. Thus, addition of the 1-lithio-1-methoxycyclopropane *136* (*vide supra*, Sect. 4.6.3, Eq. (44)) to *287*,

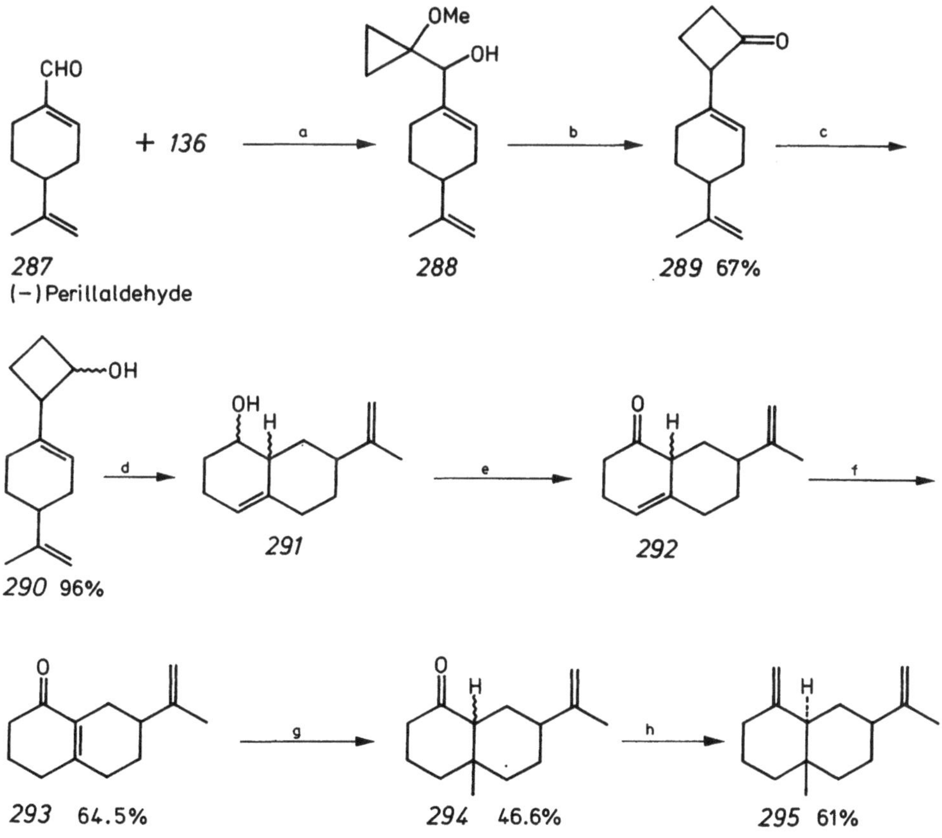

Scheme 5. Synthesis of (—)-β-selinene [153]
Reagents: a) THF, —78 °C; b) 48% aq. HBF_4, THF, 25 °C, 15 mn, 67%; c) $LiAlH_4$, Et_2O, 0 °C, 100%; d) KH, THF, reflux, 1 hr; e) MeCOMe, CrO_3; f) Al_2O_3, EtOAc-Hexane (3:97), 64.5% from *294*; g) CuI, MeLi, Et_2O, BF_3-Et_2O, —70 °C, 46%; h) NaH, DMSO, $Ph_3\overset{+}{P}CH_3$, Br^-, 80 °C, 70 hr, 61%.

followed by exposure of the crude product *288* to 5% HBF_4 in THF for 10 min, yielded the 2-vinylcyclobutanone *289*. It is noteworthy that this type of rearrangement could be conducted under conditions that did not affect the sensitive isopropenyl group. Reduction of *289* with $LiAlH_4$ yielded the 2-vinylcyclobutanol *290* as a 18:82 cis, trans mixture, which upon treatment with potassium hydride in refluxing THF for 1 hr produced the cyclohexenol *291* by a base-induced $C_4 \rightarrow C_6$ ring expansion. Oxidation of *291* with Jones reagent yielded the nonconjugated enone *292* which was converted to the conjugated enone *293* in 64.5% yield from *290* by rapid passage through a short column of basic alumina. Conjugate addition of a methyl group, accomplished by the use of CH_3CuBF_3 [154] yielded the ketone *294* (as a mixture of two stereoisomers with one highly predominating), in 46.6% yield. Finally, Wittig olefination in DMSO according to a procedure known to convert a cis-trans mixture of such decalones to the trans epimer [155], converted the ketone *294* into the optically active ($[\alpha]_D$ = —49.5° [c = 6.55, hexane]) (—)-β-selinene *295* contaminated with only about 5% of an isomer as shown by capillary G.C., (Scheme 5) [153].

This type of rearrangement was even capable of counteracting the aromaticity of the furan ring. Thus, the adduct *296* of 1-methoxycyclopropyllithium *136* to furfural was rearranged with 5% HBF_4 in THF without much destruction of the sensitive furan ring. The resulting 2-(2-furfuryl)cyclobutanone *297* obtained in 54% yield from *136*, was reduced with $LiAlH_4$ to the cyclobutanol *298a*, and this compound was treated with potassium hydride to yield the furanocyclohexenol *299*, albeit in modest yield; the tertiary alcohol *298b*, from the addition of MeLi to *297*, led, however, under identical conditions exclusively to fragmentation products, Eq. (83) [153].

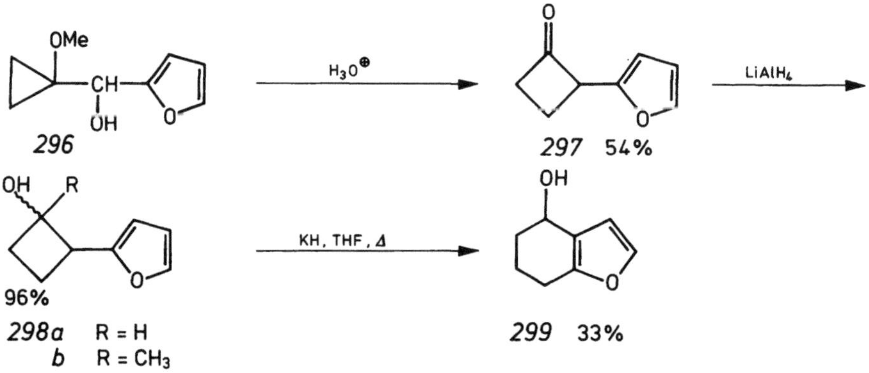

(83)

The stereoselective preparation of the decalol diene *305*, which is a model compound for an important intermediate in the synthesis of Compactin [156], the inhibitor of a key step in the cholesterol biosynthesis [157], has also been achieved following this strategy. Thus, 1,3-cyclohexadiene-2-carboxaldehyde *300* was reacted with *136* to give the cyclopropylcarbinol *301*, which was rearranged to the acid sensitive cyclobutanone *302* in dilute acid, in 52% overall yield from *300*. Then, the major reduction product obtained from *302* with $LiAlH_4$, *i.e.*, the trans-cyclobutanol *303*, underwent $C_4 \rightarrow C_6$ ring enlargement in the presence of KH in refluxing THF to give the axial and equatorial cyclohexenols *305* and *306* in a ratio of 92:8. The cis-cyclobutanol *304*, formed uncontaminated with the trans-isomer *303* by reduction with K-selectride, was reacted with KH under similar conditions to give the same cyclohexenols *305* and *306* in a ratio of 72:28. Thus, the readily available 2-(1,3-cyclohexadien-2-yl)-cyclobutanone *302*, provided rather stereoselectively and efficiently the compound *305*, a precursor to compactin (Scheme 6) [153].

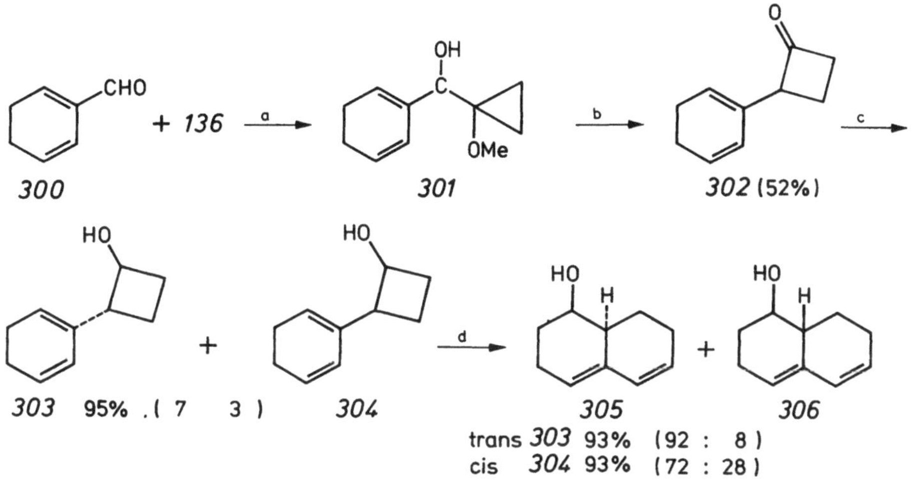

Scheme 6. Synthesis of a compactin precursor [153]
Reagents: a) THF, −78 °C; b) 48% HBF_4, THF, 0 °C, 20 min. 52%; c) $LiAlH_4$, Et_2O, 0 °C, 93%; d) KH, THF, reflux, 1 hr, 93%.

5.3.3. Thermal

Reaction of α(*n*-butylthiomethylene) cyclohexanone *307* with 1-phenylthiocyclopropyllithium *116a* did afford the desired adduct *308a* efficiently, but, the expected $C_3 \rightarrow C_4$ ring expansion into cyclobutanone *310* could not be effected cleanly, probably because the intermediate cation *309a* was more stable than its analogue without the

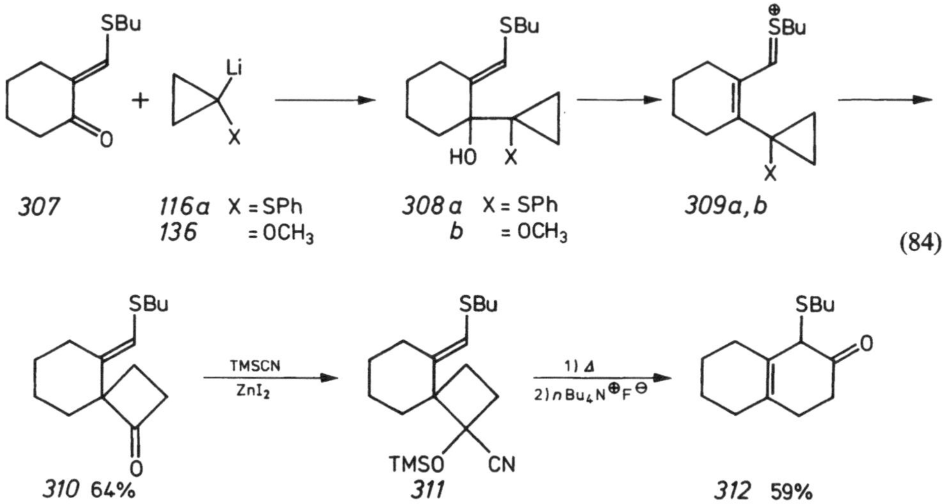

(84)

sulfur atom [61,62]. Better results were obtained by the use of 1-methoxycyclopropyllithium *136* providing *308b* which, upon treatment with a 48% aqueous HBF_4 solution in THF (1:4) [73], led via *309b* to the spirocyclobutanone *310* in 64% yield. The derivative of the cyclobutanone chosen for the $C_4 \rightarrow C_6$ ring expansion was the easily prepared, trimethylsilyl cyanohydrin *311*, which had the requisite electron withdrawing group [158]. Thus, *310*, was mixed with 1.5 equivalents of trimethylsilylcyanide (TMSCN) and a catalytic amount of ZnI_2; the resulting mixture was added to anhydrous diglyme and refluxed for 2.5 hr to give, after treatment with n-$Bu_4N^+F^-$ the cyclohexenone *312* in 59% yield (Eq. 84)) [159].

5.4 Involving $C_3 \rightarrow C_4 \rightarrow C_8$ Ring Expansions

5.4.1. Base Induced

There are several important classes of natural products, which contain eight-membered rings, but, owing to unfavorable entropic factors as well as large increases in enthalpy resulting from transannular and torsional strain, the syntheses of cyclooctanes from acyclic precursors were only marginally successful [160]. Recently a novel and versatile cyclooctane synthesis has been developed, based upon the anionic oxy-Cope rearrangement of 1,2-dialkenylcyclobutanols, now readily available from 2-vinylcyclobutanones, products of $C_3 \rightarrow C_4$ ring expansions of 1-donor substituted cyclopropane derivatives. For instance, reaction of 2-methyl 2-(2-methylpropen-1-yl)-cyclobutanone *313* [63,73] with vinyllithium afforded essentially the trans-divinylcyclobutanol *314* in nearly quantitative yield. Treatment of *314* with potassium hydride in THF at room temperature resulted in rapid rearrangement to the (Z)- and (E)-cyclooctenones *315* and *316*, in a 79/21 ratio, which were isolated in 62% overall yield from the cyclobutanone *313* (Eq. (85)) [161].

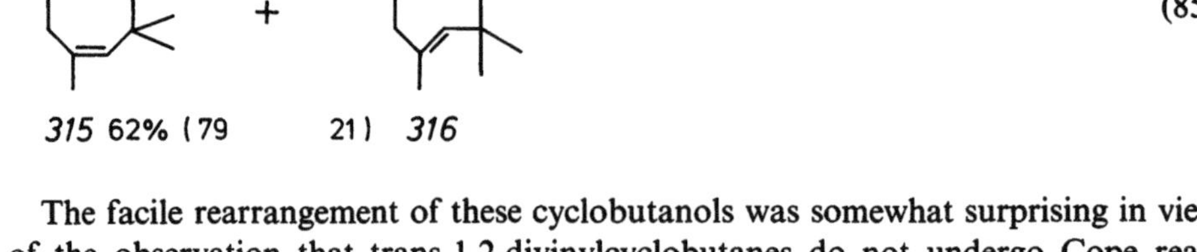

(85)

The facile rearrangement of these cyclobutanols was somewhat surprising in view of the observation that trans-1,2-divinylcyclobutanes do not undergo Cope rearrangement, but instead react predominantly via [1,3])-shift processes at elevated temperatures [162]. It is reasonable to assume that *314* is a trans-dialkenylcyclobutanol and that the rearrangement to cyclooctenones presumably occurred via initial isomerization to the cis-isomer and subsequent Cope rearrangement, since no products resulting from [1,3]-rearrangements were detected in these reactions [162]. More readily, after reaction of 5-methylenespiro[3.5]nonan-1-one *317* with vinyllithium a mixture of bicyclo[6.4.0]dodec-1(8)-en-4-one *320* and 5-(2-methyl-1-cyclohexen-1-yl)-1-penten-3-one *321* was isolated in 78% yield and in a ratio of 44/56; presumably, *320* was formed via the rapid oxy-Cope rearrangement of the intermediate cis-divinylcyclobutanol *318*, while the ring-opened ketone *321* was produced via a retro-ene reaction of the intermediate trans divinylcyclobutanol *319*, Eq. (86) [161].

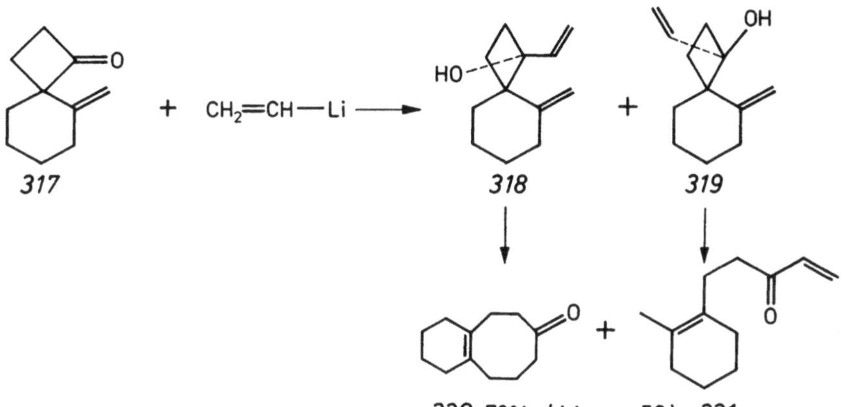

320 78% (44 56) 321

(86)

Whatever the case, the formation of the cyclooctenone *320* from the cyclobutanone *317* constituted the first example of a one-step $C_4 \rightarrow C_8$ ring expansion. In the same way, the spirocyclobutanone *266* [63,73], was treated with vinylmagnesium bromide to generate a mixture of the diastereomeric cyclobutanols *322* and *323* in 78% yield (ratio 21/79). Each diastereomer, separated by HPLC, individually subjected to KH in THF at room temperature, rearranged cleanly to bicyclo[5.3.1]undec-1(11)-en-4-one in 80% yield, Eq. (87) [161].

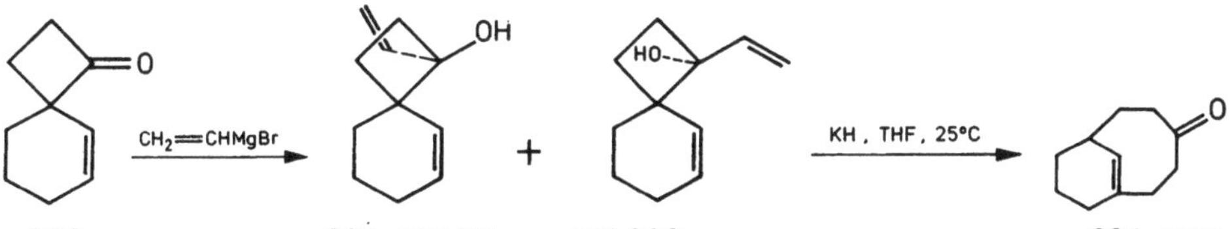

266 322 78% (21 79) 323 324 80%

(87)

5.4.2. Thermal

This methodology has been illustrated by the total synthesis of poitediol *334*, a sesquiterpene diol isolated from the red seaweed Laurencia Poitei [163]. Thus, stereoselective reduction of the cyclohexenone *325* with DIBAH in Et_2O at −100 °C and cyclopropanation [164] afforded in 60% yield a separable mixture of norcaranols *326a, b* (8:1 ratio). Oxidation of the major compound with PCC in CH_2Cl_2 [107] and addition of vinylmagnesium bromide gave the norcaranol *327* (77% yield), which underwent quantitative $C_3 \rightarrow C_4$ ring expansion to the bicyclo[3.2.0]heptanone *328* upon treatment with 1 equivalent of $BF_3 \cdot Et_2O$. Addition of lithium acetylide gave the 1-alkynyl-2-vinylcyclobutanol *329* which, underwent unprecedented oxy-Cope rearrangement on simple heating in hexane at 50 °C for 4 hr to give in 55% yield the cycloocta-4,7-dien-3-one *330*. Geminal dimethylation and carbonyl transposition were achieved by addition of methyllithium, oxidative rearrangement with PCC and reaction with dimethylcuprate to afford the enone *331* in 74% overall yield. Reduction with DIBAH in ether at −78 °C and benzylation followed by epoxidation with MCPBA afforded a 3:2 mixture of the isomeric epoxides *332*, which were treated with $LiEt_3BH$. After separation, the suitable diastereomer of the corresponding alcohol was successively protected as a SEM ether [165], debenzylated and oxidized (DMSO-$(COCl)_2$) [109] to give the cyclooctanone *333* in 54% overall yield. Introduction of the α-methylene group was accomplished in 60% yield by enolate formation with LDA, quenching with formaldehyde and dehydration of the intermediate alcohol. Finally, reduction with DIBAH produced a separable 1:1 mixture of alcohols which were deprotected to afford (±) poitediol *334* and (±)-4-epi poitediol *335*, (Scheme 7) [166].

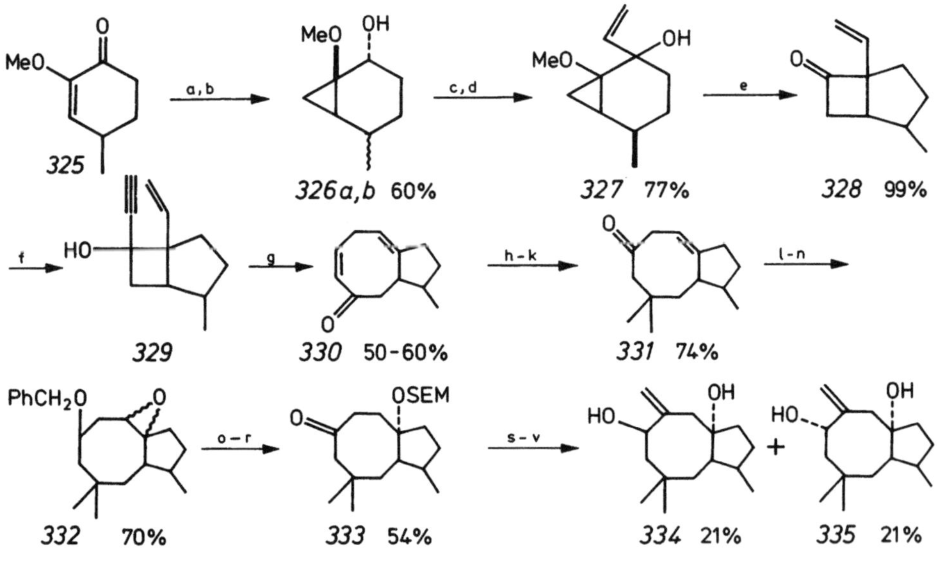

Scheme 7. Synthesis of (±) poitediol [166]
Reagents: a) DIBAH, Et_2O, −100 °C; b) Et_2Zn, CH_2I_2, toluene, 60 °C, 60%; c) PCC, CH_2Cl_2; d) $CH_2{=}CHMgBr$, THF, 77%; e) BF_3-Et_2O, 99%; f) $LiC{\equiv}CH$, THF, −30 °C, 5 min; g) hexane, 50 °C, 4 h, 50—60%; h) MeLi, Et_2O, −78 °C; i) PCC, CH_2Cl_2; k) $LiMe_2Cu$, Me_2S, Et_2O; l) DIBAH, Et_2O, −78 °C; m) KH, THF, PCH_2Br; n) MCPBA, $CHCl_3$; o) $LiEt_3BH$, THF; p) SEMCl, i-Pr_2NEt, THF, 50 °C q) Na, NH_3; r) $(COCl)_2$, Me_2SO, −78 °C, 54%; s) LDA, THF, CH_2O; t) Me_3SiCl, i-Pr_2NEt, 60%; u) DIBAH, hexane, −78 °C; v) 0.1 M HCl, MeOH.

5.5 Involving $C_3 \rightarrow C_5$ Ring Expansions

5.5.1 Introduction

The thermal vinylcyclopropane-cyclopentene rearrangement was discovered by Neureiter in 1959 [167]. After many mechanistic and theoretical studies [168], this thermal $C_3 \rightarrow C_5$ ring expansion was incorporated into many useful synthetic schemes. Various hetero analogues of the vinylcyclopropane system have also been investigated; the rearrangements of cyclopropyl ketones and cyclopropyl imines, for instance, provided useful synthetic methods to dihydrofurans and dihydropyrroles, respectively [169,170]. Photochemical and transition metal promoted rearrangements of vinylcyclopropanes have also been emphasized [3,4]. Although the thermal vinylcyclopropane rearrangement usually required a free energy of activation of 48 to 53 kcal/mol and occurred between 250–600 °C [168], dramatic acceleration has been obtained for the ring expansion of 2-alkoxy and 2-carbanion-substituted vinylcyclopropanes providing cyclopentenes with high yield and stereospecificity at 25 °C and —30 °C, respectively [171].

This section is concerned only with the thermal rearrangements of 1-trimethylsiloxy, 1-alkoxy-, 1-phenylthio- and 1-trimethylsilylvinylcyclopropanes into cyclopentene derivatives, which occurred either on heating in the liquid phase (sealed tube) at about 300 °C for 30 min, or by passing through a conditioned hot tube at 300 °C with a contact time of 4 sec or by flash thermolysis at 600 °C for 10 m sec [3].

5.5.2 Via 1-Trimethylsiloxyvinylcyclopropanes

It has been shown that 1-vinylcyclopropanols underwent thermal rearrangement into cyclobutanones (*vide supra*, Sect. 5.1.1, Eq. (68)) [41], an alternate pathway, however, was followed with a trimethylsilyl protected hydroxyl group providing the regiospecific formation of a silyl enol ether of a cyclopentanone. Moreover, the 1-trimethylsiloxy group on the cyclopropane ring appeared to facilitate the rearrangement by 5 kcal/mol [172], whereas it hampered the rearrangement by 3 kcal/mol when placed on the double bond of the vinylcyclopropane system [173]. This driving substituent effect of the siloxy group was illustrated by the thermal rearrangement of the 1-(2-cyclopropylvinyl)-1-trimethylsiloxycyclopropane *336*, which led exclusively to the 3-cyclopropylcyclopentanone silyl enol ether *337*, although two vinylcyclopropane moities could *a priori* be involved in the rearrangement, the isomer *338* was not obtained, Eq. (88) [15].

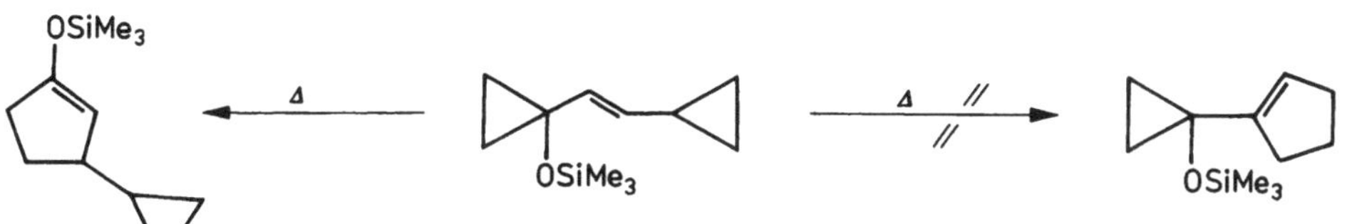

337 336 338

(88)

Acid or base hydrolysis of the thermolysis products, *i.e.*, the 1-siloxycyclopentenes, unmasked the carbonyl group and provided the corresponding cyclopentanones [15, 43, 174], while dehydrosilylation led to cyclopentenones [58, 92]. On the other hand, treatment either with methyllithium [175], n-butyllithium [176] or lithium amide in ammonia [138] allowed the generation of the corresponding lithium enolates, which could then be alkylated regiospecifically to introduce further alkyl groups leading to 2,3-disubstituted cyclopentanones. Alternatively, the cyclopentanone enol silyl ethers have also been alkylated directly in the presence of Lewis acids [177].

Because of the discovery of a growing number of naturally occurring substances of biological importance that contain the five-membered ring moiety [178], the synthesis of cyclopentanoid compounds is a subject of present interest. Indeed, among the various approaches recently investigated, the thermal vinylcyclopropane-cyclopentene rearrangement of readily available 1-siloxy-1-vinylcyclopropanes (*vide supra*, Sect. 4.1.5) constitutes an efficient three-carbon annelation process [179].

5.5.2.1. From the Cyclopropanone Hemiacetal

Aside from the ready preparation of some α,β-disubstituted cyclopentanones, the utility of the cyclopropanone hemiacetal approach has been illustrated by the total synthesis of the methyl ester of 11-deoxyprostaglandin E_2 *342* [15]. Towards this end, 1-trimethylsilylbutadiynylcyclopropanol *13*, readily available from the cyclopropanone hemiacetal *3* (*vide supra*, Sect. 2.1, Eq. (6)) was successively treated with dihydropyran in CH_2Cl_2 in the presence of 10% mol. equiv. of PPTS [101]. Desilylation by potassium fluoride in DMF [138], formation of the lithium salt with *n*-BuLi and condensation with hexanal gave the propargylic alcohol *339* in 64% overall yield. The

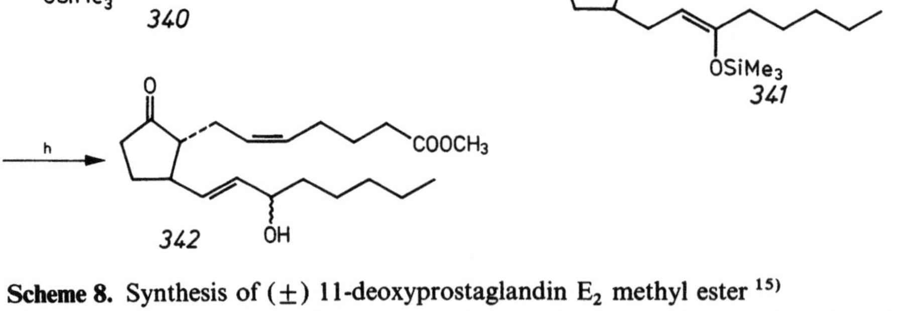

Scheme 8. Synthesis of (±) 11-deoxyprostaglandin E_2 methyl ester [15]
Reagents: a) DHP, PPTS, $HCCl_3$, 100%; b) KF, 2 H_2O, DMF, 91%; c) *n*-BuLi, THF, $CH_3(CH_2)_4CHO$, 70%; d) EtOH, PPTS, 55 °C, 95%; e) $LiAlH_4$, THF, 65 °C, 86.5%; f) $ClSiMe_3$, NEt_3, DMSO, 82,5%; g) Flash vacuum pyrolysis at 600 °C, 100%; h) NH_2Li, NH_3, methyl cis-7-bromo-5-heptenoate, 44.5%.

conversion of *339* to 1-trimethylsiloxy-1-(5-trimethylsiloxydeca-1,3 dienyl) cyclopropane *340* involved the removal of the THP protecting group [101], and lithium aluminum hydride reduction of the two triple bonds followed by double silylation in the presence of DMSO [180]. Flash thermolysis of *340* at 600 °C provided the prostaglandin precursor *341* in nearly quantitative yield (~95%); the overall yield from *13* was 44%. Finally the lithium enolate generated in liquid ammonia by reaction of *341* with lithium amide [138] was alkylated with a four-fold excess of methyl (Z)-7-bromo-5-heptenoate to yield a *ca.* 50:50 diastereomeric mixture of the 11-deoxyprostaglandin E_2 methyl esters *342* and its C_{15}-epimer, (Scheme 8) [15].

It is noteworthy that the regio- and stereoselectivity of the allylation of *341* was recently improved by using the Pd-catalyzed coupling reaction of the corresponding lithium cyclopentenolate-BEt_3 complex with the appropriate (Z)-allylic acetate, which provided *342* in 74% yield [176].

5.5.2.2 From Oxaspiropentane

Ring opening of the oxaspiropentane *343* upon treatment with sodium phenylselenide (*vide supra*, Sect. 4.5, Eq. (34)) [59] and O-silylation produce the vinylcyclopropanol trimethylsilyl ether *344* which, on flash thermolysis at 670 °C, gave the siloxycyclopentene *345* as a 2:1 mixture of epimers at $C_{(8)}$. Then, allylation of the more substituted enolate arising from *345*, opens a convenient way to the antitumor agent, aphidicolin *346* [181].

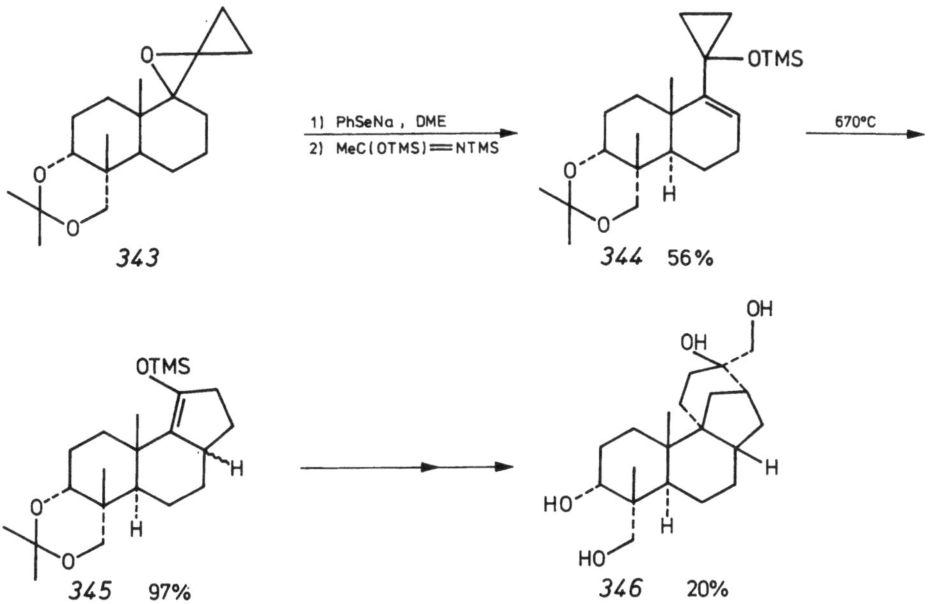

Scheme 9. Synthesis of (±) aphidicolin [181]

5.5.2.3 From 1-Hydroxycyclopropylcarbonyl Derivatives

It has been shown that 1-hydroxycyclopropanecarboxaldehyde derivatives *171* underwent Wittig olefination with cyclohexylidenephosphorane (*vide supra*, Sect. 4.7,

Eq. (55)); the O-silylated derivative [180], (contrary to the corresponding oxaspiropentane *110* (*vide supra*, Sect. 3.5, Eq. (33)) readily led to the 1-siloxyvinylcyclopropane *111*, which upon thermolysis regioselectively gave the silyl enol ether of spiro[4.5]-decan-2-one *347*, Eq. (89) [111].

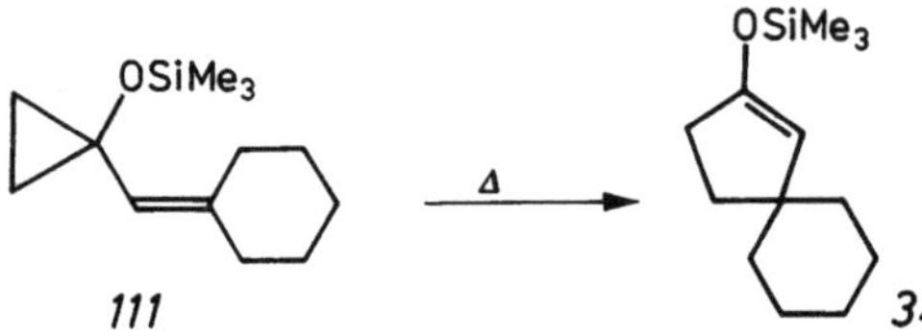

111 *347* 95% (89)

The spiro compound *347* constitutes the basic carbon framework found in sesquiterpenes of the spirovetivane and acorane class [184], which have been the target of many syntheses [185]. The thermal $C_3 \rightarrow C_5$ ring expansion *111* → *347* was of

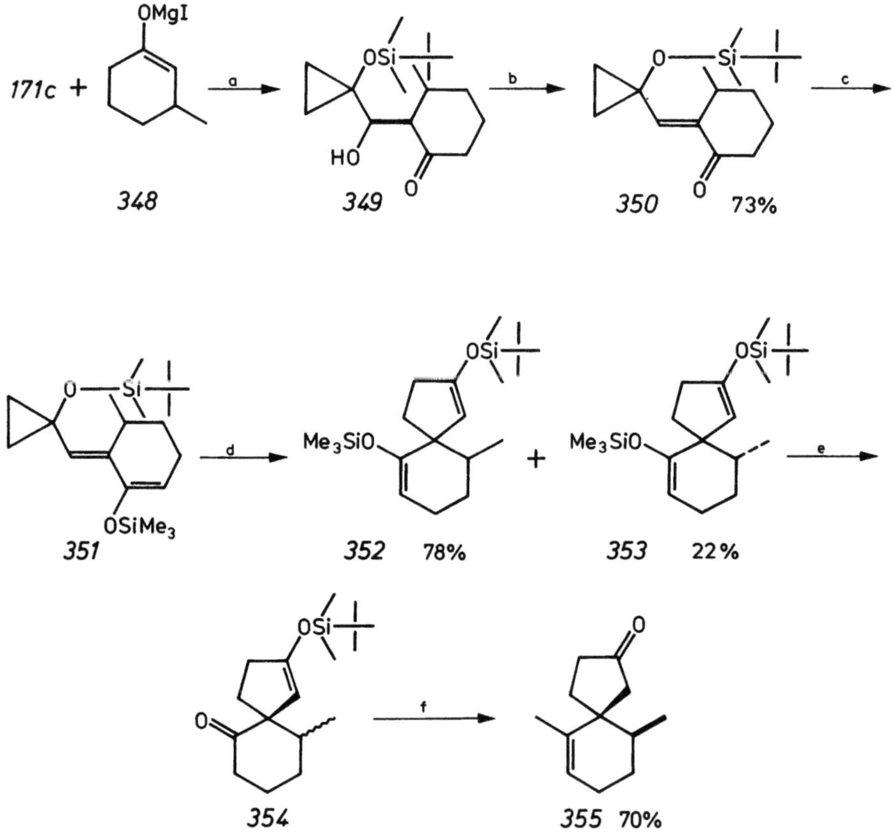

Scheme 10. Synthesis of a Spirovetivane [111]
Reagents: a) Et_2O, 0 °C; b) i, Ac_2O, pyridine, r.t.; ii, DBN, C_6H_6, 80 °C, 24 h, 73%; c) i, LDA, DMF, 0 °C; ii, $ClSiMe_3$, NEt_3, 0 °C, 100%; d) F.V.T., 600 °C; e) NEt_3, MeOH, 30 °C; f) i, $IMgCH_3$, Et_2O; ii, pTsOH, C_6H_6, 80 °C, 70%.

particular advantage in that it directly provided a spiroketone with the carbonyl group in the 2-position as found in spirovetivanes such as *355*. The latter is a constituent of Vetiver oil and plays a significant role in the reconstitution of this essential oil; in addition it constitutes a particularly convenient intermediate for the synthesis of biogenetically related spirovetivanes of economic importance such as α-vetispirene, β-vetivone, hinesol, agarospirol, etc. [186]

The stereoselective total synthesis of the challenging spiroketone *355* has been achieved from 1-(t-butyldimethylsiloxy)cyclopropanecarboxaldehyde *171c* (*vide supra*, Sect. 4.7, Eq. (55)). Thus, trapping of the enolate anion *348* resulting from the cuprous iodide-catalyzed addition of methylmagnesium iodide to 2-cyclohexen-1-one with *171c* afforded the ketol *349*, which was treated with successively acetic anhydride in pyridine and with 1,5-diazabicyclo[4.3.0]non-5-ene in refluxing benzene, to produce the enone *350* in 73% overall yield from *171c*. Ketone *350* was converted into the silyl enol ether *351* upon reaction with lithium diisopropylamide in DME and trimethylsilyl chloride in NEt_3 at 0 °C. Flash thermolysis at 600 °C quantitatively gave a mixture of the spirocyclic bissilyl enol ethers *352* (78%) with the desired stereochemistry and *353* (22%), resulting from the preferential rearrangement to the less sterically shielded side of the six-membered ring of *351*. On the other hand, flash thermolysis of the enone *350* produced a substantial amount of desilylated derivatives of *352* and *353*. Highly chemoselective desilylation to the epimeric cyclohexanones *354* was achieved upon treatment with 0.1 M triethylamine in methanol [187] at 30 °C for 48 hr. Finally addition of methylmagnesium iodide, followed by dehydration of the corresponding epimeric alcohols and desilylation on treatment with p-toluenesulfonic acid in refluxing benzene provided a mixture of spiroenones, from which the expected spiroenone *355* was isolated in 70% yield, (Scheme 10) [111].

The accessibility to 2,3-disubstituted — and 4,5-disubstituted 2-cyclopentanones by this thermal $C_3 \rightarrow C_5$ ring expansion has been illustrated by the syntheses of dihydrojasmone [92], cis-Jasmone [92] and dicranenone A [188].

Jasmonoïds, important raw materials in the perfume industry, are among the best known and most often synthetized members of the cyclopentanoïd class, because these simple compounds incorporate the 2,3-dialkylated cyclopentanone and cyclo-

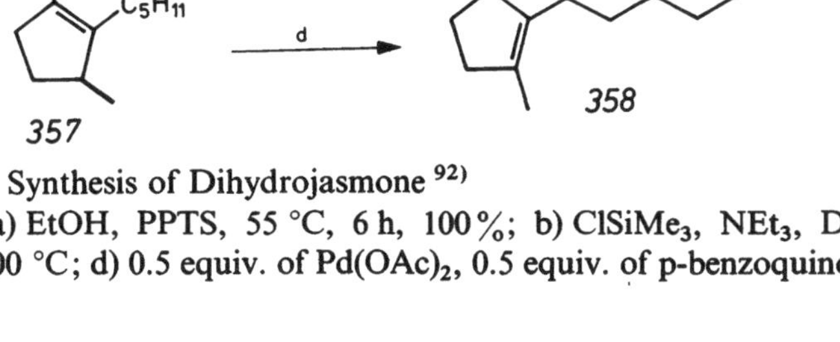

Scheme 11. Synthesis of Dihydrojasmone [92]
Reagents: a) EtOH, PPTS, 55 °C, 6 h, 100%; b) $ClSiMe_3$, NEt_3, DMSO, 96%; c) F.V.T. 600 °C; d) 0.5 equiv. of $Pd(OAc)_2$, 0.5 equiv. of p-benzoquinone, CH_3CN, 92%.

pentenone units found in a large number of biologically active natural products. Dihydrojasmone for example, was also prepared from the readily available 1-hydroxy vinylcyclopropane derivative *177* (*vide supra*, Sect. 4.7, Eq. (57)). Thus, removal of the THP protecting group [101] and 0-silylation [180] gave a E,Z mixture of the disubstituted 1-trimethylsiloxy 1-vinylcyclopropane *356*. Flash thermolysis at 600 °C of the isomeric olefins mixture produced exclusively the cyclopentanone silyl enol ether *357* which, on treatment with palladium acetate and p-benzoquinone yielded dihydrojasmone *358* in 88% overall yield from *177b*, (Scheme 11) [92].

The synthesis of cis-jasmone from the aldehyde *171b* was realized as shown in Scheme 12. First of all, *171b* was treated with ethylidenetriphenylphosphorane to produce the (E,Z)-olefin *359a*, cleavage of the THP group [101] and O-silylation [180] led to the 1-trimethylsiloxy-1-vinylcyclopropane *359b* in 92% overall yield. Flash thermolysis gave the 3-methyl-1-trimethylsiloxycyclopentene *360* and addition of phenylselenenyl bromide [189] the 3-methyl-2-phenylselenocyclopentanone *361*. Treatment of crude *361* with LDA in THF containing 2 equivalents of HMPA followed by alkylation with cis 2-pentenyl bromide resulted in the formation of cyclopentanone *362* in 89% overall yield. Unfortunately, oxidative elimination of the α-selenocyclopentanone *362* gave a mixture of endo and exocyclic enones; however, this defect was easily corrected by the procedure of Liotta which allows the isomerization *362* → *363* upon treatment with LDA in THF-HMPA at −78 °C [190]. Elimination by a two phase system containing methylene chloride and 30% hydrogen peroxide gave the cyclopentenone *364* which was isomerized to jasmone *365* by sodium methoxide in methanol, (Scheme 12) [92].

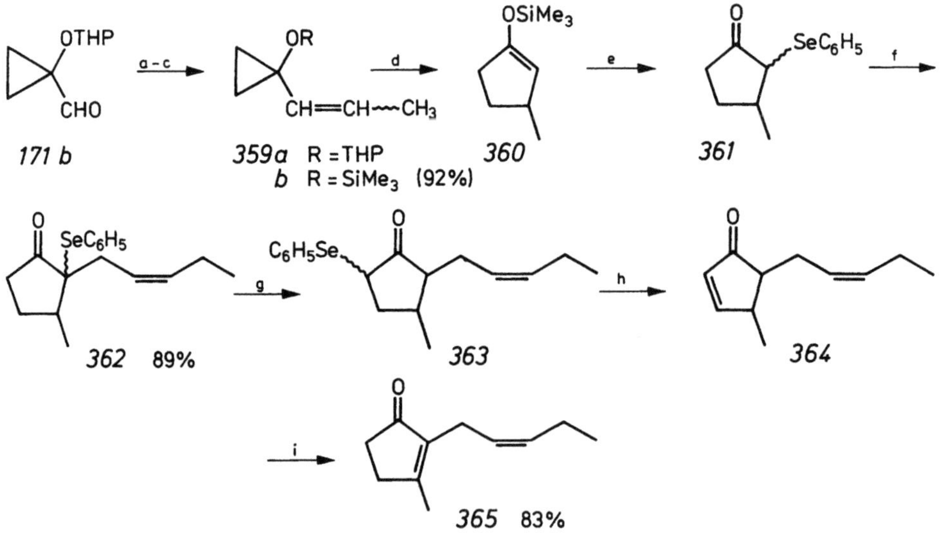

Scheme 12. Syntheses of cis-jasmone [92]
Reagents: a) $CH_3CH{=}P(C_6H_5)_3$; THF, reflux, 18 h, 84%; b) EtOH, PPTS, 55 °C; c) $ClSiMe_3$, NEt_3, DMSO, 92%; d) FVT, 600 °C; e) C_6H_5SeBr, Et_2O, −78 °C; f) LDA, THF, HMPA, *cis* 2-pentanyl bromide, −78 °C, 2 hr, r.t., 15 hr, 89%; g) LDA, THF, HMPA, −78 °C; h) 30% H_2O_2, CH_2Cl_2; i) $NaOCH_3$, CH_3OH, r.t., 83%.

The dicranenones, which constitute a new class of fatty acids containing a cyclopentenone ring with a structural similarity to prostanoids and jasmonoids, have recently been isolated from Japanese mosses, and possess antimicrobial activities [191]. With respect to the synthesis of dicranenone A, the ethyl-E-3-(1-tetrahydropyranyloxycyclopropyl)prop-2-enoate *219*, readily available from the aldehyde *171b* (*vide supra*, Sect. 5.11, Scheme 1), was successively deprotected (EtOH, PPTS [101]) and 0-trimethylsilylated ($ClSiMe_3$, NEt_3, DMSO [180]). Reduction of the ester group with DIBAH and silylation of the resulting allylic alcohol led to the (E)-disiloxyvinylcyclopropane *366* in 78% overall yield from *219*. Upon flash thermolysis at 600 °C, *366* underwent quantitative, regiospecific ring expansion into the cyclopentenol silyl ether *367*. Alkylation of *367* with (Z)-pent-2-enylbromide in the presence of zinc bromide [191] gave the expected 2-allylcyclopentanone in only 20—40% yield. In an alternative approach, however, the copper salt-catalyzed cyclopropanation of *367* with ethyl diazoacetate [193] in refluxing benzene gave the ethyl cyclopropanecarboxylate

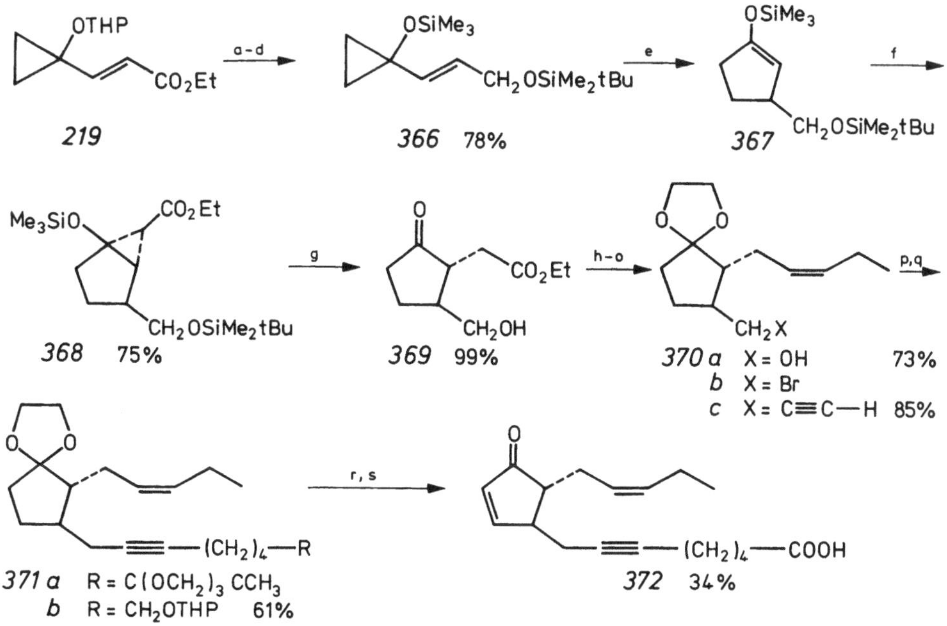

Scheme 13. Synthesis of (±) dicranenone A [188]

Reagents: a) EtOH, PPTS, 55 °C, 95%; b) Me_3SiCl, NEt_3, DMSO, 98%; c) iBu_2AlH, Toluene, —60 °C; d) ClSit-$BuMe_2$, imidazole, DMF, 20 h, 93%; e) 600 °C; f) $Cu(MeCOCH_2COMe)_2$, C_6H_6, 80 °C, ethyl diazoacetate, 75%; g) EtOH, $ClSiMe_3$, r.t. 12 hr, 99%; h) $HOCH_2CH_2OH$, C_6H_6, pTsOH, 80 °C; i) $ClSitBuMe_2$; j) $LiAlH_4$, THF; j) $LiAlH_4$, THF, 65 °C, 1 h; h) DMSO, $(COCl)_2$; l) $EtCH{=}PPh_3$, 73%; m) pTsCl, pyridine, 0 °C, 82%; n) LiBr, Me_2CO reflux, 12 hr, 92%; o) $LiC{\equiv}CH{-}NH_2CH_2CH_2NH_2$, (1.5 equiv.) DMSO, 8 °C; p) *n*-BuLi, THF, 0 °C, I-$(CH_2)_4R$, 23%; q) : i, *n*-BuLi, THF, 0 °C; ii, $B[(CH_2)_5OTHP]_3$, THF; iii, I_2, ether —78 °C, 61%; r) : i, $CH_3COOH/H_2O/THF$ 6.5:2.5:1.0, 45 °C; ii, DHP, pTsOH, CH_2Cl_2, r.t., 80%; s) i, LDA, THF, —78 °C; ii, PhSeCl; iii, H_2O_2, pyridine, CH_2Cl_2, r.t., 55%; t) i, p-TsOH, CH_3OH, r.t., 82%; ii, CrO_3-H_2SO_4, acetone, —20 °C, 94%.

368 in 75% yield. On simple addition to ethanol acidified by a drop of $ClSiMe_3$, *368* underwent desilylation and regiospecific opening of the cyclopropane ring to afford the γ-keto ester *369*. Acetalization with ethylene glycol in the presence of PPTS [101], followed by reduction and oxidation gave, after Wittig olefination of the corresponding aldehyde with salt-free *n*-propylidenetriphenylphosphorane [194], the acetal *370a* in 73% overall yield from *369*. Tosylation and treatment with lithium bromide in acetone, followed by acetylenation with the lithium acetylide ethylenediamine complex in DMSO [195] led to the propyne derivative *370c* in 85% yield. Alkylation of the terminal acetylenic carbon with the ortho-ester prepared from 5-iodopentanoic acid chloride and 3-methyl-3-hydroxymethyloxetane [196] gave the orthoester *371a* in 23% yield [188], while alkylation with tris[5-(tetrahydropyran-2-yloxy)pentyl]borane followed by oxidation with iodine gave *371b* in 61% yield [191]. Final steps toward the dicranenone A *372* involved introduction of the double bond at the required position in the five-membered ring; thus, for instance the cyclopentanone obtained after deacetalization of *371b* was treated with LDA at —78 °C and the corresponding kinetic enolate [197] was trapped with PhSeCl and oxidized with hydrogen peroxide [198]. After removal of the THP group (p-TsOH, CH_3OH, r.t.) and oxidation with Jones reagent, the racemic dicranenone A *372* was obtained in 34% overall yield from *371b*, (Scheme 13) [188].

5.5.3 Via 1-Phenylthiovinylcyclopropanes

As shown in Eq. (35) (*vide supra*, Sect. 4.6.1) 1-arylthiocyclopropyllithium *116* reacted with carbonyl compounds to provide, after dehydration, 1-arylthiovinylcyclopropane derivatives which were also able to undergo thermal $C_3 \rightarrow C_5$ ring expansion to cyclopentenol thioethers [62]. In principle, these enol thioethers can be desulfurized to yield regiospecifically generated cyclopentenes [199], hydrolyzed to give regiospecifically generated cyclopentenones [200] or hydrogenated to cyclopentanes. The regioselectivity of this annelation process which parallels that based upon the oxaspiropentane intermediate (*vide supra*, Sect. 5.5.2.2) and its high stereoselectivity which enhanced the utility of this approach have been discussed in a recent review on this topic [63b]. The thermal rearrangement of β-(1-phenylthio)cyclopropylenones, however, was shown to afford a mixture of cyclopentenoïd products. For instance, all attempts to effect the thermal ring expansion of *373*, a precursor of the challenging spiro[4.5]decane skeleton (*vide supra*, Sect. 5.5.2.3) gave at best a 9% yield of the incompletely characterized phenylthiocyclopentene *374*, Eq. (90) [64]. (For comparison see Eq. (89) and Scheme 10).

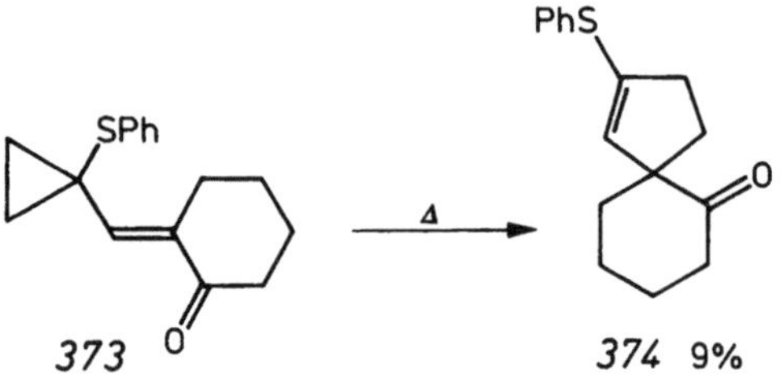

373 374 9%

(90)

5.5.4 Via 1-Trimethylsilylcyclopropanes

Contrary to the accelerating effect of the trimethylsiloxy group (*vide supra*, Sect. 5.5.2, Eq. (88), the trimethylsilyl group retarded the thermal vinylcyclopropane rearrangement, as it exerts a destabilizing effect on the diradical formed after rupture of the cyclopropane ring [201]. Consequently, on the pyrolysis of 1-cyclopropyl-1-(1-trimethylsilylcyclopropyl)ethylene *375*, the opening of the unsubstituted three-membered ring was kinetically favored to initially lead to *376* and ultimately to *377*; the intermediate vinylsilane *378* and allylsilane *379* involving initial rearrangement of the silylcyclopropane moiety were not observed, indicating clearly that $k_H \gg k_{Si}$, Eq. 91) [81,202].

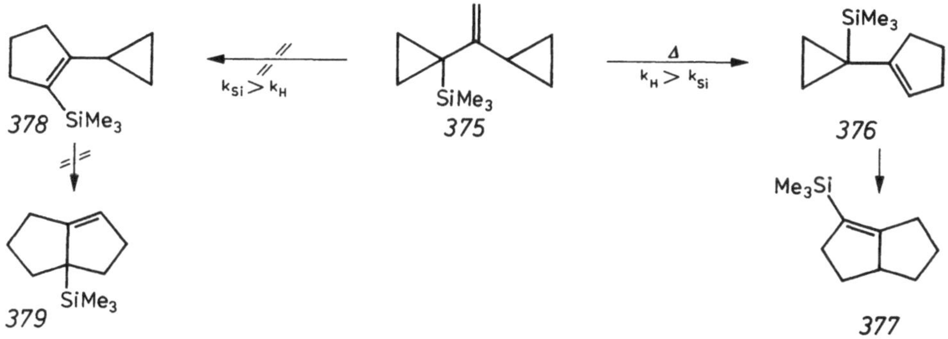

(91)

Under controlled conditions, vinylsilanes such as *377* undergo regioselective electrophilic substitutions. Thus, acetylation by means of acetyl chloride and aluminum chloride in CH_2Cl_2 at −78 °C afforded *380*, while bromination led to the vinyl bromide *381*. Epoxidation with m-chloroperbenzoic acid efficiently transformed *377* into a 60:40 mixture of the epimers of *382*, Eq. (92) [82].

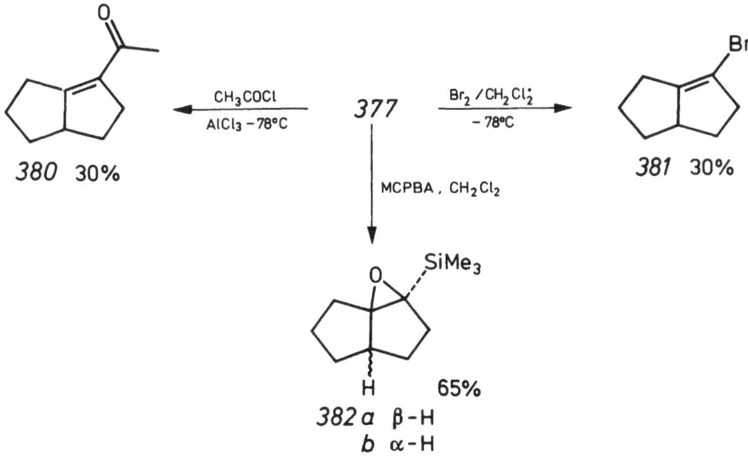

(92)

Furthermore, such vinylcyclopropanes have also successfully been used for the stereoselective construction of quaternary carbon centers such as those which occur in many spirovetivane-type-sesquiterpenes [184]. Thus, the (Z)-1-[1-(trimethylsilyl-cyclopropyl)methylene]-2,6-dimethyl-2-cyclohexene *160*, readily available from the 1-trimethylsilylcyclopropanecarboxaldehyde *156* (*vide supra*, Sect. 4.6.4, Eq. (51)), underwent thermal bond reorganization upon heating to 560 °C in a tube packed with quartz chip to yield a 4:1 mixture of the spirocyclic vinylsilanes *383* and *384*. The predominance of *383* signalled again preferential rearrangement to the less sterically shielded side of the six-membered ring of *160* [89,91] (*vide supra*, Sect. 5.5.2.3, Scheme 10). Various attempts to acylate *383* chemospecifically at its vinylsilane center under Friedel-Crafts conditions failed; similarly exposure to bromine, iodine, cyanogen bromide and dichloromethyl methylether-$TiCl_4$ [203] led only to formation of tarry products [91]. Since α, β-epoxysilanes can usually be converted into carbonyl compounds [204], the synthesis of *386* was pursued as a possible route to the challenging spirovetivane *355*. To this end, epoxidation of *383* and *384* with 3 equivalents of m-chloroperbenzoic acid in CH_2Cl_2 provided the diepoxide *385*. Selective cleavage of the cyclohexane-annellated oxirane ring was accomplished by treating *385* with diphenyl diselenide and sodium borohydride in ethanol [205], the resulting β-hydroxy selenide then underwent reductive elimination [206] to afford *386* in 45% yield. Unfortunately, a variety of protic and Lewis acid catalysts known for their efficiency in transforming α, β epoxysilanes to carbonyl compounds [204] did not give the expected spirocyclopentanone *355* (For comparison see Scheme 10). In related studies, copper-mediated reactions of isopropenyl magnesium bromide and isopropenyllithium with *386* gave no evidence of coupling product, Eq. (93) [91].

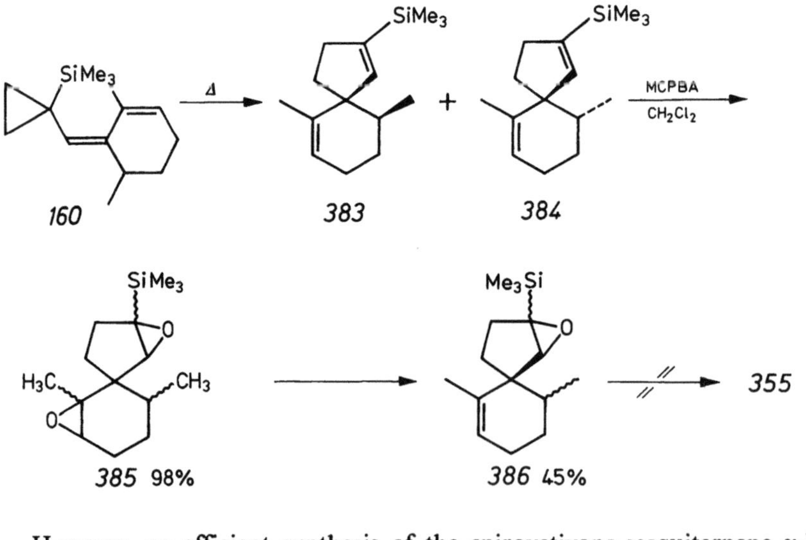

(93)

However, an efficient synthesis of the spirovetivane-sesquiterpene α-Vetispirene *392* was achieved initiating the cleavage of the silicon-cyclopropane carbon bond of 160 by fluoride ion. This was effected with anhydrous tetra-*n*-butyl ammonium fluoride in THF solution containing acetone at reflux temperature for 10 hr. The pentadienyl anion *387* thus generated experienced highly regioselective addition of acetone at the cyclopropyl carbon atom, thereby avoiding the development of methylenecyclopropane moiety and providing the isomerically pure alcohol *388* in 90% yield. Subsequent conversion to the methyl ether *389* with sodium hydride and methyl iodide, thermal $C_3 \rightarrow C_5$ ring expansion at 440 °C produced a 5:1 mixture of the cyclopentenes *390* and *391* in quantitative yield, which upon exposure to p-toluenesulfonic acid in benzene for 25–30 min at 5–20 °C yielded a mixture of the (±)-α-vetispirene *392* and its C_{10}-epimer (ratio 5:1), (Scheme 14) [89,91].

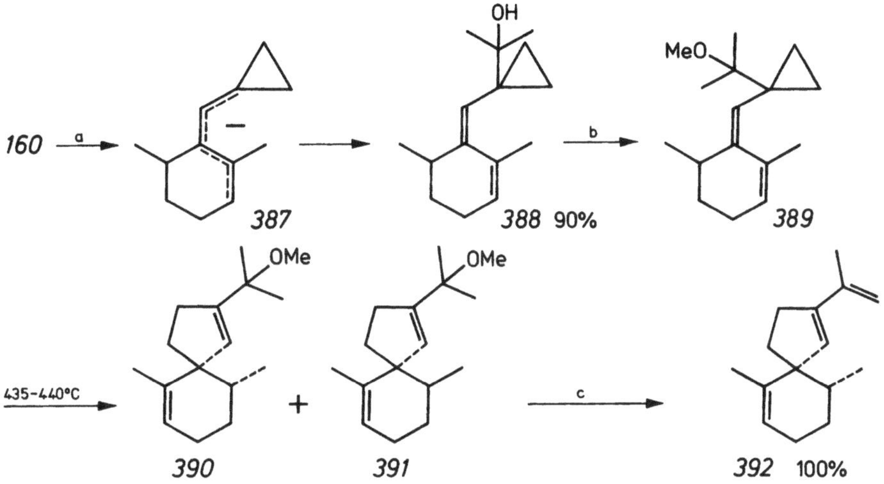

Scheme 14. Synthesis of (±)-α-vetispirene
Reagents: a) $Bu_4N^+F^-$, CH_3COCH_3, THF, reflux, 10 hr, 90%; b) :i, NaH, THF, HMPT, −20 °C; ii, CH_3I, 0 °C, 100%; c) pTsOH, C_6H_6, 10 °C, 5 hr, 100%.

5.5.5 Miscellaneous

Metal-promoted vinylcyclopropane $C_3 \rightarrow C_5$ ring expansions have been reported. Thus, ethyl 2-methoxy-2-vinylcyclopropanecarboxylate *393* rearranged to a 2:1 mixture of 3-methoxy-2-cyclopentenecarboxylate *396* and 3-methoxy-3-cyclopentenecarboxylate *397* on heating at 160 °C in the presence of catalytic amounts of copper bronze or copper(I) chloride; in contrast, platinum and rhodium complexes catalyzed

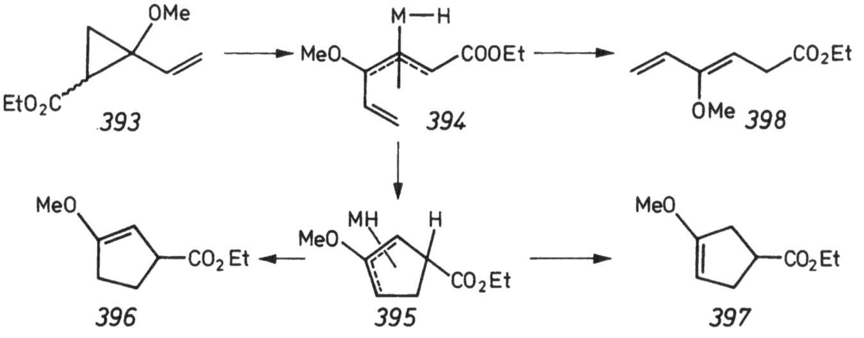

(94)

the ring opening with subsequent H-migration to yield the ethyl 4-methoxy-3,5-hexadienoate *398*. This dichotomy and the production of both *396* and *397*, rather than only the vinylcyclopropane-cyclopentene rearrangement product *397* was consistent with the generation of the intermediate *395* from the initially formed η^3-allyl metal hydride complex *394* [207] (Eq. (94)) [208].

The dicyclopropyl ketimine *198* prepared from the O,N-cyclopropanone hemiacetal *194* (*vide supra*, Sect. 4.9, Eq. (62)), heated in xylene with ammonium chloride for 4 hr underwent ring expansion exclusively to the enamine *399* followed by isomerization to the cyclopropyl pyrroline *400*. Although further ring expansion was not observed on prolonged heating, *400* was converted to the hydrobromide *401* with anhydrous HBr [209] which upon heating to 140 °C for 10 min experienced a second cyclopropyl imine rearrangement to provide the pyrrolizidone *403* in 51% yield, most probably via the HBr adduct *401* by cyclization to the pyrroline *402* followed by acid-induced hydrolysis, Eq. (95) [129].

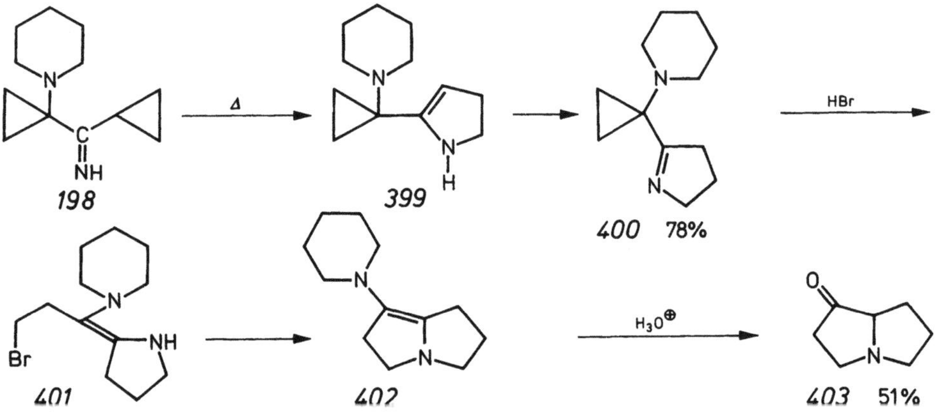

(95)

This sequence opened a route to pyrrolizidines that are of obvious interest in natural product synthesis. The effects of substituents in determining the preferred opening (A versus B) of the cyclopropyl rings of the dicyclopropyl ketimines *404* have been examined. Thus, on heating *404a, b* in refluxing xylene to 140 °C for 1.5–5 hr, in the presence of ammonium chloride, the pyrrolines *405a, b* were obtained exclusively through the rearrangement of the less substituted or thiophenyl-substituted cyclo-

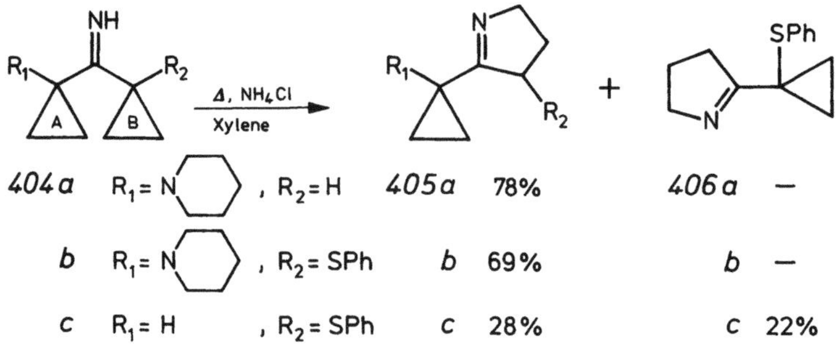

(96)

propane rings (B), while with the substrate *404c* a combination of steric and electronic factors yielded a mixture of products *405c* and *406c* (Eq. (96))[129b].

A completely different result was obtained when the dicyclopropyl ketimines were subjected to acid-catalyzed thermolysis in the presence of a non-nucleophilic counterion. Thus, on heating *404a, b* in xylene with the dimethyl ether complex of fluoroboric acid, both substrates underwent rearrangement to pyrrole derivatives *407a, b* resulting from ring-opening of the cyclopropyl ring A containing the electron-releasing piperidino group, in disagreement with previous reports[209,210], Eq. (97)[129b].

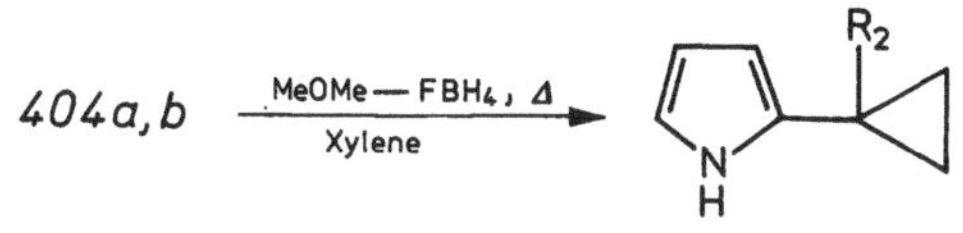

407a = 32%
b = 42%

(97)

5.6 Involving Ring Openings

5.6.1 From Cyclopropyl Derivatives

It has been shown that the regioselective cyclopropanation of enol silyl ethers of α-enones provided 1-donor substituted-vinylcyclopropanes which underwent base-induced ring opening to provide specific α or α′-monomethylation of conjugated cycloalkenones (*vide supra*, Sect. 4.3, Eq. (25)). On the other hand, epoxidation (H_2O_2) and acid-induced (HCO_2H) ring opening of vinylcyclopropane derivatives have been used for the introduction of the allylic alcohol substituent of the prostaglandins PGE_1 and PGF_1[211]. Other useful synthetic applications of cyclopropane ring openings have been reviewed[3]. It is noteworthy that 1-ethoxy-1-trimethylsiloxycyclopropane *2* can act as a synthetic equivalent of the β-anion of ethyl propionate[99]. Thus, addition of *2* to carbonyl compounds was readily achieved with the aid of one equivalent of titanium tetrachloride to provide γ-lactones in high yields[212]. Zinc homoenolates of alkylpropionate *408* were also available from the reaction of *2* with zinc chloride

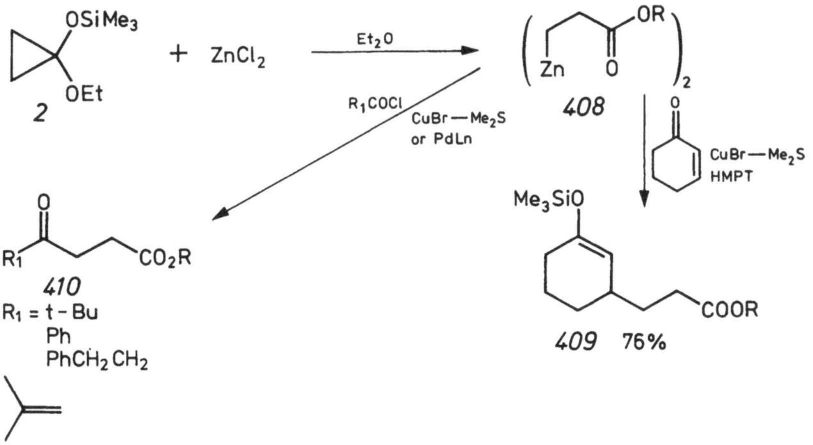

(98)

in ether; successive addition of a catalytic amount (5%) of CuBr—Me_2S complex, hexamethylphosphoric triamide (HMPT) and an unsaturated carbonyl compound at 0 °C led to the regiospecific silyl enol ether (>99%) of 6-oxo esters, such as *409*. The addition of acyl chlorides under the same conditions led to 4-oxo esters *410* [213]. On the other hand, transmetalation of *408* with a catalytic amount of palladium (5% of $PdCl_2(o\text{-}Tol_3P)_2$) and coupling reactions with aryl-, vinyl- or acyl halides in THF led to products *410* of homoenolate arylation, vinylation or acylation respectively, Eq. (98) [214].

Although *2* was inert to tributyltin chloride [215], reaction of the readily available 1-vinylcyclopropanols *69a, b* (*vide supra*, Sect. 4.1) with trimethyl- or tributylstannyl chloride in ether in the presence of triethylamine and DMSO, resulted in ring opening and β-C-stannylation to provide the enone *411* Eq. (99) [217]. Under the same conditions, Me_3SiCl led to O-silylation exclusively [180] (*vide supra*, Sect. 4.5.2.1, Scheme 8). Alkylation and acylation of these intermediates involving trans-metalation [216] have been investigated.

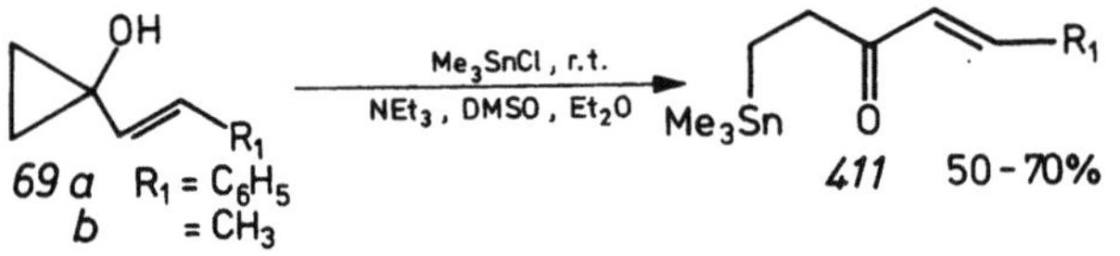

(99)

5.6.2 From Cyclobutyl Derivatives

The utility of the ring opening of cyclobutanone derivatives has been demonstrated by the total synthesis of the sex pheromone grandisol [139] (*vide supra*, Sect. 5.1.2, Scheme 2). In addition to uses discussed above, Baeyer-Villiger oxidation of cyclobutanones led to γ-butyrolactones [218] which also constituted important precursor to cyclopentenones [219]. Anion-stabilizing groups such as phenyl, bromine, sulfur etc. at the α-position of a cyclobutanone facilitated the ring cleavage by nucleophiles including hydroxide, methoxide and methyllithium. This method allowed net replacement of the carbon-oxygen bonds of a carbonyl group by either C—H or C—R bonds (reductive alkylation) or by two C—R bonds (geminal bis-alkylation) in a highly stereoselective fashion [220, 221]. It has been utilized for instance to generate the methyl deoxypodocarpate *414*, an important structural unit in many natural products. The key step of this synthesis involved the base induced cleavage of the ring of the α,α-dithiocyclobutanone *412*, prepared by spiroannelation of the suitable tricyclic ketone with the diphenylsulfonium cyclopropylide *103* (*vide infra*, Sect. 4.5, Eq. (30) followed by α-thioacetalisation [223]. Then, hydrolysis of the dithiane *413* and decarbonylation

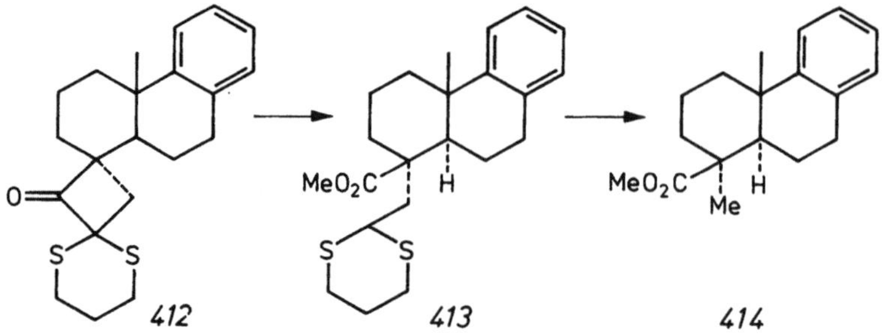

(100)

with Wilkinson's catalyst [224] gave, in a highly stereoselective fashion, methyl deoxypodocarpate *414*, Eq. (100) [221].

Addition of nucleophiles to 2-vinylcyclobutanone followed by epoxidation provided systems which underwent base-induced ring-opening with hydride, alkyl and aryl organometallic reagents as well as with ester enolates [63]. Among the various possible applications, this functional carbon chain elongation procedure has been used in cyclopentane synthesis. Thus, the aldol product of 1-arylthiocyclopropanecarboxaldehyde *119* with 3-(benzyloxy)octan-2-one was dehydrated to provide the vinylcyclopropane *415* (*vide supra*, Sect. 4.6.1, Eq. (36)). After reduction with DIBAH and *n*-BuLi the allylic alcohol *416* then underwent acid-induced $C_3 \rightarrow C_4$ ring expansion upon treatment with 48% HBF_4 to the 2-vinylcyclobutanone *417* (*vide supra*, Sect. 5.1.2, Eq. (74)). Addition of the ester enolate prepared from methyl acetate and LDA followed by epoxidation with m-chloroperbenzoic acid gave the epoxide *418* which was directly fragmented upon treatment with methanolic magnesium methylate to give a 1:1 mixture of Z and E allylic alcohols *419* [63]. Finally, direct reaction of *419* with boron trifluoride etherate in CH_2Cl_2 at 0 °C led to the O-alkylation product *420* which underwent smooth isomerization with bis(1,2-diphenylphosphinoethane)-palladium in refluxing dioxane, in the presence of O,N-bis(trimethylsilyl) acetamide to avoid decarbomethoxylation [225,226] to provide cyclopentanone *421*, the Roussel-Uclaf intermediate to prostaglandins PGA_2 [227] and PGE [228], opening a formal synthetic entry into PG$'_s$ (for a comparable entry to prostanoids *vide supra*, Scheme 8)

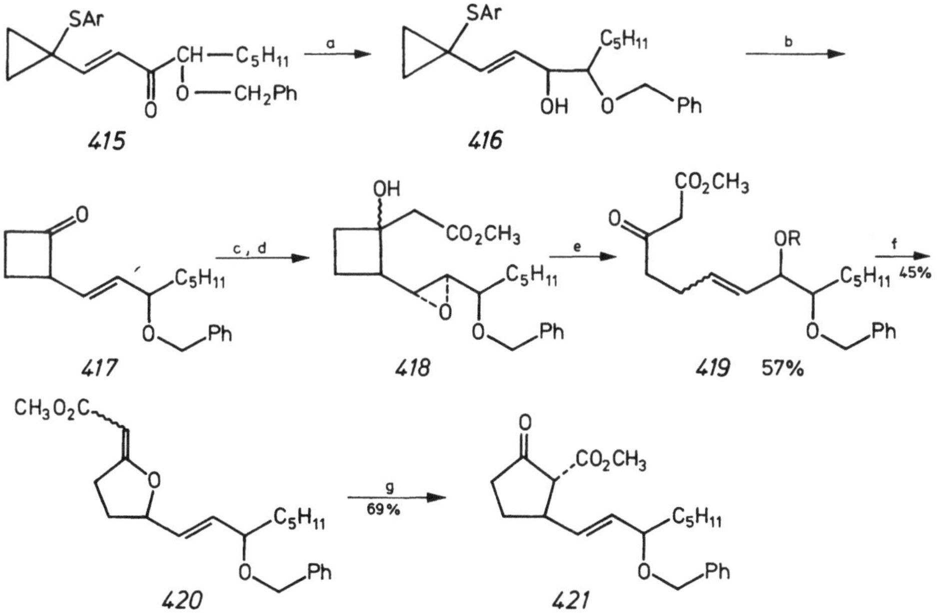

415 416 417 418 419 420 421

Scheme 15. Synthesis of prostaglandin precursors [63]
Reagents: a) DIBAH, n-BuLi, THF, r.t., 98%; b) 48% HBF_4, Et_2O, 41%; c) LDA, MeOAc, Et_2O, −78 °C, 85%; d) MCPBA, CH_2Cl_2, r.t.; e) $Mg(OCH_3)_2$, MeOH, 2°C, 72 hr; f) BF_3—Et_2O, CH_2Cl_2, −78 °C, 0 °C, 45%; g) $(DIPHOS)_2Pd$, $Me(Me_3SiO)C = SiMe_3$, dioxane, reflux, 5 hr, 69%.

[15]. Furthermore, a lactone related to *421* has been found to have hypotensive activity [229] (Scheme 15) [63].

6 Conclusion

1-Donor substituted ethynylcyclopropanes have been prepared from the readily available cyclopropanone hemiacetal or from various sources of chlorovinyl-, chlorovinylidene- and ethynylcarbenes as well as from the thermal rearrangement of ethynyl vinyl oxiranes. Surprisingly unreactive towards acids, they undergo $C_3 \rightarrow C_4$ ring expansion only with *t*-butylhypochlorite or with metachloroperbenzoic acid; their thermal $C_3 \rightarrow C_5$ (C_6 or C_7) ring expansions required the presence of a 2-methyl or 2-vinyl substituent on the three-membered ring. The ring opening of the cyclopropyl cation could be completely avoided by delocalization of the positive charge through the adjacent triple bond.

1-Donor substituted vinylcyclopropanes appears to be much more attractive building blocks owing to their exceptional reactivity. Various routes have been developed towards these challenging compounds involving the simple nucleophilic addition to the cyclopropanone hemiacetal of vinylic- or acetylenic organometallic reagents followed by stereoselective hydride reduction, the iron induced cyclization of 1,3-dichloroacetone adducts, the chemoselective Simmons-Smith cyclopropanation of α-enone silyl enol ethers, the dye sensitized photooxygenation of alkylidenecyclopropanes, the regioselective base induced ring opening of oxaspiropentanes, the nucleophilic addition of 1-heterosubstituted lithiocyclopropanes to carbonyl compounds followed by acid-induced dehydration, as well as the simple Wittig olefination with 1-hydroxycyclopropylcarbonyl compounds. While 1-vinylcyclopropanol and vinylogous derivatives undergo thermal or acid induced $C_3 \rightarrow C_4$ ring expansion leading to cyclobutanone derivatives under mild conditions, the rearrangement of 1-phenylthio-, 1-methyl (or phenyl) seleno- and 1-methoxyvinylcyclopropanes required more drastic conditions. It is worth noting that 2-vinylcyclobutanones are readily formed in this way and that these are efficient precursors of the C_5, C_6 or C_8 homologous rings, which are formed by subsequent thermal, acid- and base-induced as well as photolytic rearrangements, as illustrated by the total synthesis of dihydrojasmone and a methylenomycin B precursor (C_5), of (—)-β-selinene and a compactin precursor (C_6) and of (—) poitediol (C_8). From the synthetic point of view, the thermal vinylcyclopropane — cyclopentene ring expansion is also very useful; thus the accelerating substituent effect of the trimethylsiloxy group of O-silylated 1-vinylcyclopropanol derivatives allowed the chemo-, regio- and stereoselective formation of cyclopentenol silyl ethers leading to cyclopentanone, cyclopentenone and spiroketone derivatives which encompass an important class of biologically active substances, as illustrated by the total synthesis of (±) 11-deoxyprostaglandin E_2 methyl ester, of (±) aphidicolin, of a spirovetivane, of dihydrojasmone and cis-jasmone and of dicranenone A. Although the 1-arylthio-, and 1-trimethylsilylvinylcyclopropane derivatives, in general, also underwent the thermal $C_3 \rightarrow C_5$ ring expansion providing cyclopentenol thioethers and 1-trimethylsilylcyclopentene respectively, the final generation of the carbonyl group was more difficult compared to the simple methanolysis of a silyl enol ether.

The usefulness of the ring opening of these C_3 or C_4 derivatives has also been demonstrated by the regioselective α- or α′-monomethylation of the α-enones and by the highly stereoselective preparations of (±) grandisol and of (±) methyl deoxypodocarpate. The ring opening of these cyclopropanol derivatives induced by Lewis acids such as $TiCl_4$, $ZnCl_2$, Me_3SnCl, etc. allowing subsequent transmetalation also appears to be very promising. Now easily available, the 1-donor substituted vinylcyclopropanes therefore constitute highly recommendable building blocks.

7 Acknowledgement

I wish to thank my co-workers for their contributions to the work mentioned in this review. I express my gratitude to Prof. J. K. Crandall and to Drs. J. P. Barnier, A. Fadel, G. Rousseau and J. Seyden for reading and commenting the manuscript, and the Centre National de la Recherche Scientifique for financial support.

8 References

1. Perkin, W. H.: Chem. Ber. *17*, 54 (1884); Röder, F.: Ann. Chem. *227*, 13 (1885); Fittig, R.: Ann. Chem. *227*, 25 (1885)
2. a) de Meijere, A.: Angew. Chem. *91*, 867 (1979); Angew. Chem. Int. Ed. Engl. *18*, 809 (1979); b) de Meijere, A.: Chemie in unserer Zeit *16*, 13 (1982) Weinheim, Verlag Chemie 1982; c) de Meijere, A.: Bull. Soc. Chim. Belges *93*, 241 (1984)
3. Salaün, J.: The chemistry of the cyclopropyl group, Chap. 12 (Rappoport, Z., ed.), London—New York—Sydney, Interscience Publishers, in press
4. Hudlicky, T., Kutchan, T. M., Naqvi, S. M.: Organic Reactions, Vol. 33, p. 247, New York—Chichester—Brisbane—Toronto—Singapore, John Wiley and sons, inc. 1985
5. Wiberg, K. B., Hess Jr, B. A., Ashe III, A. J.: Carbonium Ions, Vol. 3, p. 1295 (Olah, G. A., Schleyer, P. v. R., ed.), New York, Wiley Interscience 1972
6. Turro, N. J.: Acc. Chem. Res. *25*, 2 (1969); Wasserman, H. H., Clark, G. M., Turley, P. C.: Top. Curr. Chem. *47*, 73 (1974)
7. Salaün, J., Marguerite, J.: Org. Synth. *63*, 47 (1984)
8. Salaün, J.: Chem. Rev. *83*, 619 (1983)
9. Lipp, P., Koster, R.: Liebigs Ann. Chem. *499*, 1 (1932); Chem. Ber. *64*, 2823 (1931)
10. Wasserman, H. H., Cochoy, R. E., Baird, M. S.: J. Am. Chem. Soc. *91*, 2375 (1969)
11. Rousseau, G., Slougui, N.: Tetrahedron Lett. *24*, 1251 (1983)
12. Wasserman, H. H., Clagett, D. C.: J. Am. Chem. Soc. *88*, 5368 (1966)
13. Salaün, J.: J. Org. Chem. *41*, 1237 (1976); J. Org. Chem. *42*, 28 (1977)
14. Cochoy, R. E.: Ph. D. Dissertation, Yale University 1969
15. Salaün, J., Ollivier, J.: Nouv. J. Chem. *5*, 587 (1981)
16. Salaün, J., Bennani, F., Compain, J. C., Fadel, A., Ollivier, J.: J. Org. Chem. *45*, 4129 (1980)
17. Brown, H. C., Rao, C. G.: ibid. *43*, 3602 (1978)
18. Holmes, A. B., Jennings-White, C. L. D., Schulthess, A. H.: J. Chem. Soc. Chem. Comm. 840 (1979)
19. Tobey, S. W., West, R.: J. Am. Chem. Soc. *88*, 2478, 2481 (1966)
20. Weber, W., de Meijere, A.: Angew. Chem. *92*, 135 (1980); Angew. Chem. Int. Ed. Engl. *19*, 138 (1980); Chem. Ber. *118*, 2450 (1985); Kostikov, R., de Meijere, A.: J. Chem. Soc. Chem. Commun. *1984*, 1528
21. Liese, T., de Meijere, A.: Angew. Chem. *94*, 65 (1982); Angew. Chem. Int. Ed. Engl. *21*, 65 (1982); Liese, T., Splettstoesser, G., de Meijere, A.: Tetrahedron Lett. *23*, 3341 (1982); Liese, T., de Meijere, A.: Chem. Ber. *119*, 2995 (1986); Bengtson, G., Keyaniyan, S., de Meijere, A.: Chem. Ber. *119*, 3607 (1986)
22. Keyaniyan, S., Göthling, W., de Meijere, A.: Tetrahedron Lett. *25*, 4105 (1984); Chem. Ber.

120, 395 (1987); Keyaniyan, S., Apel, M., Richmond, J. P., de Meijere, A.: Angew. Chem. *97*, 763 (1985); Angew. Chem. Int. Ed. Engl. *24*, 770 (1985)
23. Franck-Neumann, M., Geoffroy, P., Lohmann, J. J.: Tetrahedron Lett. *24*, 1775 (1983); Franck-Neumann, M., Geoffroy, P.: Tetrahedron Lett. *24*, 1779 (1983)
24. Hauptmann, H.: Tetrahedron *32*, 1293 (1976)
25. Bourelle-Wargnier, F., Vincent, M., Chuche, J.: J. Org. Chem. *45*, 428 (1980)
26. Dalacker, V., Hopf, H.: Tetrahedron Lett. 15 (1974)
27. Dolbier, Jr., W. R., Garza, O. T., Al-Sader, B. H.: J. Am. Chem. Soc. *97*, 5038 (1975); Tetrahedron Lett. 887 (1976)
28. Salaün, J., Ollivier, J.: unpublished
29. Roberts, J. D., Chambers, V. C.: J. Am. Chem. Soc. *73*, 5034 (1951); Brown, H. C., Rao, C., Ravindanathan, M.: J. Am. Chem. Soc. *99*, 7663 (1977) and references cited therein
30. Berson, J. A., Olin, S. S.: ibid. *92*, 1086 (1970)
31. Howell, B. A., Jewett, J. G.: ibid. *93*, 798 (1971)
32. Landgrebe, J. A., Becker, L. W.: ibid. *89*, 2505 (1967)
33. Grunwald, E., Winstein, S.: ibid. *70*, 846 (1948)
34. Deppuy, C. H., Schnack, L. G., Hausser, J. W.: ibid. *88*, 3343 (1966)
35. Schleyer, P. v. R., Fry, J. L., Lam, L. K. M., Lancelot, C. J.: ibid. *92*, 2542 (1970)
36. Radom, L., Hariharan, P. C., Pople, J. A., Schleyer, P. v. R.: ibid. *95*, 6531 (1973)
37. Salaün, J.: J. Org. Chem. *43*, 2809 (1978)
38. Sato, F., Ishikawa, H., Sato, M.: Tetrahedron Lett. *21*, 365 (1980)
39. Ollivier, J.: Ph. D. Dissertation, University of Paris XI, Orsay 1986
40. Deppuy, C. H., Dappen, G. M., Eilers, K. L., Klein, R. A.: J. Org. Chem. *29*, 2813 (1964); Gibson, D. H., De Puy, C. H.: Chem. Rev. *74*, 605 (1974)
41. Salaün, J., Conia, J. M.: Tetrahedron Lett. 2849 (1972); Salaün, J., Garnier, B., Conia, J. M.: Tetrahedron *30*, 1413 (1974)
42. Denis, J. M., Girard, C., Conia, J. M.: Synthesis 549 (1972); Conia, J. M., Girard, C.: Tetrahedron Lett. 2767 (1973); Girard, C., Conia, J. M.: Tetrahedron Lett. 3333 (1974); Girard, C., Amice, P., Barnier, J. P., Conia, J. M.: Tetrahedron Lett. *37*, 3329 (1974); Conia, J. M.: Pure Appl. Chem. *43*, 317 (1975)
43. a) Wenkert, E., Berges, D. A., Golob, N. F.: J. Am. Chem. Soc. *100*, 1263 (1978); b) Wenkert, E.: Acc. Chem. Res. *13*, 27 (1980) and references cited therein
44. Ultimoto, K., Tamura, M., Sisido, K.: Tetrahedron *29*, 1169 (1973)
45. Sousa, J. A., Bluhm, A. L.: J. Org. Chem. *25*, 108 (1960)
46. Salaün, J., Conia, J. M.: J. Chem. Soc. Chem. Commun. 1579 (1971); Salaün, J., Champion, J., Conia, J. M.: Org. Synth. *57*, 36 (1977)
47. Bartlett, P. D., Ho, M. S.: J. Am. Chem. Soc. *96*, 627 (1974)
48. Rousseau, G., Le Perchec, P., Conia, J. M.: Tetrahedron *32*, 2533 (1976); Tetrahedron *34*, 3475 (1978)
49. Crandall, J. K., Paulson, D. R.: J. Org. Chem. *33*, 991 and 3291 (1968)
50. Wiseman, J. R., Chan, H. F.: J. Am. Chem. Soc. *92*, 4749 (1970)
51. Aue, D. H., Meshishnek, M. J., Shellhamer, D. F.: Tetrahedron Lett. 4799 (1973); Erden, I., de Meijere, A., Rousseau, G., Conia, J. M.: Tetrahedron Lett. *21*, 2501 (1980)
52. Makosza, M., Wawrzyniewicz, M.: ibid. 4659 (1969)
53. Braun, M., Seebach, D.: Angew. Chem. *86*, 279 (1974); Angew. Chem. Int. Ed. Engl. *13*, 277 (1974); Braun, M., Dammann, R., Seebach, D.: Chem. Ber. *108*, 2368 (1975)
54. Dammann, R., Braun, M., Seebach, D.: Helv. Chim. Acta *59*, 2821 (1976); Hiyama, T., Takehara, S., Kitatani, K., Nozaki, H.: Tetrahedron Lett. 3295 (1974)
55. Wiechert, R.: Angew. Chem. *82*, 219 (1970); Angew. Chem. Int. Ed. Engl. *9*, 237 (1970)
56. Johnson, C. R., Katekar, G. F., Huxol, R. F., Janiga, E. R.: J. Am. Chem. Soc. *93*, 371 (1971)
57. Bogdanowicz, M. J., Trost, B. M.: Tetrahedron Lett. 887 (1972); Trost, B. M., Bogdanowicz, M. J.: J. Am. Chem. Soc. *95*, 289, 5311 (1973)
58. Trost, B. M., Kurozumi, S.: Tetrahedron Lett. 1929 (1974)
59. Trost, B. M., Scudder, P. H.: J. Am. Chem. Soc. *99*, 7601 (1977)
60. Truce, W. E., Hollister, K. R., Lindy, L. B., Parr, J. E.: J. Org. Chem. *33*, 43 (1968)
61. Trost, B. M., Keeley, D., Bogdanowicz, M. J.: J. Am. Chem. Soc. *95*, 3068 (1973); Trost, B. M.,

Keeley, D. E., Arndt, H. C., Rigby, J. M., Bogdanowicz, M. J.: J. Am. Chem. Soc. *99*, 3080 (1977)
62. Trost, B. M., Keeley, D. E.: ibid. *98*, 248 (1976)
63. a) Trost, B. M., Jungheim, L. N.: J. Am. Chem. Soc. *102*, 7910 (1980); b) B. M. Trost: Top. Curr. Chem. *133*, 3 (1986)
64. Byers, J. H., Spencer, T. A.: Tetrahedron Lett. *26*, 717 (1985)
65. Gadwood, R. C.: ibid. *25*, 5851 (1984)
66. Halazy, S., Lucchetti, J., Krief, A.: ibid. 3971 (1978)
67. Burgess, E. M., Penton, H. R., Taylor, E. A.: J. Org. Chem. *38*, 26 (1973)
68. Halazy, S., Krief, A.: Tetrahedron Lett. *22*, 1829 (1981)
69. Masuyama, Y., Ueno, Y., Okawara, M.: Chem. Lett. 835 (1977)
70. a) Halazy S., Krief, A.: Tetrahedron Lett. *22*, 1833 (1981); b) Krief, A.: Top. Curr. Chem. *135*, 1 (1987)
71. Cohen, T., Matz, J. R.: J. Am. Chem. Soc. *102*, 6900 (1980)
72. Braun, M., Seebach, D.: Chem. Ber. *109*, 669 (1976)
73. Cohen, T., Matz, J. R.: Tetrahedron Lett. *22*, 2455 (1981)
74. Gadwood, R. C., Rubino, M. R., Nagarajan, S. C., Michel, S. T.: J. Org. Chem. *50*, 3255 (1985)
75. Colvin, E.: Silicon in Organic Synthesis, London—Boston—Sydney—Willington—Durban—Toronto, Butterworths 1980
76. Seyferth, D., Cohen, H. M.: Inorg. Chem. *1*, 913 (1962); Sakurai, H., Hosomi, A., Kumada, M.: Tetrahedron Lett. 2469 (1968)
77. Cohen, T., Sherbine, J. P., Matz, J. R., Hutchins, R. R., Mc Henry, B. M., Willey, P. R.: J. Am. Chem. Soc. *106*, 3245 (1984)
78. Tanaka, K., Uneme, H., Matsui, S., Tanikaga, R., Kaji, A.: Chem. Lett. 287 (1980)
79. Paquette, L. A., Horn, K. A., Wells, G. J.: Tetrahedron Lett. *23*, 259 (1982)
80. Halazy, S., Dumont, W., Krief, A.: ibid. 4737 (1981)
81. Paquette, L. A., Wells, G. J., Horn, K. A., Yan, T. H.: Tetrahedron *39*, 913 (1983)
82. Warner, P. M., Le, D.: J. Org. Chem. *47*, 893 (1982); Ainsworth, C., Kuo, Y. N.: J. Organometal. Chem. *46*, 73 (1972)
83. Gröbel, B. T., Seebach, D.: Angew. Chem. *86*, 102 (1974); Angew. Chem. Int. Ed. Engl. *13*, 83 (1974); Chan, T. H., Mychajlowskij, W.: Tetrahedron Lett. 3479 (1974)
84. Chan, T. H., Mychajlowskij, W., Ong, B. S., Harpp, D. N.: J. Org. Chem. *43*, 1526 (1978)
85. Sawada, S., Inouye, Y.: Bull. Chem. Soc. Japan *42*, 2669 (1969)
86. Wells, G. J., Yan, T. H., Paquette, L. A.: J. Org. Chem. *49*, 3604 (1984)
87. Ojima, I., Kogure, T., Nagai, Y.: Chem. Lett. 985 (1975); Noyori, R., Umeda, I., Ishigami, T.: J. Org. Chem. *37*, 1542 (1972)
88. Ohkata, K., Paquette, L. A.: J. Am. Chem. Soc. *102*, 1082 (1980)
89. Yan, T. H., Paquette, L. A.: Tetrahedron Lett. *23*, 3227 (1982)
90. Trost, B. M., Salzmann, T. N., Hiroi, K. J.: J. Am. Chem. Soc. *98*, 4887 (1976)
91. Paquette, L. A., Yan, T. H., Wells, G. J.: J. Org. Chem. *49*, 3610 (1984); Paquette, L. A., Wells, G. J., Wickham, G.: J. Org. Chem. *49*, 3618 (1984)
92. Salaün, J., Almirantis, Y.: Tetrahedron *39*, 2421 (1983)
93. Rühlmann, K.: Synthesis 236 (1971); Bloomfield, J. J., Owsley, D. C., Nelke, J. M.: Org. Reactions *23*, 259 (1976); Bloomfield, J. J., Nelke, J. M.: Org. Synth. *57*, 1 (1977)
94. Heine, H. G.: Chem. Ber. *104*, 2869 (1971); Denis, J. M., Conia, J. M.: Tetrahedron Lett. 2845 (1971); Barnier, J. P., Denis, J. M., Salaün, J., Conia, J. M.: Tetrahedron *30*, 1405 (1974)
95. Heine, H. G., Wendisch, D.: Liebigs Ann. Chem. 463 (1976)
96. Ollivier, J., Salaün, J.: Tetrahedron Lett. *25*, 1269 (1984)
97. Warner, P. M., Le, D.: J. Org. Chem. *47*, 893 (1982)
98. Konen, D. A., Silbert, L. S., Pfeffer, P. E.: ibid. *40*, 3253 (1975)
99. Werstiuk, N. H.: Tetrahedron *39*, 205 (1983)
100. Sauers, R. R., Schlosberg, S. B., Pfeffer, P. E.: J. Org. Chem. *33*, 2175 (1968); De Puy, C. H., Breitbeil, F. W., de Bruin, K. R.: J. Am. Chem. Soc. *88*, 3347 (1966)
101. Miyashita, N., Yoshikoshi, A., Grieco, P. A.: J. Org. Chem. *42*, 3772 (1977)
102. Brook, P. R., Harrison, J. M.: J. Chem. Soc. Chem. Commun. 997 (1972)

103. Corey, E. J., Gras, J. L., Ulrich, P.: Tetrahedron Lett. 809 (1976)
104. Corey, E. J., Wollenberg, R. H.: ibid. 4705 (1976)
105. Corey, E. J., Venkateswarlu, A.: J. Am. Chem. Soc. *94*, 6190 (1972)
106. Sugasawa, S., Zasski, J.: J. Pharm. Soc. Japan **550** 1050 (1927)
107. Corey, E. J., Suggs, J. W.: Tetrahedron Lett. **2647** (1975)
108. Corey, E. J., Schmidt, G.: ibid. 399 (1979)
109. Mancuso, A. J., Huang, S. L., Swern, D.: J. Org. Chem. *43*, 2480 (1978)
110. Salaün, J., Fadel, A., Conia, J. M.: Tetrahedron Lett. 1429 (1979)
111. Barnier, J. P., Salaün, J.: ibid. *25*, 1273 (1984) and references cited therein
112. Tilborg, W. J. M., Schaafsma, S. E. Steinberg, H., de Boer, T. J.: Rec. Trav. Chim. Pays-Bas *86*, 419 (1967)
113. Kato, K.: Tetrahedron Lett. *21*, 4925 (1980)
114. Stork, G., Depezay, J. C., d'Angelo, J.: ibid. 389 (1975)
115. Schlosser, M., Brich, Z.: Helv. Chim. Acta *61*, 1903 (1978)
116. Beckmann, S., Geiger, H.: Methoden Org. Chem. (Houben-Weyl) *4*, 445 (1971)
117. Otto, H. H.: Dtsch. Apoth. Ztg. *115*, 89 (1975)
118. Tarchini, C., Rohmer, M., Djerassi, C.: Helv. Chim. Acta *62*, 1210 (1979)
119. Karpf, M., Djerassi, C.: J. Am. Chem. Soc. *103*, 302 (1981)
120. Pettus, J. A. Jr., Moore, R. E.: J. Chem. Soc. Chem. Commun. 1093 (1970); Moore, R. E., Pettus, J. A. Jr., Mistysyn, J.: J. Org. Chem. *39*, 2201 (1974); Müller, D. G., Clayton, M. N., Gassmann, G., Boland, W., Marner, F. J., Jaenicke, L.: Experientia *40*, 211 (1984); Müller, D. G., Clayton, M. N., Gassmann, G.: Boland, W., Marner, F. J., Schotten, T., Jaenicke, L.: Naturwiss. *72*, 97 (1985)
121. Schotten, T., Boland, W., Jaenicke, L.: Helv. Chim. Acta *68*, 1186 (1985)
122. Dorsch, D., Kunz, E., Helmchen, G.: Tetrahedron Lett. *26*, 3319 (1985)
123. Colobert, F., Genet, J. P.: Tetrahedron Lett. *26*, 2779 (1985)
124. Guibe-Jampel, E., Rousseau, G., Salaün, J.: J. Chem. Soc. Chem. Commun. 1080 (1987)
125. Salaün, J., Karkour, B.: Tetrahedron Lett., 1080 1987
126. Sullivan, G. R.: Topics in Stereochem. *10*, 287 (1978); Fraser, R. R.: Asymmetric Syntheisis (Morrison, J. D., ed.) Vol. 1, p. 173, New York, Academic Press 1983
127. a) Piers, E., Banville, J., Lau, C. K., Nagakura, I.: Can. J. Chem. *60*, 2965 (1982); b) Piers, E., Lau, C. K., Nagakura, I.: Can. J. Chem. *61*, 288 (1983)
128. Wasserman, H. H., Dion, R. P.: Tetrahedron Lett. *23*, 785 (1982)
129. a) Wasserman, H. H., Dion, R. P.: ibid. *23*, 1413 (1982); b) *24*, 3409 (1983)
130. Schmidt, A., Köbrich, G.: ibid. 2561 (1974)
131. Richey Jr., H. G.: Cyclopropylcarbonium Ions, in Carbonium ions, Vol. III, Chap. 25, p. 1201 (ed. Olah, G. A., Schleyer, P. v. R.) New York, Wiley-Interscience 1972
132. Conia, J. M., Robson, M. J.: Angew. Chem. *87*, 505 (1975); Angew. Chem. Int. Ed. Engl. *14*, 473 (1975) and references cited therein
133. Wasserman, H. H., Hearn, M. J., Cochoy, R. E.: J. Org. Chem. *45*, 2874 (1980)
134. Crandall, J. K., Conover, W. W.: ibid. *43*, 3533 (1978)
135. Woodward, R. B., Hoffman, R.: Angew. Chem. *81*, 797 (1969); Angew. Chem. Int. Ed. Engl. *8*, 781 (1969)
136. Pittman, C. V., Olah, G. A.: J. Am. Chem. Soc. *87*, 2998 and 5123 (1965); Olah, G. A., Jeuell, C. L., Kelly, D. P., Porter, R. D.: J. Am. Chem. Soc. *94*, 146 (1972)
137. Yamakado, Y., Ishiguro, M., Ikeda, N., Yamamoto, H.: ibid. *103*, 5568 (1981)
138. Patterson Jr., J. W., Fried, J. H.: J. Org. Chem. *39*, 2506 (1974)
139. Trost, B. M., Keeley, D. E., Arndt, H. C., Bogdanowicz, M. J.: J. Am. Chem. Soc. *99*, 3088 (1977)
140. See references 29 in ref. 139
141. Halazy, S., Zutterman, F., Krief, A.: Tetrahedron Lett. *23*, 4385 (1982)
142. Danheiser, R. L., Martinez-Davila, C., Sard, H.: Tetrahedron *37*, 3943 (1981)
143. Jackson, D. A., Rey, M., Dreiding, A. S.: Helv. Chim. Acta *66*, 2330 (1983); Tetrahedron Lett. *24*, 4817 (1983)
144. Eaton, P. E., Carlson, G. R., Lee, J. T.: J. Org. Chem. *38*, 4071 (1973)
145. Matz, J. R., Cohen, T.: Tetrahedron Lett. *22*, 2459 (1981)
146. Jemow, J., Tautz, W., Rosen, P., Williams, T. H.: J. Org. Chem. *44*, 4212 (1979); Mikolajczyk,

M.: Current Trends in Organic Synthesis (Nozaki, H., ed.) New York, Oxford, Pergamon Press, 347 (1983)
147. Barnier, J. P., Karkour, B., Salaün, J.: J. Chem. Soc. Chem. Commun. 1270 (1985)
148. Gadwood, R. C.: J. Org. Chem. *48*, 2098 (1983)
149. Houk, K. N.: Chem. Rev. *76*, 1 (1976); Dauben, W. G., Lobder, DG. Ipaktschi: Fortschr. Chem. Forsch. *54*, 73 (1975)
150. Lyle, T. A., Mereyala, H. B., Pascual, A., Frei, B.: Helv. Chim. Acta *67*, 774 (1984)
151. Brown, H. C., Krishnamurthy, S.: J. Am. Chem. Soc. *94*, 7159 (1972)
152. Mitsui, S., Ito, M., Naubu, A., Senda, Y.: J. Catal. *36*, 119 (1975)
153. Cohen, T., Bhupathy, M., Matz, J. R.: J. Am. Chem. Soc. *105*, 520 (1983)
154. Yamamoto, Y., Yamamoto, S., Yatagai, H., Ishihara, Y., Maruyama, K.: J. Org. Chem. *47*, 119 (1982)
155. Mc Kenzie, B. D., Angelo, M. M., Wolinsky, J. J.: ibid. *44*, 4042 (1979)
156. Wang, N. Y., Hsu, C. T., Sih, C. J.: J. Am. Chem. Soc. *103*, 6538 (1981); Funk, R. L., Zeller, W. E.: J. Org. Chem. *47*, 180 (1982)
157. Brown, A. G., Smale, T. C., King, T. J., Hasenkamp, R., Thompson, R. H.: J. Chem. Soc. Perkin Trans 1, 1165 (1976); Brown, M. S., Faust, J. R., Goldstein, J. L., Kaneko, I., Endo, A.: J. Biol. Chem. *253*, 1121 (1978) and references cited therein
158. Evans, D. A., Truesdale, L. K., Caroll, G. L.: J. Chem. Soc. Chem. Commun. 55 (1973); Deuchert, K., Hertenstein, U., Hünig, S.: Synthesis 777, 1973; Deuchert, K., Hertenstein, U., Hünig, S., Wehner, G.: Chem. Ber. *112*, 2045 (1979)
159. Byers, J. H., Spencer, T. A.: Tetrahedron Lett. *26*, 713 (1985)
160. Illuminati, G., Mandolini, L.: Acc. Chem. Res. *14*, 95 (1981)
161. Gadwood, R. C., Lett, R. M.: J. Org. Chem. *47*, 2268 (1982)
162. Berson, J. A., Dervan, P. B.: J. Am. Chem. Soc. *95*, 267, 269 (1973) and references cited therein
163. Fenical, W., Schulte, G. R., Finer, J., Clardy, J.: J. Org. Chem. *43*, 3628 (1978)
164. Furukawa, J., Kawabata, N., Nishimura, J.: Tetrahedron Lett. 3353 (1966); Tetrahedron *24*, 53 (1968)
165. Lipschutz, B. H., Pegram, J. J.: Tetrahedron Lett. 3343 (1980)
166. Gadwood, R. C., Lett, R. M., Wissinger, J. E.: J. Am. Chem. Soc. *106*, 3869 (1984)
167. Neureiter, N. P.: J. Org. Chem. *24*, 2044 (1959)
168. Milvitskaya, E. M., Tarakanova, A. V., Plate, A. F.: Russ. Chem. Rev. *45*, 469 (1976); Frey, H. M., Walsh, R.: Chem. Rev. *69*, 103 (1969); Frey, H. M.: Adv. Phys. Org. Chem. *4*. 147 (1966); Gutsche, C. D., Redmore, D.: Adv. Alicyclic Chem. *2*, 161 (1968); Willcott, M. R., Cargill, R. L., Sears, A. B.: Prog. Phys. Org. Chem. *9*, 25 (1972); Hudlicky, T., Kutchan, T. M., Naqvi, S. M.: Org. Reactions *33*, 246 (1985)
169. Alonso, M., Morales, A.: J. Org. Chem. *45*, 4530 (1980)
170. Stevens, R. V.: The total synthesis of Natural Products (ed. ApSimon, J.), Vol. 3, p. 439, New York, John Wiley and Sons 1977
171. Danheiser, R. L., Martinez-Davila, C., Morin, J. M.: J. Org. Chem. *45*, 1340 (1980); Danheiser, R. L., Martinez-Davila, C., Auchus, R. J., Kadonaga, J. T.: J. Am. Chem. Soc. *103*, 2443 (1981); Danheiser, R. L., Bronson, J. J., Okano, K.: J. Am. Chem. Soc. *107*, 4579 (1985)
172. Trost, B. M., Scudder, P. H.: J. Org. Chem. *46*, 506 (1981)
173. Monti, S. A., Cowherd, F. G., Mc Aninch, T. W.: ibid. *40*, 858 (1975)
174. Trost, B. M.: Top. Curr. Chem. *41*, 1 (1973); Acc. Chem. Res. *7*, 85 (1974); Pure Appl. Chem. *43*, 565 (1975)
175. Stork, G., Hudrlik, P. F.: J. Am. Chem. Soc. *90*, 4462, 4464 (1968); House, H. O., Czuba, L. J., Gall, M., Olmstead, H. D.: J. Org. Chem. *34*, 2324 (1969); House, H. O., Gall, M., Olmstead, H. D.: J. Org. Chem. *36*, 2361 (1971)
176. Luo, F. T., Negishi, E.: Tetrahedron Lett. *26*, 2177 (1985); J. Org. Chem. *50*, 4762 (1985)
177. For a recent review on Lewis acid induced α-alkylation of carbonyl compounds, see: Reetz, M. T.: Angew. Chem. Int. Ed. Engl. *21*, 96 (1982). See also: Colvin, E. in "Silicon in Organic Synthesis", p. 221, London, Butterworths 1981
178. Mikołajczylk, M., Grzejszczak, S., Lyzwa, P.: Tetrahedron Lett. *23*, 2237 (1982) and references cited therein
179. Trost, B. M.: Chem. Soc. Rev. *11*, 141 (1982) and references cited therein

180. Visser, R. G., Bos, H. J. T., Brandsma, L.: Rec. Trav. Chim. Pays-Bas *99*, 70 (1980)
181. Trost, B. M., Nishimura, Y., Yamamoto, K., Mc Elvain, S. S.: J. Am. Chem. Soc. *101*, 1328 (1979)
182. Ito, Y., Hirao, T., Saegusa, T. J.: J. Org. Chem. *43*, 1011 (1978)
183. Dalziel, W., Hesp, B., Stevenson, K. M., Jarvis, J. A.: J. Chem. Soc. Perkin Trans I, 2841 (1973)
184. Marshall, J. A., Brady, S. F., Anderson, N. H.: Fortsch. Chem. Org. Naturst. *31*, 283 (1974)
185. Subrahamanian, K. P., Reusch, W.: Tetrahedron Lett. 3789 (1978); Ibuka, T., Hayashi, K., Minakata, H., Inubushi, Y.: Tetrahedron Lett. 159 (1979) and references cited therein
186. Caine, D., Boucugnani, A. A., Pennington, W. R.: J. Org. Chem. *41*, 3632 (1976); Büchi, G., Berthet, D., Decorzant, R., Grieder, A., Hauser, A.: J. Org. Chem. *41*, 3208 (1976)
187. Creary, X., Rollin, A. J.: ibid. *44*, 1017 (1979)
188. Ollivier, J., Salaün, J.: J. Chem. Soc. Chem. Commun. 1269 (1985)
189. Ryu, I., Murai, S., Niwa, I., Sonoda, N.: Synthesis 874 (1977)
190. Liotta, D., Barnum, C. S., Saindane, M.: J. Org. Chem. *46*, 4301 (1981)
191. Ichikawa, T., Namikawa, M., Yamada, K., Sakai, K., Kondo, K.: Tetrahedron Lett. *24*, 337 (1983); Sakai, K., Fujimoto, T., Yamashita, M., Kondo, K.: Tetrahedron Lett. *26*, 2089 (1985)
192. Paterson, I.: ibid 1519 (1979)
193. Reissig, H. U., Hirsch, E.: Angew. Chem. *92*, 839 (1980); Angew. Chem. Int. Ed. Engl. *19*, 813 (1980)
194. Tanaka, H., Torii, S.: J. Org. Chem. *40*, 462 (1975)
195. Smith, W. N., Beumel, Jr., O. F.: Synth. Commun. 441 (1974)
196. Corey, E. J., Raju, N.: Tetrahedron Lett. *24*, 5571 (1983)
197. Corey, E. J., Gross, A. W.: ibid. *25*, 495 (1984)
198. Reich, H. J., Renga, J. M., Reich, I. L.: J. Am. Chem. Soc. *97*, 5434 (1975)
199. Boar, R. B., Hawkins, D. W., McGhie, J. F., Barton, D. H. R.: J. Chem. Soc. Perkin Trans *1*, 654 (1973)
200. Geiss, K., Seuring, B., Pieter, R., Seebach, D.: Angew. Chem. *86*, 484 (1974); Angew. Chem. Int. Ed. Engl. *13*, 479 (1974)
201. An α-trimethylsilyl group has been shown to destabilize a radical. cf. Sommer, L. H., Dorfman, F., Goldberg, G. M., Whitmore, F. G.: J. Am. Chem. Soc. *68*, 488 (1946)
202. Paquette, L. A., Wells, G. J., Horn, K. A., Yan, T. H.: Tetrahedron Lett. *23*, 263 (1982)
203. Yamamoto, K., Yoshitake, J., Qui, N. J., Tsuji, J.: Chem. Lett. 859 (1978)
204. Stork, G., Colvin, E.: J. Am. Chem. Soc. *93*, 2080 (1971); Hudrlik, P. F., Hudrlik, A. M., Rona, R. J., Misra, R. N., Withers, G. P.: J. Am. Chem. Soc. *99*, 1993 (1977); Hudrlik, P. F., Misra, R. N., Withers, G. P., Hudrlik, A. M., Rona, R. J., Arcoleo, J. P.: Tetrahedron Lett. 1453 (1976)
205. Sharpless, K. B., Lauer, R. F.: J. Am. Chem. Soc. *95*, 2697 (1973)
206. Reich, H. J., Rusek, J. J., Olson, R. E.: ibid. *101*, 2225 (1979)
207. Tulip, T. H., Ibers, J. A.: ibid. *100*, 3252 (1978); *101*, 4201 (1979)
208. Doyle, M. P., Van Leusen, D.: ibid. *103*, 5917 (1981)
209. Stevens, R. V.: Acc. Chem. Res. *10*, 193 (1977)
210. Stevens, R. V., Sheu, J. T.: J. Chem. Soc. Chem. Commun. 682 (1975)
211. Schneider, W. P., Axen, U., Lincoln, F. H., Pike, J. E., Thompson, J. L.: J. Am. Chem. Soc. *90*, 5895 (1968); Axen, U., Lincoln, F. H., Thompson, J. L.: J. Chem. Soc. Chem. Commun. 303 (1969)
212. Nakamura, E., Kuwajima, I.: J. Am. Chem. Soc. *99*, 7360 (1977)
213. Nukamura, E., Kuwajima, I.: ibid. *106*, 3368 (1984); Kuwajima, I.: Top. Curr. Chem., this volume
214. Nakamura, E., Kuwajima, I.: Tetrahedron Lett. *27*, 83 (1986)
215. Nakamura, E., Shimada, J., Kuwajima, I.: Organometallics *4*, 641 (1985)
216. Milstein, D., Stille, J. K.: J. Org. Chem. *44*, 1613 (1979); Labadie, J. W., Stille, J. K.: J. Am. Chem. Soc. *105*, 6129 (1983); Labadie, J. W., Tueting, D., Stille, J. K.: J. Org. Chem. *48*, 4634 (1983)
217. Salaün, J., Barnier, J. P.: unpublished
218. Bogdanowicz, M. J., Ambelang, T., Trost, B. M.: Tetrahedron Lett. 923 (1973); Trost, B. M., Bogdanowicz, M. J.: J. Am. Chem. Soc. *95*, 5321 (1973)

219. Ansell, M. F., Palmer, M. H.: Quart. Rev. *18*, 211 (1964); Eaton, P. E., Cooper, G. F., Johnson, R. C., Mueller, R. H.: J. Org. Chem. *37*, 1947 (1972)
220. Trost, B. M., Bogdanowicz, M. C., Kern, J.: J. Am. Chem. Soc. *97*, 2218 (1975)
221. Trost, B. M., Preckel, M., Leichter, L. M.: ibid. *97*, 2224 (1975)
222. Stork, G., Burgstahler, A.: ibid. *73*, 3544 (1951)
223. Bredereck, H., Effenberger, F., Simchen, G.: Chem. Ber. *96*, 1350 (1963); *98*, 1078 (1965); Bredereck, H., Simchen, G., Rebsdat, S., Kantlehner, W., Horn, P., Wahl, R., Hoffmann, H., Grieshaber, P.: Chem. Ber. *101*, 41 (1968)
224. Osborn, J. A., Jardine, F. H., Young, J. F., Wilkinson, G.: J. Chem. Soc. A. 1711 (1966); Ohno, K., Tsuji, J.: J. Am. Chem. Soc. *90*, 99 (1968); Walborsky, H. M., Allen, L. E.: J. Am. Chem. Soc. *93*, 5465 (1971)
225. Trost, B. M., Runge, T. A., Jungheim, L. N.: J. Am. Chem. Soc. *102*, 2840 (1980)
226. Tsuji, J., Kobayashi, Y., Kataoka, H., Takahashi, T.: Tetrahedron Lett. 1475 (1980)
227. Martel, J., Blade-Font, A., Marie, C., Vivat, M., Toromanoff, E., Buenida, J.: Bull. Chim. Soc. Fr. II, 131 (1978); Buenida, J., Nierat, J., Vivat, M.: Bull. Soc. Chim. Fr. II, 614 (1979)
228. Taber, D. F.: J. Am. Chem. Soc. *99*, 3513 (1977); Toru, T., Kurozumi, S., Tanaka, T., Muira, S.: Tetrahedron Lett. 4087 (1976); Konda, K., Umemoto, T., Takahatake, Y., Tunemoto, D.: Tetrahedron Lett. 113 (1977)
229. Roussel-Uclaf Fr. Demande 2,383,936; Chem. Abstr. *91*, 389,98 (1979)

Donor-Acceptor-Substituted Cyclopropanes: Versatile Building Blocks in Organic Synthesis

Hans-Ulrich Reißig

Institut für Organische Chemie der Technischen Hochschule Darmstadt, Petersenstr. 22, D-6100 Darmstadt, FRG

Syntheses and ring opening reactions of vicinally donor-acceptor-substituted cyclopropanes are treated in this review. Special attention is directed to methyl 2-siloxy cyclopropanecarboxylates which display an particularly rich and versatile chemistry leading to synthesis of polyfunctional compounds, certain carbocycles, and heterocyclic systems. Other classes of donor-acceptor-substituted cyclopropanes are also regarded emphasizing new developments. Although preparative aspects are strongly accentuated, interesting mechanistic features, characteristic for this type of three-membered rings, will also be discussed.

1 Introduction

Reactions of cyclopropane derivatives activated by one type of functional group have been well understood and applied in organic synthesis for quite some time [1]. The far-reaching analogy between reactivity of olefins and cyclopropanes can be explained by the π-type orbitals of strained three membered carbocycles and their interaction with the activating substituents [2, 3].

Therefore acceptor cyclopropanes *1* will be ring opened by nucleophiles N^- to provide products like *2* (homo Michael addition) as depicted in Eq. 1. On the other hand, electrophiles E^+ cleave donor activated cyclopropanes *3* affording adducts *4* or *5* which demonstrates that the cyclopropane serves as a homoenolate equivalent in this sequence (Eq. 2). Seebach consequently classified these methods as umpolung with the cyclopropane "trick" [4].

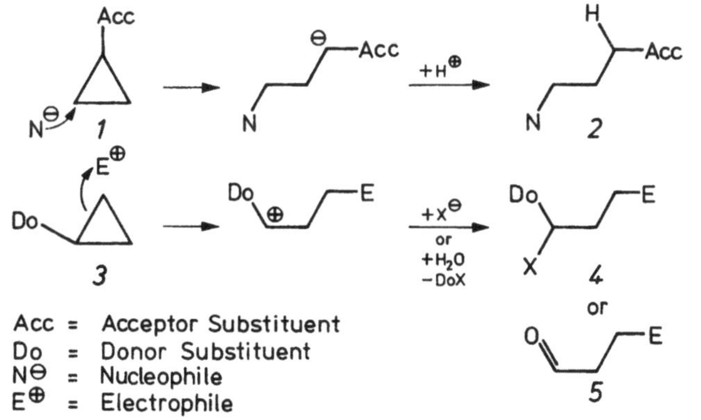

(1)

(2)

Chemistry of cyclopropane derivatives, which combine different sorts of activating substituents, has less definitely been investigated and their synthetic potentials have only recently been explored. Whereas geminally donor-acceptor-substituted systems *6* were treated in preceeding contributions to this series [5, 6, 7], this review will be confined to vicinally activated compounds *7*.

Their reactivity should reflect charge distribution as shown in the mesomeric formula *8*. This intuitive view is supported by MNDO calculations [8]. It is therefore to be expected that nucleophiles and/or electrophiles add to *7* affording products with the general structure *9* or derivatives thereof. Thus *7* should combine the features of a homo Michael system and those of a homoenolate equivalent.

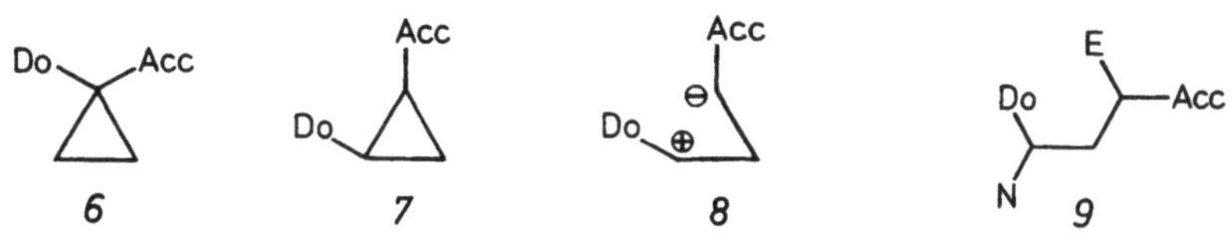

An application of donor-acceptor-substituted derivatives such as *7* not only effects cyclopropane cleavage under milder conditions, but also generates products having at least two functional groups. This is of special value with regard to further synthetic manipulations.

The review will be restricted to systems carrying strong π-acceptors such as carbonyl, cyano, or sulfonyl groups; halide, phenyl, vinyl, or related groups will not be regarded as activating substituents. With respect to the donor function oxy-, amino-, and thio-cyclopropanes will be considered. The trimethylsilylmethyl unit is the weakest donor dealt with, whereas processes as illustrated in Eq. 3 are beyond the scope of this article.

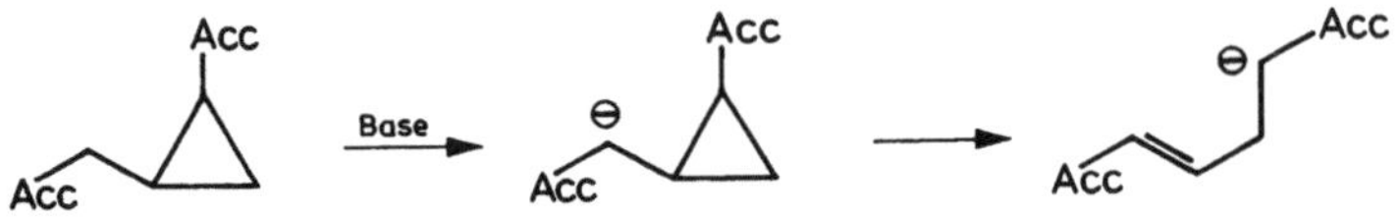

(3)

This review will concentrate on recent developments in the eighties and to synthetic applications. The literature coverage includes 1986.

2 Donor-Acceptor-Substituted Cyclopropanes — Principles of Synthesis and Ring Cleavage

2.1 Synthesis

Not surprisingly, cyclopropanes became synthetic tools only after methods of their preparation had been improved. This is mainly due to progress in carbene chemistry, and several modes to construct donor-acceptor-substituted cyclopropanes also benefit from these advantages. Scheme 1 is an attempt to summarize the most important possibilities of how to build up the systems under discussion and it shows that four of six paths use a carbene or its equivalent.

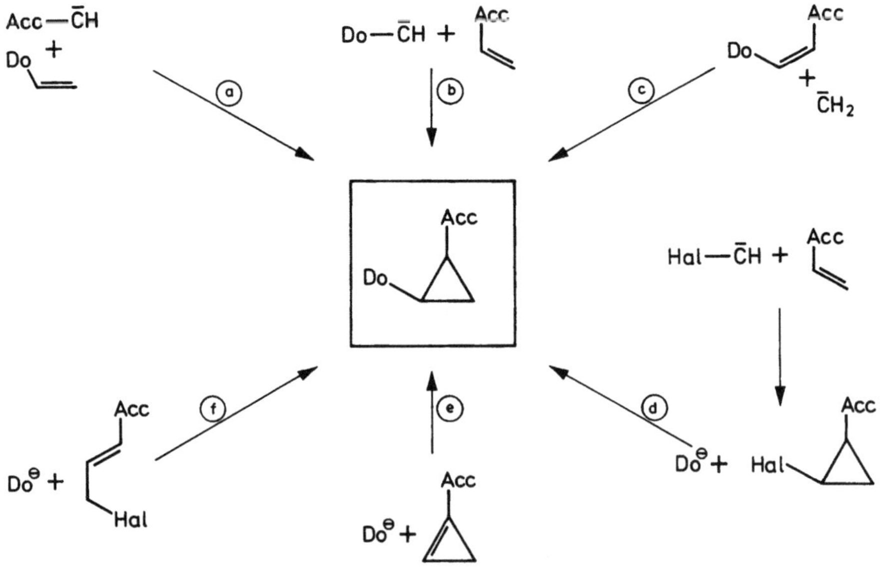

Scheme 1. Possible pathways for preparations of donor-acceptor-substituted cyclopropanes

From the three direct [2 + 1]-cycloaddition routes, path a employing electronrich olefins and acceptor-substituted carbenes is the most efficient one, since the alkenes can be synthesized from carbonyl compounds or other precursors and the carbenes are produced from easily available diazo alkanes. Therefore this very flexible mode to construct donor-acceptor substituted cyclopropanes is by far the most frequently used route.

It is rather difficult to generate donor-substituted carbenes (path b), and methylenation of push-pull-olefins (path c) is not very efficient due to the low reactivity of these alkenes. Therefore these two alternative [2 + 1]-cycloadditions have been of relatively low importance so far. This is also true for the addition of suitable nucleophiles to cyclopropenes activated by electronwithdrawing substituents (path e).

However, a respectable alternative to the direct [2 + 1]-routes (a)–(c) is the variant using halo- or dihalocyclopropanes as precursors for the desired target molecules (path d). The cyclopropane ring is formed by addition of halocarbenes to the olefin and subsequent change of functionalities is achieved by treatment with nucleophiles. It is very unlikely that a direct substitution incorporates the donor-substituent. Instead an elimination/addition sequence with the intermediacy of a cyclopropene has to be assumed.

The very first synthesis [9] of a donor-acceptor-substituted cyclopropane had been performed along route (f). Michael addition of a suitable nucleophile is followed by an intramolecular substitution (elimination of Hal^-) to create the three membered ring. Since all carbon atoms are already assembled in the starting material, and the reaction is not very general, it did not receive too much attention.

2.2 Ring Cleavage

Concerted cycloadditions and radical processes do not play a prominent role in the transformations of donor-acceptor-substituted cyclopropanes. Due to their equipment with functionalities able to stabilize charges, reactions via polar intermediates are the most frequent ones. Although the facility of ring cleavage is strongly influenced by the kind and number of activating substituents involved, a general reactivity pattern can be drawn and will hopefully help to understand the following paragraphs (Scheme 2).

In certain systems — especially in those carrying more than one donor or acceptor group — a ring expansion to five-membered heterocycles *11* occurs at relatively low temperatures. This process can be classified as a 1,3-sigmatropic shift and is most conveniently explained by the intermediacy of the 1,3-zwitterion *10*.

The nucleophilic character of the unsubstituted cyclopropane is usually still predominating in donor-acceptor-activated derivatives. Thus, most ring openings start with the attack of an electrophile E^+ and give products *13* passing through intermediates of type *12* (Scheme 2). In contrast, additions of nucleophiles N^- giving *15* via *14* are so far restricted to exceptional cases with enhanced activity.

Conversion of the acceptor substituent to a function containing a good leaving group LG allows transformation of *16* to the olefin *18* with the homoallyl cation *17* as an intermediate.

Structures *11*, *13*, *15*, and *18* are often not the isolated products, since further transformations can proceed under the reaction conditions employed to bring about opening. Thus in protic media very often carbonyl groups are generated according to Eq. 4.

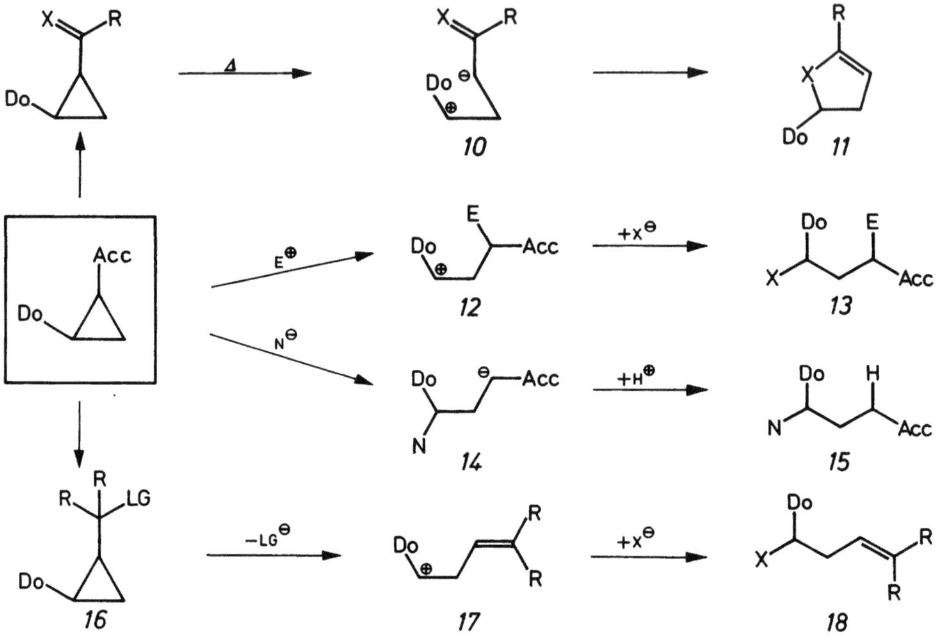

Scheme 2. General reactivity pattern of donor-acceptor-substituted cyclopropanes

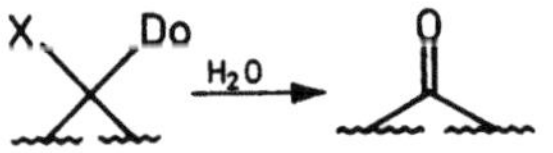

Do = RO, RS, R_2N ... (4)

3 Alkoxy Groups as Donor-Substituents

As early as 1938 Rambaud reported the first synthesis of a donor-acceptor-substituted cyclopropane — obtained by the addition/elimination path (f) (Scheme 1) — and he also recognized that these cyclopropanes are prone to ring cleavage providing 1,4-dicarbonyl compounds. After saponification *19* opens to *20* which is oxidized by air to the isolated succinic acid *21* [9].

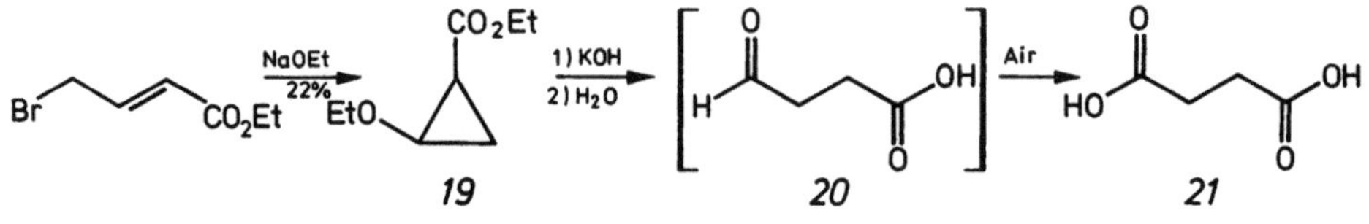

(5)

Later Julia and coworkers developed another important mode of ring breaking: addition of Grignard compounds or reduction converts the ester *22* to the corresponding alcohols which are startingpoints for a cyclopropylcarbinyl/homoallyl cation rearrangement. After acid treatment β,γ-unsaturated carbonyl compounds (e.g. *23*) can be isolated, which sometimes isomerize to the α,β-unsaturated systems [10].

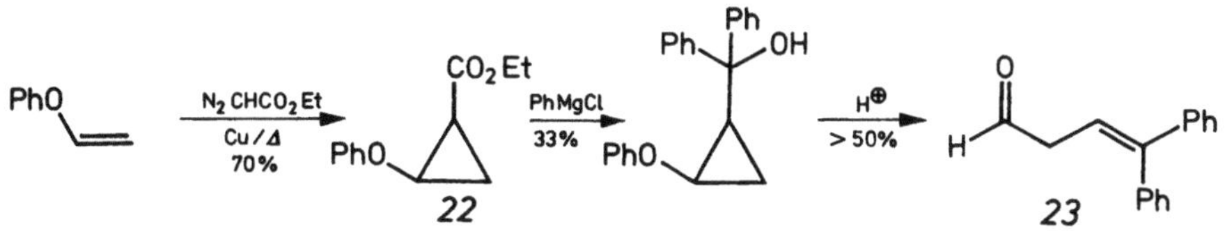

(6)

Mainly these two variants of cyclopropane cleavage or modifications thereof were explored for preparative purposes and for natural product syntheses since 1970. Not too long ago Wenkert reviewed his very impressive contributions in this field [11]. Therefore only principles and typical applications will be repeated here and supplemented by more recent examples.

3.1 Simple Enol Ethers as Olefins

Enol ethers react with diazo ketones in the presence of Cu-catalysts to give cyclopropanes such as *24*. Ring cleavage with acid and subsequent intramolecular aldol condensation constitutes a flexible route to cyclopentenones (Eq. 7) [12, 13]. This procedure has also been applied to a synthesis of *cis*-jasmone employing isopropenyl acetate as a donor olefin [14].

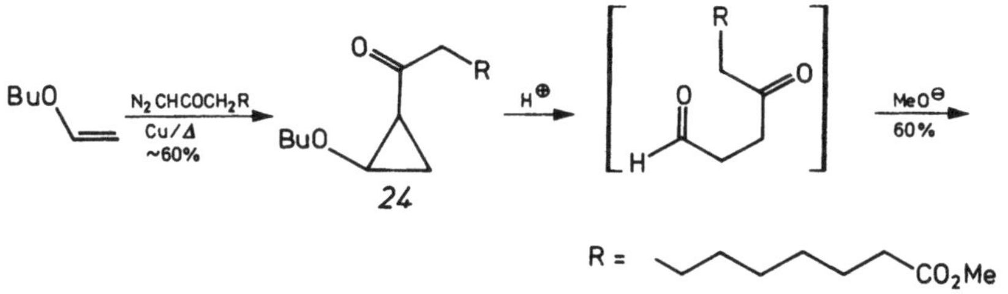

(7)

As can be expected, use of ethyl diazoacetate procides γ-oxoesters [15] or γ-oxocarboxylic acids [16] from enol ethers. Emploging the Julia method with *25* leads to the β,γ-unsaturated aldehyde *26*. Thus, this sequence establishes an overall α-vinylation of a given aldehyde [12, 17, 18].

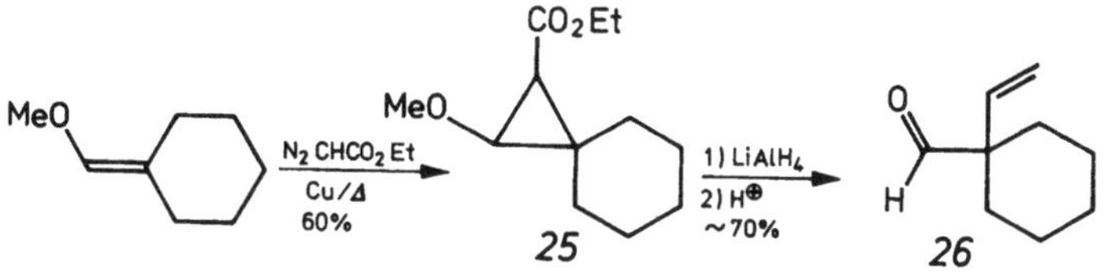

(8)

The two modes of ring cleavage have been joined together in the synthesis of bicyclic lactone *27* — a precursor for certain alkaloids [19].

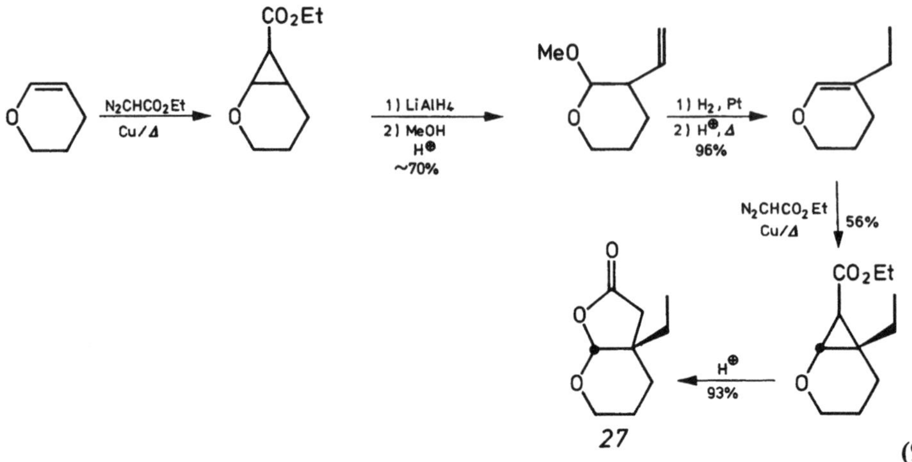

(9)

In attempts to prepare alkoxy-substituted cyclopropane carbaldehydes like *29* only ring expanded compounds *30* have been isolated. The two-step route via carbinol *28* [20] and the selective reduction [21] with DIBAI both afford the 2-alkoxy-dihydrofuran *30*.

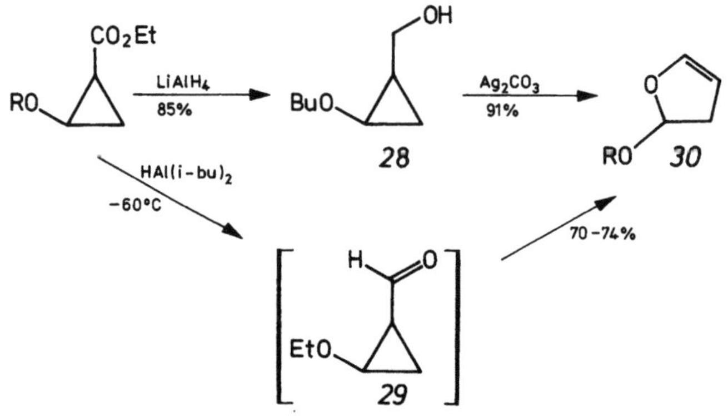

(10)

A very intriguing system *31* with donor-acceptor substitution pattern can be obtained by an intramolecular cyclopropanation [22]. The tricyclic product *31* permits several ring cleavage reactions (Scheme 3), some of which are not known for less strained cyclopropanes. All methods lead to oxabicyclo[2.2.2]octane derivatives. Thus, hydrogenolysis proceeds under relatively mild conditions breaking the bond between the donor and the acceptor substituted carbons exclusively.

Cuprates add directly to the carbonyl group; however, promoted by the Lewis acid BF_3, ring cleavage is facilitated and a 1,5-addition affords the bicyclic compound *32*. This mode of cyclopropane opening is a key step in the short preparation of the monoterpene eucalyptol [22].

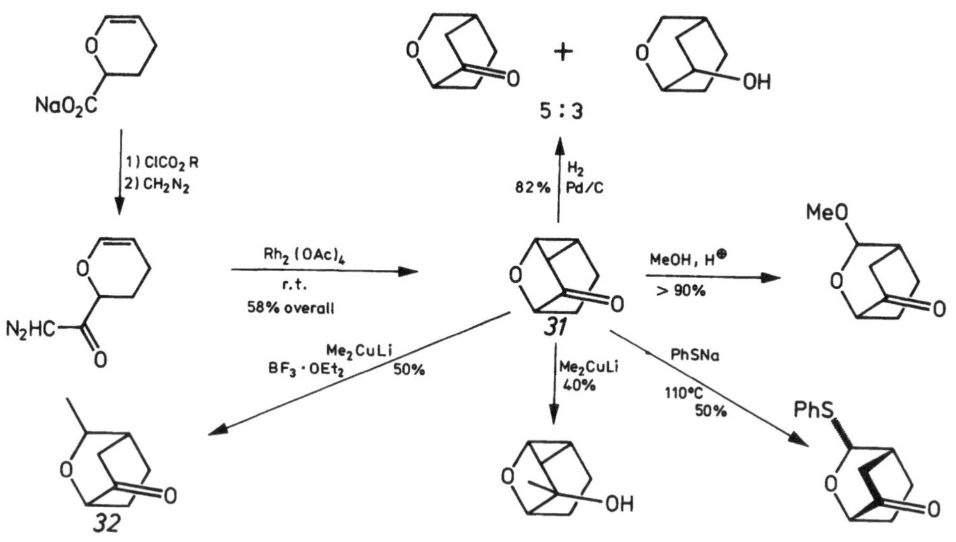

Scheme 3. Ring opening reactions of cyclopropane *31* obtained by intramolecular cyclopropanation

3.2 Furan as Donor Substrate

Cyclopropanation of dienol ethers is not regioselective and therefore of low preparative importance [23]. However, furan gives the bicyclic compound *33* (homofuran derivative) which can also be classified as a donor-acceptor-substituted cyclopropane. The low yield obtained upon photolytic generation of the carbene [24] has impressively been improved by $Rh_2(OAc)_4$-catalysis under thermal conditions [25]. Heating of *33* results in electrocyclic ring opening to afford the *trans, cis*-muconic acid semialdehyde *34*, whereas employment of strong acids brings about cleavage to the more stable *trans,trans*-compound *35* [25]. Similar experiments have been performed with 2,5-dimethyl furan [26].

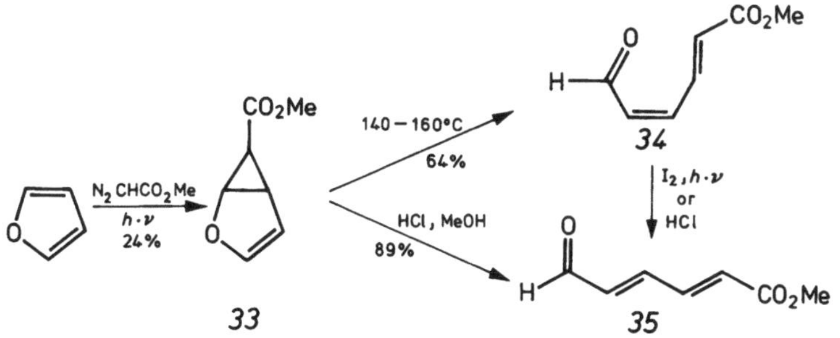

(11)

This strategy to prepare *trans,cis*-configurated functionalized dienes like *34* has elegantly been exploited for syntheses of HETEs (hydroxyeicosatetraenoic acids) and leukotrienes [27]. These metabolites of arachidonic acid have received much attention due to their biological activity. Syntheses of HETEs, for instance, follow the principle outlined in Eq. 12 with the acid catalysed ring opening of homofuran derivatives *36* to *37* as the stereoselective key step.

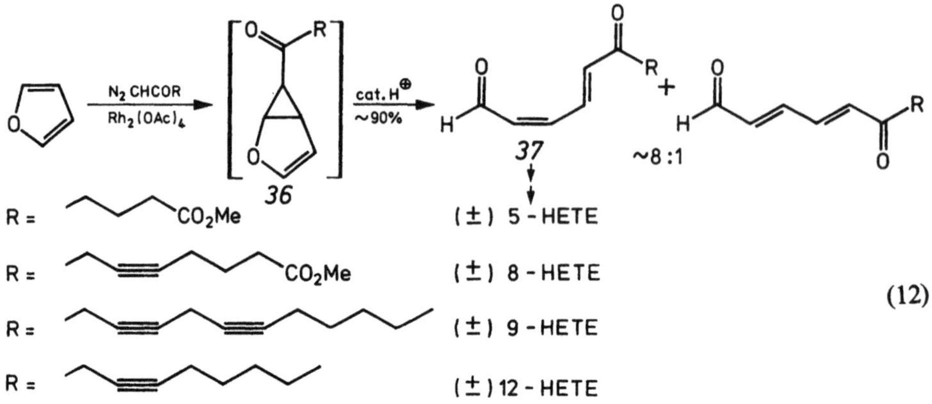
N2CHCOR
Rh2(OAc)4
cat. H⊕
~90%
36
37
~8:1
R =
CO2Me
(±) 5-HETE
R =
CO2Me
(±) 8-HETE
R =
(±) 9-HETE
R =
(±) 12-HETE

(12)

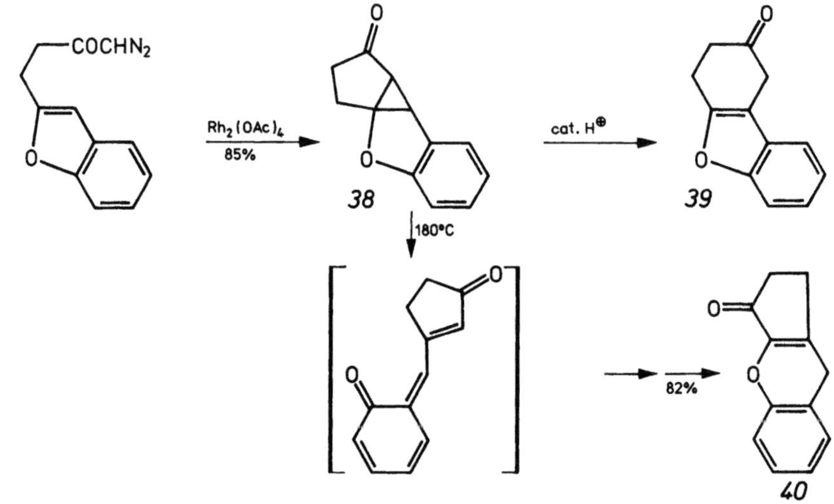
COCHN2
Rh2(OAc)4
85%
38
cat. H⊕
39
180°C
82%
40

(13)

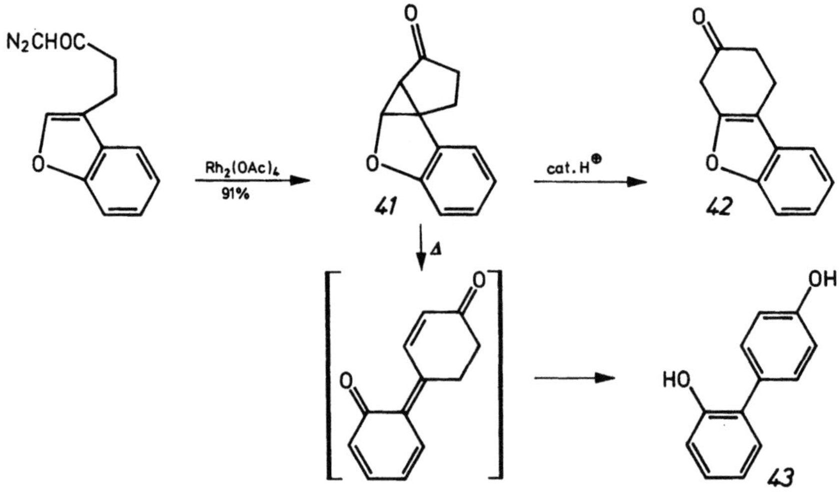
N2CHOC
Rh2(OAc)4
91%
41
cat. H⊕
42
Δ
OH
HO
43

(14)

A report dealing with the intramolecular mode of this cyclopropanation uncovers a remarkable dependency of the ring cleavage on the substitution pattern and the conditions employed [28]. Two examples with regioisomeric benzofuran systems (Eq. 13 and 14) demonstrate that thermolysis of *38* or *41*, respectively, causes electrocyclic reactions which finally give the benzopyran derivative *40* or the biphenyl compound *43*. Traces of acid completely change the cleavage behaviour and give "normal" products *39* or *42*. For the formation of these tricyclic benzofuran derivatives breakage of the internal cyclopropane bond is necessary to give the annulated cyclohexenone systems instead of the conceivable five-ring spiro compounds. Simpler furan derivatives without a condensed benzo ring give similar results [28].

3.3 Diketene as Donor Olefin

The enol acetate moiety in diketene can be utilized for cyclopropane formation. Unfortunately, with most diazo compounds, yields are rather moderate [29], and therefore the synthetic value of methods developed on this basis is restricted. As exemplified by the ethyl diazoacetate adduct *44* (Scheme 4) the ring opening of this masked tricarbonyl compound can lead to different classes of acyclic or cyclic products. The outcome of these reactions depends on the conditions employed. They simultaneously transform the β-ketoester unit present in *44* [29b].

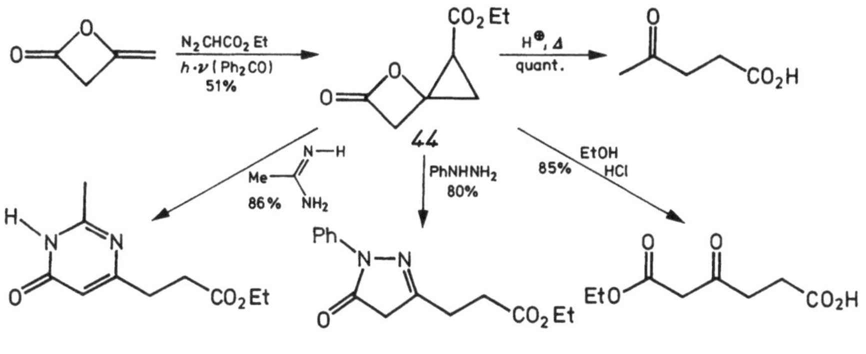

Scheme 4. Ring cleavage reactions of diketene adduct *44*

Addition of a suitably substituted diazo ketone to diketene has served as a key step — albeit with low efficiency — in yet another synthesis of *cis*-jasmone [30].

3.4 Derivatives with Two Donor or Two Acceptor Groups

1,2-Dialkoxyalkenes such as *45* behave very similarly to enol ethers and yield cyclopropanes like *46* with reasonable efficiency (Eq. 15). Ring cleavage to formyl ketone *47* with acid is followed by intramolecular aldol addition and elimination of methanol to give the cyclopentenone *48*. This can be isomerized to *49* — an intermediate for prostaglandin PGE_1. The same strategy has been employed to prepare tetrahydropyrethrolone methyl ether *50* [31].

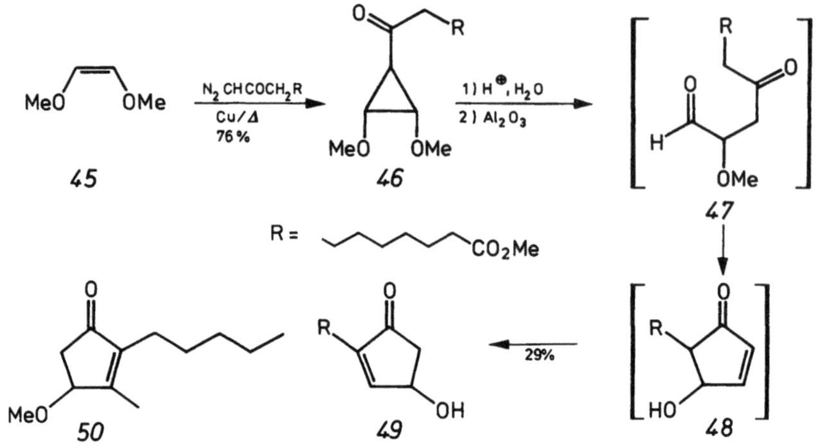

(15)

O,O-ketene acetals *51* have also been reacted with diazo ketones under copper catalysis. However, not the expected cyclopropanes but the corresponding dihydrofuran derivatives *52* are obtained as products [32]. Very likely the three-membered precursors of *52* are unstable under the cyclopropanation conditions (Cf. Scheme 5). Hydrolysis of *52* yields the γ-oxocarboxylic acids *53*.

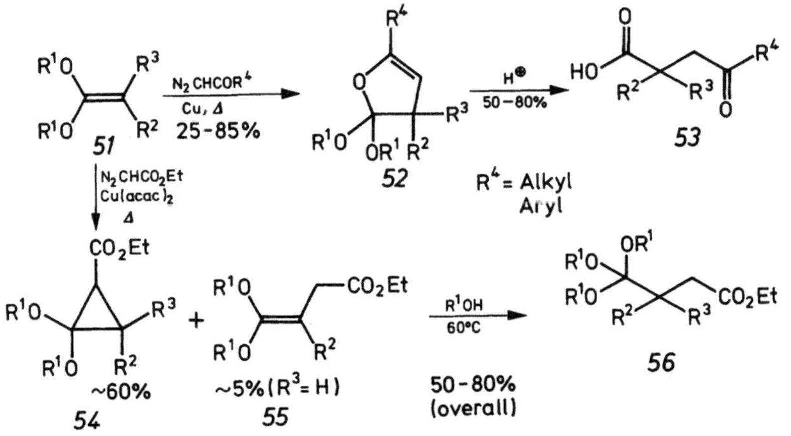

(16)

If the acceptor strength of the carbene substituent is slightly reduced, bisdonor substituted cyclopropanes become stable enough to be isolated. Thus the esters *54* are formed predominantly with their acyclic isomers *55* as side products [33] (Cf. rearrangement in Eq. 68). Cyclopropanes *54* are rather sensitive to moisture and usually converted to the orthoesters *56* without purification in reasonable overall yield [34]. A similar dependency of the resulting products on the starting diazo compound has been found for other ketene acetals [35] and also in reactions using the more active Rh-catalyst [36].

The combination of one donor substituent and two acceptor groups also enhances the cyclopropane activity. Although compounds *57* can be synthesized without special precautions, they readily expand at room temperature to the isomeric dihydrofuran derivatives if dissolved in dimethyl sulfoxide[37]. By assistance of the highly polar solvent the charges might be stabilized in an intermediate 1,3-zwitterion formed by heterolytic cleavage of the cyclopropane bond.

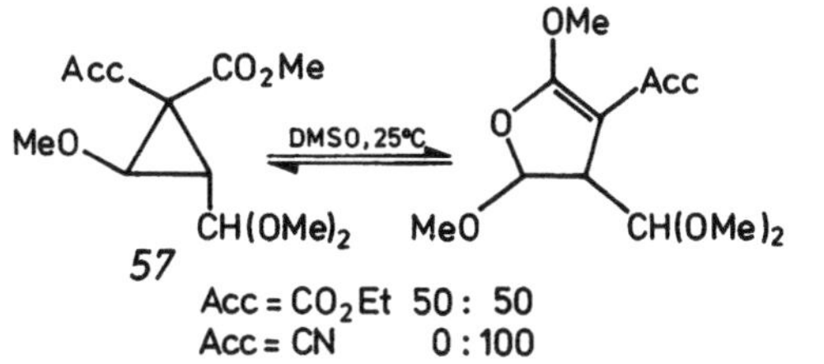

(17)

Reaction of the enol ether *58* with dimethyl diazomalonate provides the spiro compound *59* in high yield. Reduction and acid catalyzed cyclopropane cleavage gives the unsaturated γ-lactol *60* which can be oxidized to β-methylene γ-butyrolactone *61*[20].

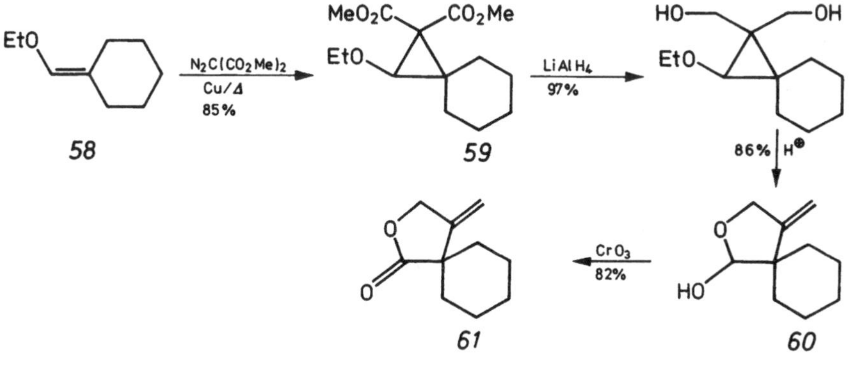

(18)

Under similar conditions diazo acetoacetate does not afford cyclopropanes but dihydrofurans as *62* which can be aromatized (e.g. to *63*)[20]. A different furan derivative *64* is obtained from ethyl diazopyruvate as outlined in Eq. 19[38]. Possibly cyclopropanes are intermediates in these reactions, which rearrange to the five-membered heterocycles under the conditions employed.

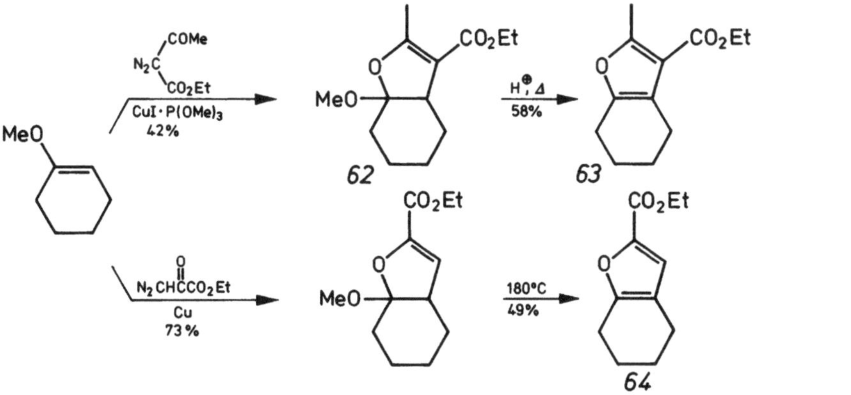

(19)

The results in this section impressively illustrate that a well balanced strength of the donor and the acceptor substituents is essential for the stability of cyclopropanes. Otherwise the cyclopropane bond between these functional groups is too fragile and is eventually broken to provide ring expanded or ring opened products.

3.5 Other Methods for the Synthesis of Acceptor Activated Oxycyclopropanes and Their Ring Cleavage Reactions

3.5.1 Reaction of Electrondeficient Olefins with Donor-Carbene-Equivalents

One interesting application of Fischer-type carbene complexes in organic synthesis is their addition to acceptor olefins affording methoxy substituted cyclopropanes *65* (Eq. 20).

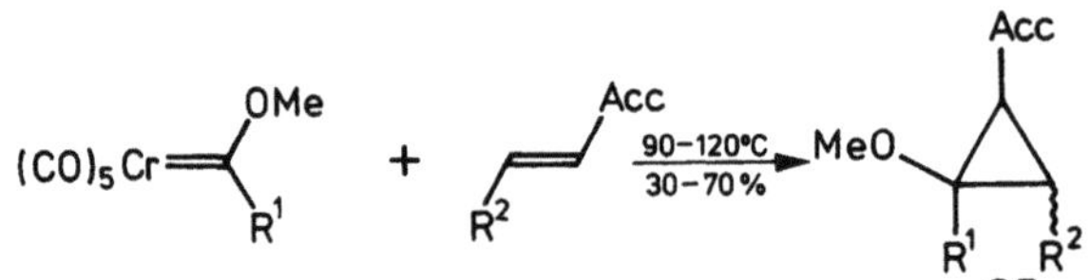

65

(20)

The initial reports [39] only contain a few examples with unsaturated esters used as the olefinic component in large excess (entries 1–3, Table 1). Recent investigations [40], however, underline that this complementary approach to donor-acceptor-substituted cyclopropanes is rather general. Since equimolar amounts of olefins and carbene complexes are sufficient to give good results (entries 4–8), this method might be of preparative value.

With α,β-unsaturated ketones the expected cyclopropanes could not be isolated, but acrylonitrile derivatives can also be used as acceptor olefins [40]. They provide cyanocyclopropanes in good yield (entries 6, 7), which might be interesting precursors for other cyclopropanes or ring opened compounds because of the synthetic versatility of the nitrile group.

There are limitations when the olefin is substituted with bulky groups however, flexibility, is guaranteed by R^1, which may be an aryl or an alkyl group (entry 8).

Table 1. Synthesis of Donor-Acceptor-Substituted Cyclopropanes with Fischer-Carbene-Complexes

Entry	R^1	R^2	Acc	*cis/trans*	Yield	Ref.
1	Ph	Ph	CO_2Me	1:3.9	34%	[39]
2	Ph	CO_2Et	CO_2Et	—	39%	[39]
3	Ph	Me	CO_2Me [a]	1:2.5	60%	[39]
4	Ph	Me	CO_2Me [b]	1:1.5	59%	[40]
5	Ph	H	CO_2Me	1:1.3	57%	[40]
6	Ph	H	CN	1:1	72%	[40]
7	Ph	Me	CN [c]	[d]	53%	[40]
8	n-Bu	H	CO_2Me	~1:1	30%	[40]

[a] 10 equivalents of olefin; [b] Equimolar amounts of olefin; [c] *cis/trans* mixture of olefins; [d] 4 isomers

Since alkoxysubstituted methyl cyclopropanecarboxylates like *66* undergo ring opening upon acid treatment to provide the γ-oxoester *67*, the cyclopropanation/ring cleavage sequence establishes an overall nucleophilic acylation of α,β-unsaturated esters [40].

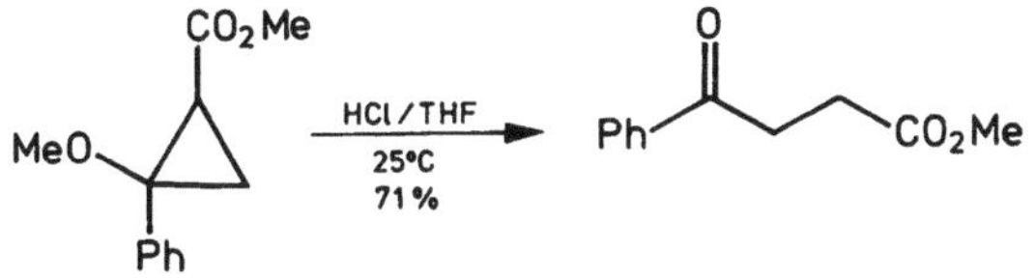

66 *67* (21)

Catalytic hydrogenolysis of the nitrile *68* is regioselective and affords γ-methoxy γ-phenyl butyronitrile *69* as the single product [40].

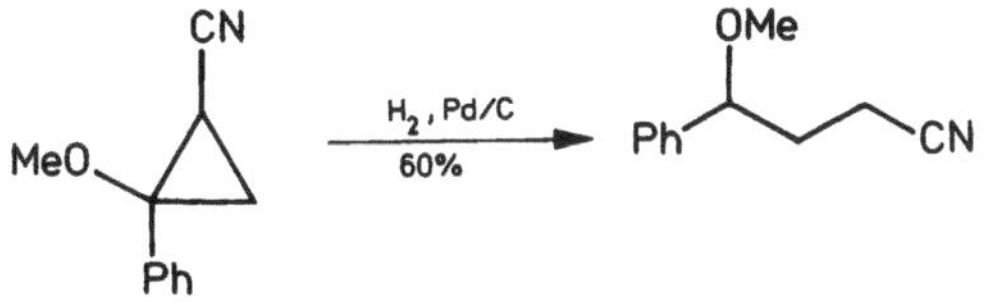

68 *69* (22)

3.5.2 Methylenation of Donor-Acceptor-Substituted Olefins

Vicinally donor-acceptor-substituted olefins usually are rather unreactive species. Nevertheless cyclopropanation of 2,2-dimethyl-3(2 H)-furanone *70* could be executed with dimethyl oxosulfonium methylide as a methylene source. The bicyclic compound *71* is formed in modest yield accompanied by the spiro epoxide as a second product in almost equal amounts. Carbinols *72* derived from *71* by alkyl lithium addition can be nitrosated and photolyzed to suffer a Barton fragmentation. The resulting γ-oxoaldehydes are directly cyclized to afford the 2-substituted cyclopentenones *73* in good yield [41].

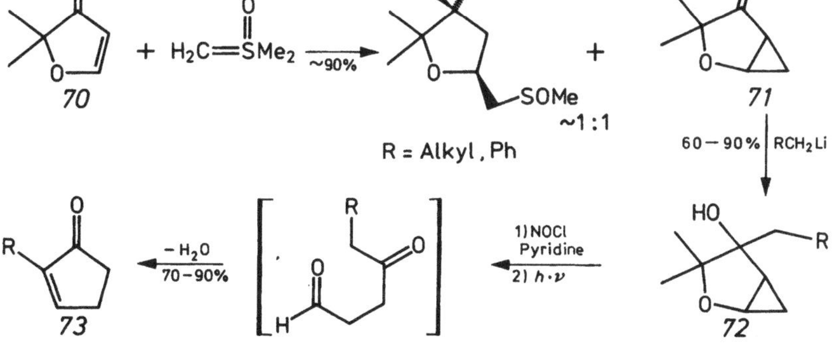

(23)

Reactions of benzo-4-pyrones with the methylide have been reported earlier. After hydrolysis 1,4-dicarbonyl compounds could be isolated [42].

3.5.3 Halocyclopropanes as Precursors

The efficient formation of dihalocyclopropanes could be extended to allyl alcohol adducts *74*, which are oxidized to *75*. The resulting ketones are very useful synthetic building blocks. Reaction with alkoxide, for instance, affords dihydrofuran derivatives *76* via alkoxysubstituted cyclopropyl ketones [43]. This ring enlargement might be a purely thermal process (1,3-sigmatropic shift), but other mechanistic possibilities could also be conceived (cf. Scheme 2).

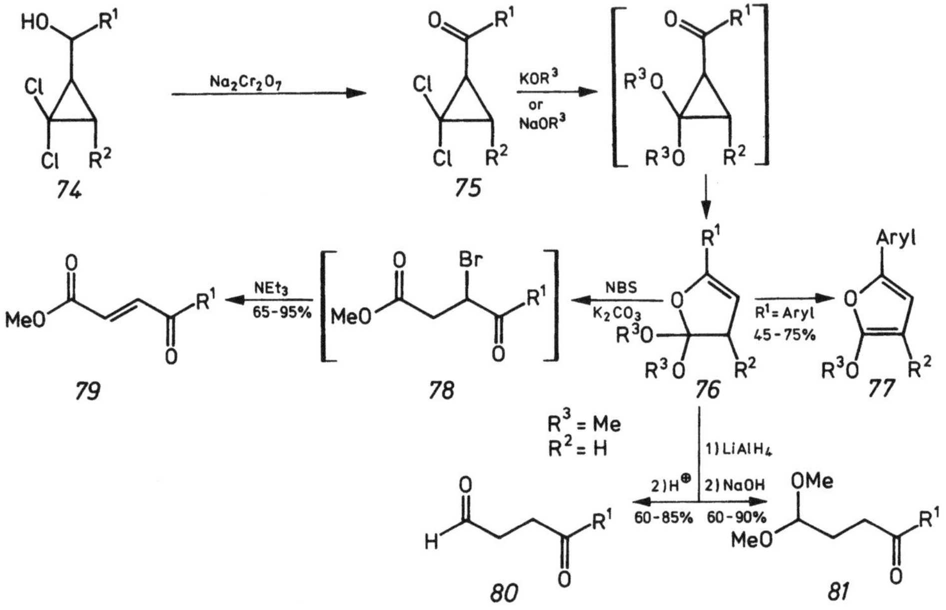

Scheme 5. Reactions with dichlorocyclopropylketone *75* as key starting material

The cyclic enol ethers *76* either form 1-alkoxy furans *77* by elimination [44] or, more interestingly, they are oxidized by bromosuccinimide to brominated intermediates *78* which give α,β-unsaturated γ-oxoesters *79* after base treatment [45]. Reaction of the cyclic orthoesters *76* with $LiAlH_4$ leads to γ-oxoaldehydes *80* or their acetals *81* depending on the work-up procedure [46].

As to be expected, monochloro cyclopropyl ketones *82* can be ring opened to γ-oxoaldehydes *83* by treatment with sodium methoxide without preceeding reduction [47].

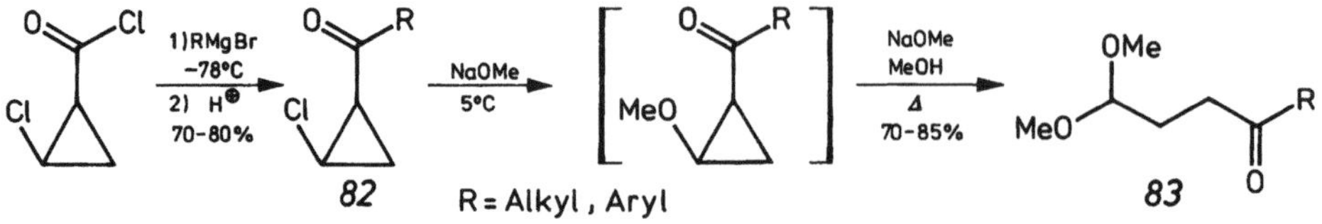

(24)

Analogously, several dichloro- or difluoro-cyclopropyl ketones provide γ-oxo-esters under similar conditions [48,49], whereas dichloro-cyclopropyl sulfones give γ-sulfonyl orthoesters [50].

3.5.4 Other Compounds as Starting Materials

Another method to construct methoxysubstituted cyclopropyl ketones is outlined in Eq. 25. Here acid chlorides and allyl chloride are combined to give *84* which could be cyclized to the key compound *85* after addition of methanol. Standard methodology brings about preparation of γ-oxoaldehydes *86* or β,γ-unsaturated acetals *87* [51].

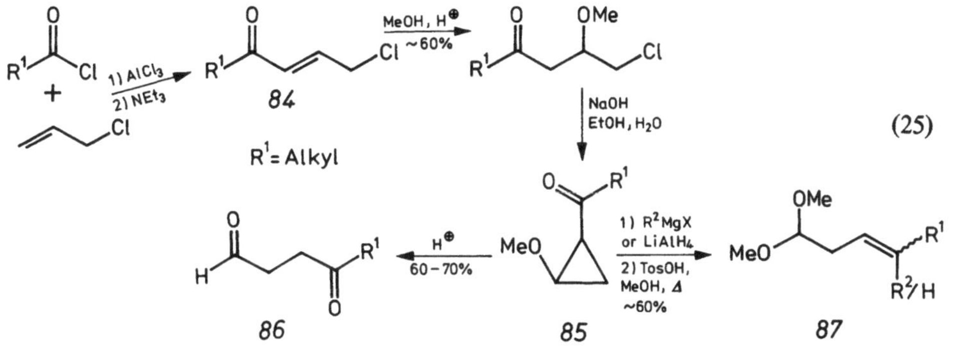

(25)

Deprotonation/alkylation of suitably functionalized cyclopropanes opens the way to a new manifold of substituted compounds. Thus cyclopropyl sulfone *88*, carrying a MEM-group as donor function, can easily be converted to the cyclopropyl anion *89* which smoothly reacts with alkyl halides yielding *90* or with ketones providing hydroxyalkylated cyclopropyl sulfones, respectively. The latter could directly be transformed to unsaturated sulfones (e.g. *92* → *93*) by acid treatment, but alkylated products *90* have to be cleaved in a two step procedure. Deprotection of the MEM-group with HBF_4 allows subsequent base induced ring opening of the resulting cyclopropanols and after elimination of sulfinate α,β-unsaturated aldehydes *91* are formed [52].

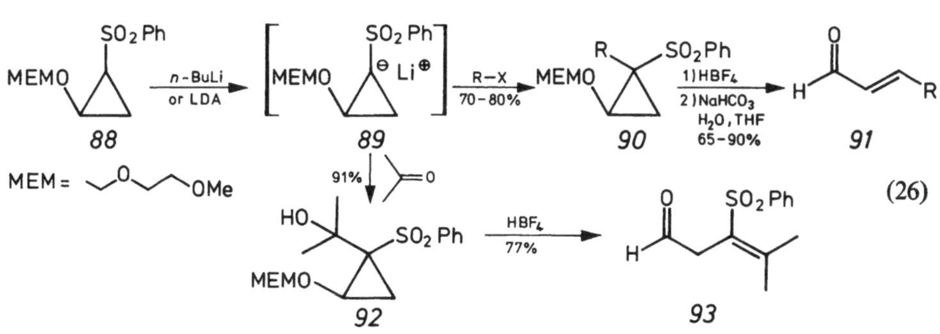

(26)

The stereochemistry of the alkylation of a carbanion related to *89* has recently been studied by another group [53].

An exceptional case of ring formation and cleavage deals with a system incorporating two cyclopropane units with donor-acceptor pattern. Photochemical synthesis of *94* and electrocyclic reaction afford the unusual bridged hexanooxepin *95* [54].

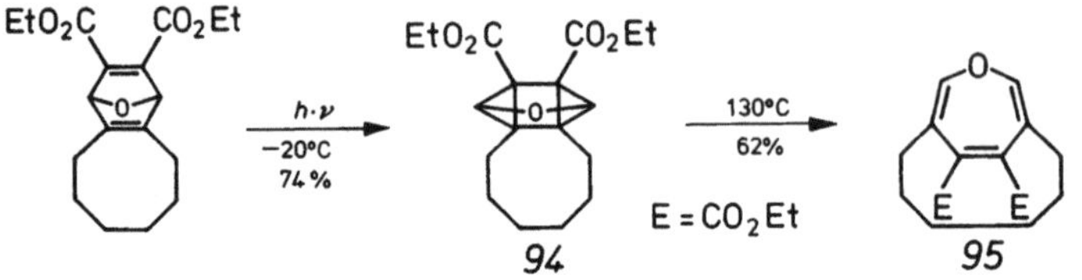

94 95 (27)

4 Trialkylsiloxy Groups as Donorsubstituents

As demonstrated in the preceeding paragraph alkoxysubstituted cyclopropanes provide access to a variety of different 1,4-dicarbonyl derivatives and are therefore valuable and versatile building blocks for organic synthesis.

Since most methods are based on alkyl enol ethers as starting materials, however, they are faced with severe drawbacks as far as selectivity and flexibility are concerned. Preparations of alkoxyalkenes are sometimes very tedious to put into practice and usually not chemo-, regio-, or stereoselective. In addition, the cyclopropanation step is frequently inefficient due to purification problems or the need of using a large excess of olefin. But most importantly cyclopropane cleavage requires rather harsh conditions (strong acid or base) because of the strength of the C—O-bond that has to be broken. Therefore these methods of ring opening are not compatible with many functional groups.

In contrast, the related silyl enol ethers are available by mild selective transformations from carbonyl compounds or other precursors [55]. Their stability and that of products derived from these alkenes can easily be regulated by choosing suitable substituents at silicon. Selective cleavage of a Si—O-bond is possible with fluoride reagents under very mild conditions, and this is why cyclopropane ring opening can now be performed with high chemoselectivity.

Exploiting these major advantages, siloxysubstituted cyclopropanes have become extremely versatile building blocks for many preparative purposes in the last decade. Besides these important practical aspects, they also have provided a deeper insight into many fascinating mechanistic features of donor-acceptor-substituted cyclopropanes.

4.1 Synthesis of Methyl 2-Siloxycyclopropanecarboxylates by Carbenoid-Addition and Their Ring Cleavage to γ-Oxoesters

A large variety of silyl enol ethers *96* has been transformed to the corresponding cyclopropanes *97* by reaction with methyl diazoacetate in the presence of copper catalysts (Eq. 28). Although at first the isolation of mainly ring-opened products had been reported [56], the preparation of methyl 2-siloxycyclopropanecarboxylates proceeds generally in very good yields (Table 2) [57].

Table 2. Synthesis of methyl 2-siloxycyclopropanecarboxylates *97* from methyl diazoacetate and silyl enol ethers *96* according to Eq. 28 and ring cleavage to γ-oxoesters *98*

Entry	R_3	R^1	R^2	R^3	Cyclopropane *97*			γ-Oxoester *98*		
					cis: *trans*	Yield	Ref.	Method[a]	Yield	Ref.
1	Me_3	H	H	H	18:72	58%	[57]	A	63%	[62]
2	Me_3	Me	H	H	35:65	70%	[57]	A	71%	[61]
3	Me_3	n-C_5H_{11}	H	H	39:61	81%	[57]	A	59%	[61]
4	Me_3	$CHMe_2$	H	H	42:58	80%	[57]	A	89%	[62]
5	Me_3	CMe_3	H	H	48:52	87%	[57]	A	90%	[62]
6	Me_3	Ph	H	H	48:52	78%	[57]	A	78%	[62]
7	Me_3	$CH=CH_2$	H	H	36:64	73%	[83]	A	75%	[62]
8	Me_3	$CMe=CH_2$	H	H	49:51	65%	[84]	—		
9	Me_3	H	H	Me	22:78	73%	[57]	B	92%	[65]
10	Me_3	CMe_3	H	Me	35:75	67%	[65]	C	99%	[65]
11	Me_3	Ph	H	Me	45:55	71%	[57]	A	80%	[62]
12	Me_3	$CH=CH_2$	H/Me	Me/H[b]	4 isomers	48%	[84]	—		
13	Me_3	H	Me	Me	25:75	79%	[57]	B	86%	[62]
14	Me_3	Me	Me	Me	32:68	86%	[57]	A	93%	[62]
15	Me_3	Ph	Me	Me	46:54	80%	[57]	A	82%	[61]
16	Me_3	H	$-(CH_2)_4-$		25:75	58%	[57]	B	80%	[62]
17	Me_3	H	$-(CH_2)_5-$		26:76	75%	[57]	B	93%	[62]
18	Me_3	$-(CH_2)_3-$		H	46:54	72%	[57]	A	91%	[62]
19	Me_3	$-(CH_2)_4-$		H	58:42	75%	[57]	A	97%	[62]
20	Me_3	$-(CH_2)_5-$		H	47:53	77%	[57]	A	93%	[62]
21	Me_3	$-(CH_2)_8-$		H	30:70	73%	[77]	—		
22	Me_3	$-CH=CH-CH_2-CH_2-$		H	64:36	65%	[83]	—		
23	Me_3	$-CHMe-(CH_2)_3-$		H	4 isomers	74%	[57]	A	91%	[62]
24	Me_3	$-(CH_2)_3-CHMe-$		H	60:40	62%	[57]	A	88%	[62]
25	Me_3	$-(CH_2)_2-CH(tBu)-CH_2-$		H	4 isomers	74%	[57]	A	83%	[61]
26	Me_3	$-(CH_2)_4-$		$OSiMe_3$	>95% *trans*	66%	[57]	D	87%	[62]
27	$tBuMe_2$	H	Me	Me	23:77	77%	[57]	—		
28	$tBuMe_2$	Ph	H	H	40:60	78%	[57]	E (F)	68% (72%)	[62]
29	$tBuMe_2$	CMe_3	H	H	42:58	80%	[57]	—		

[a] Method A: $NEt_3 \cdot 2\text{–}3\,HF$; Method B: $NEt_3 \cdot HF$; Method C: cat. $PdCl_2(MeCN)_2$, Me_2CO, H_2O; Method D: MeOH, cat. K_2CO_3, 80 °C; Method E: n-Bu_4NF; Method F: 2 N HCl; [b] Mixture of *E*/*Z*-isomers

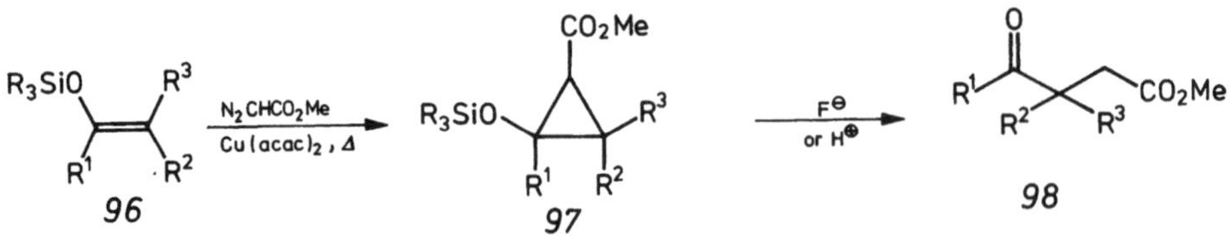

(28)

Nearly equimolar amounts of the components are sufficient to obtain satisfying results even in large-scale runs. $Cu(acac)_2$ is the most suitable catalyst; $Rh_2(OAc)_4$, which is effective even at room temperature, might be preferable in exceptional cases. As expected, the cyclopropanation step occurs regioselectively (entries 4, 14) and stereospecifically (entries 9–11), thereby retaining the olefin configuration completely in the product. On the other hand, there is only a very moderate stereoselection as far as the methoxy carbonyl group is concerned, and usually *cis/trans* mixtures are obtained [57].

If the olefin is chiral (entries 23–25) high diastereoselectivity has been observed, when the center of asymmetry is at C-3 of cyclic silyl enol ethers (entry 24). Cyclopropanation then occurs *trans* to the substituent at this carbon [57] exclusively, and due to the very mild cleavage conditions this *trans*-relationship is preserved in the subsequent ring opening (vide infra). This protocol has been applied to introduce a side chain during the stereoselective synthesis of a prostaglandin [58] and of dicranenone A [59].

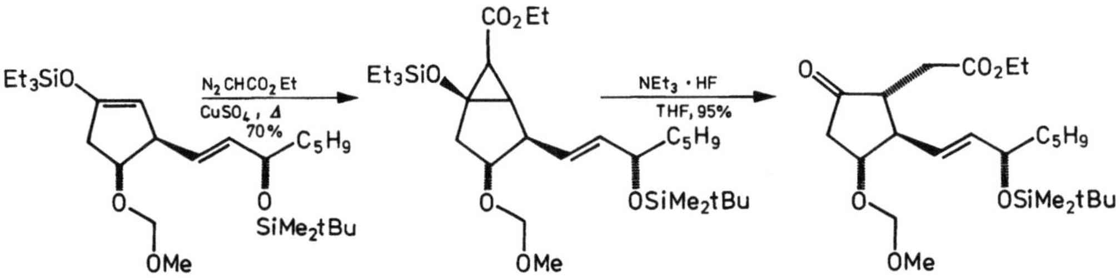

(29)

Interestingly, the cyclopropanation is completely nonstereoselective if the starting material is a chiral siloxy cyclohexene derivative with alkyl groups at C-4 or C-6 (entries 23, 25).

The prostaglandin approach above (Eq. 29) also shows that the reaction with the carbenoid is compatible with further functions and even a second olefinic unit. However, this second double bond is left unattacked only because of its deactivation by the allylic siloxy group. Competition experiments have demonstrated that simple olefins like styrene or cyclohexene react with methyl diazoacetate under copper-catalysis in rates comparable to those of silyl enol ethers [57].

Reactions of trimethylsilyl enol ethers with diazo ketones give cyclopropanes contaminated by ring opened compounds [60,61]. Use of the more stable tert-$BuMe_2Si$-derivatives or of $Rh_2(OAc)_4$ as a catalyst might eventually improve the situation. O-Silylated ketene acetals and O,S-ketene acetals, respectively, did not provide products with cyclopropane structure [61].

Methyl 2-siloxycyclopropanecarboxylates *97* are thermally stable compounds, which can nevertheless be cleaved under very mild conditions to give γ-oxoesters *98* (Eq. 28) [61]. The Me_3Si-derivatives are ring opened by acids or strong bases, but they can selectively be cleaved by fluoride reagents. This retro aldol type reaction via cyclopropanol intermediates is most effectively achieved by the easily accessible $NEt_3 \cdot xHF$ ($x = 1$—3) [63], which generates *98* in almost quantitative yield (table 2). With catalytic amounts of acid or base methanol is another operating medium for desilylation (entry 26, lit. [59]). $PdCl_2(MeCN)_2$-catalysis allows smooth ring opening in wet acetone (entry 10) [65]. When switching to the more stable *tert*-butyldimethylsiloxy compounds n-Bu_4NF is the reagent of choice for ring cleavage (entry 28). The opening of the three-membered ring completes the combination of an enolate equivalent (silyl enol ether) with an enolium equivalent (methyl diazoacetate) for the construction of the 1,4-dicarbonyl compound via a cyclopropane.

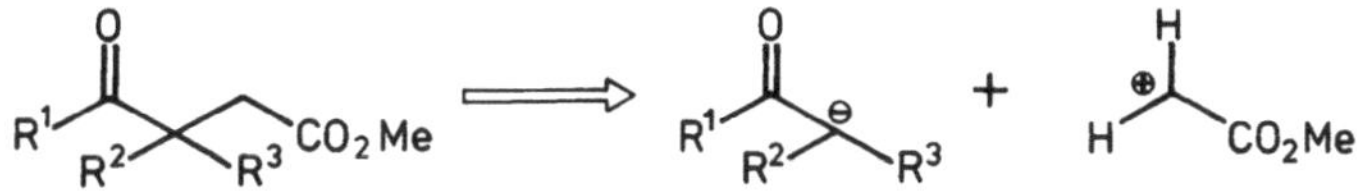

Although alkylations of enolates with α-halo ester compounds are quite effective in singular cases, these reactions often proceed with poor yield and selectivity. Therefore the siloxycyclopropane route is to be considered even for large scale preparations of relatively simple γ-oxoesters. Synthesis of rather sensitive formyl esters (entries 9, 13, 16, 17) or the stereoselective generation of *trans*-substituted cyclic γ-oxoesters as mentioned above can hardly be achieved with comparable efficiency by other methods.

The mild cleavage conditions with $NEt_3 \cdot HF$, which do not cause epimerization at centers α to the carbonyl group, are essential for an enantioselective synthesis of γ-oxoesters using optically active catalysts [64] in the cyclopropanation step. Up to 50% ee have been obtained so far [65]. Improvements should be possible, if the *trans/cis*-ratio of the siloxycyclopropane can be increased. Formylesters of type *99* are promising building blocks for further transformations (e.g. synthesis of γ-butyrolactones).

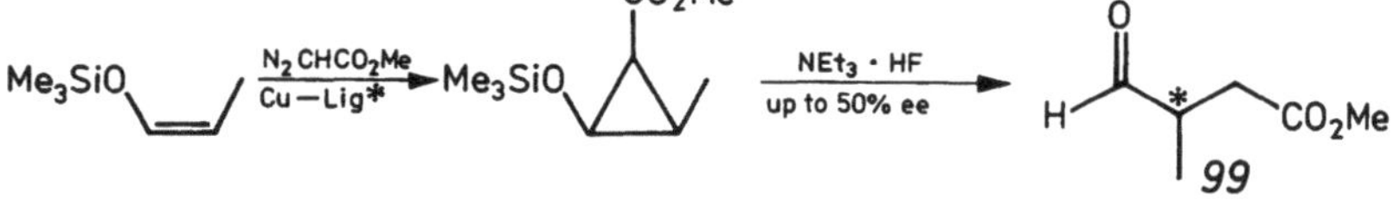

(30)

With the methodology presented so far only compounds with a $-CH_2CO_2R$ moiety can be synthesized. However, specifically mono- or dideuterated γ-oxoesters are obtainable as illustrated in Eq. 31 [57].

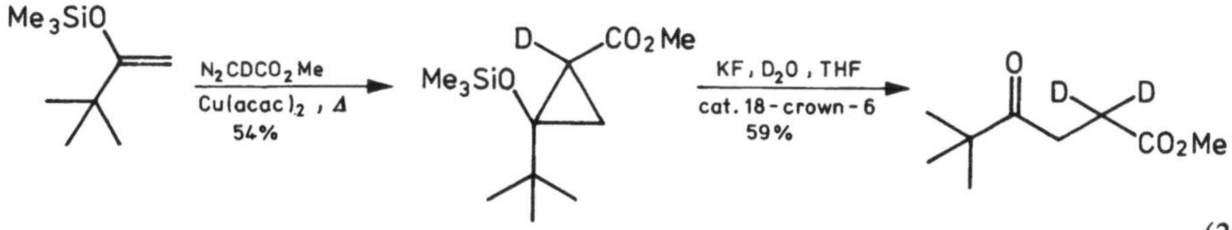

(31)

According to Scheme 1 methyl 2-siloxycyclopropanecarboxylates should also be available from donor-acceptor-substituted olefins like *100*, which are easily synthesized by silylation of the corresponding 1,3-dicarbonyl compounds. Cyclopropanation of *100* with methyl diazoacetate or diazomethane could be realized in the presence of Cu(II)-catalysts, but due to the relatively low reactivity of the olefins a large excess of diazoalkanes had to be employed. This makes the isolation of *101* troublesome and therefore direct hydrolysis with acid to give 1,4-dicarbonyl compounds *102* is advantageous (Eq. 32) [66].

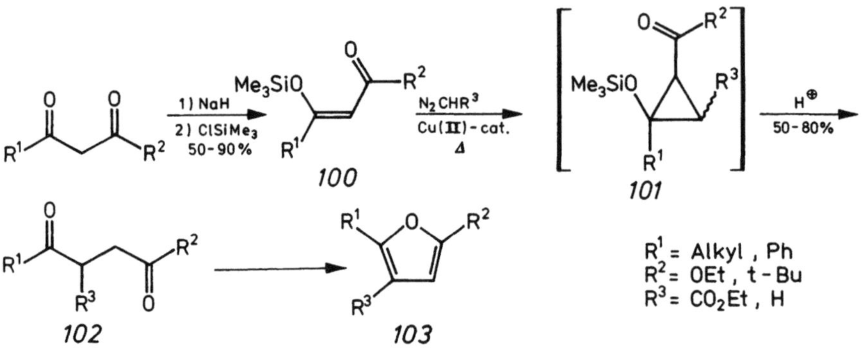

(32)

With R^2 = OEt, however, some condensation to the furan derivative *103* cannot be avoided. Homologisation of ethyl benzoylacetate via the silyl compound *100* with diazomethane to the γ-ketoester *102* is incomplete, and hydrolyzed starting material is recovered.

Photolysis of acylsilanes produces siloxysubstituted carbenes, which could be trapped by an excess of electrondeficient olefins. This route to the resulting alkyl 2-siloxycyclopropanecarboxylates seems not to be of preparative value, however [67].

Since methyl 2-siloxycyclopropanecarboxylates *97* are masked γ-oxoesters, the protected carbonyl group cannot disturb reactions modifying the ester function. This great advantage is demonstrated in the following paragraphs and especially in the deprotonation/alkylation sequence, which allows introduction of a substituent at C-1 of the cyclopropane and therefore leads to compounds containing a —CHR′CO_2R unit after ring cleavage.

4.2 Synthesis of Methyl 2-Siloxycyclopropanecarboxylates by Deprotonation/Alkylation

4.2.1 Preparative Aspects

For the introduction of further substituents at C-1 of methyl 2-siloxycyclopropanecarboxylates *97* the deprotonation and smooth alkylation of the resulting enolates is a very feasible route (Eq. 33) [68]. It opens the way to a large variety of cyclopropanes without the necessity of preparing a new diazo compound for every desired substituent at C-1.

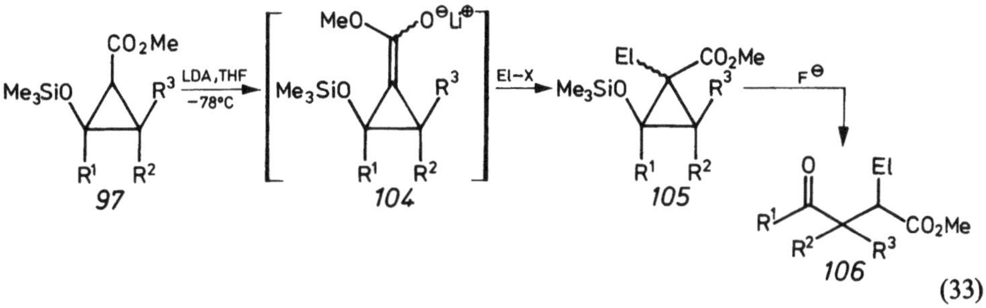

(33)

The enolates *104* can be generated in tetrahydrofuran at —78 °C with 1.5 equivalents of lithium diisopropylamide (LDA) and trapped by electrophiles at this temperature or under "warm up" conditions. Usually, very high yields of alkylation products *105* are attained (Table 3) [68].

With its low yield of 40% entry 1 is exceptional, since in this case very likely missing steric hindrance causes considerable self condensation during enolate generation [69]. This process has been reported as the exclusive reaction in attempts to deprotonate the unsubstituted ethyl cyclopropanecarboxylate [70]. Decreased CH-acidity of the starting material and increased reactivity of the corresponding enolate — both caused by I-strain in the intermediate [71] — should be responsible for this self condensation. It proceeds in the deprotonation phase, if not prevented by additional substituents as in almost all other cases in Table 3.

It should be mentioned that the existence of a carbanion located β to a potential leaving group is not self-explanatory. Since the elimination product would be a highly strained cyclopropene derivative, this process does not occur at least at low temperatures.

The examples in Table 3 demonstrate that many S_N2-active alkyl halides can be employed as electrophiles with good success [68]. Secondary alkyl halides like 2-propyl iodide, however, are not reactive enough in the low temperature ranges required for obtaining clean addition without decomposition of the enolate.

The alkylated products *105* can be transformed to the corresponding γ-oxoesters *106* in high yield by the usual ring cleavage with fluoride reagents (Eq. 33) as shown in Table 3 for some representative cases. With the cyclopropanation/alkylation/ring opening sequence one of the most flexible and efficient routes to specifically substituted γ-oxoesters *106* has been established. Many of the alkylated products shown in Table 3 are starting materials for further synthetic transformations as decribed in upcoming paragraphs.

Heteroelectrophiles like dimethyl disulfide (entries 9, 18, 29) lead to trifunctional compounds. A second methoxycarbonyl group can be introduced with methyl chloroformiate (entries 38, 42). However, the trimethylsiloxy compounds are extremely sensitive and undergo very fast ring opening to malonic ester derivatives [61].

Not surprisingly, carbonyl compounds are excellent reaction partners for the ester enolates *104*. Since the primary adducts are only isolable in singular cases and the subsequent ring opened products (γ-lactols) are very versatile precursors for synthesis of heterocycles, these addition reactions as well as those of electrophiles with a S=C unit will be discussed in a separate paragraph (see section 4.6).

Table 3. Synthesis of 1-substituted 2-siloxycyclopropanecarboxylates *105* by deprotonation of *97* and ring cleavage to γ-oxoesters *106* according to Eq. 33

Entry	R_3	Cyclopropane *97*				Cyclopropane *105*			γ-Oxoester *106*		
		R^1	R^2	R^3	El-X	*trans*:*cis*	Yield	Ref.	Method[a]	Yield	Ref.
1	Me_3	H	H	H	Me—I	90:10	40%	[68]	—		
2	Me_3	Me	H	H	Me—I	90:10	85%	[68]	A	81%	[62]
3	Me_3	Me	H	H	Bu—I	>95:5	71%	[68]	A	98%	[62]
4	Me_3	Me	H	H	$CH_2=CHCH_2-I$	90:10	77%	[68]	A	77%	[61]
5	Me_3	Me	H	H	$PhCH_2-Br$	75:25	67%	[68]	—		
6	Me_3	CMe_3	H	H	Me—I	>97:3	92%	[68]	A	90%	[62]
7	Me_3	CMe_3	H	H	Bu—I	>95:5	91%	[68]	A	90%	[62]
8	Me_3	CMe_3	H	H	$CH_2=CHCH_2-Br$	>97:3	98%	[68]	A	86%	[62]
9	Me_3	CMe_3	H	H	MeS—SMe	>95:5	96%	[68]	A	95%	[62]
10	Me_3	$CH=CH_2$	H	H	Me—I	>95:5	64%	[68]	A	70%	[62]
11	Me_3	$CH=CH_2$	H	H	$CH_2=CHCH_2-Br$	>95:5	62%	[68]	A	70%	[62]
12	Me_3	$CH=CH_2$	H	H	5-bromo-1,3-pentadiene	>95:5	80%	[86]	see section 4.4.2		
13	Me_3	Ph	H	H	Me—I	>97:3	84%	[68]	A	99%	[62]
14	Me_3	Ph	H	H	Bu—I	>97:3	74%	[68]	A	90%	[61]
15	Me_3	Ph	H	H	$CH_2=CHCH_2-Br$	>98:2	80%	[68]	—		
16	Me_3	Ph	H	H	$Me_2C=CHCH_2-Br$	>95:5	78%	[68]	—		
17	Me_3	Ph	H	H	$PhCH_2-Br$	>95:5	89%	[68]	—		
18	Me_3	Ph	H	H	MeS—SMe	>97:3	80%	[68]	—		
19	Me_3	H	H	Me	Me—I	85:15	47%	[68]	B	61%[b]	[61]
20	Me_3	Ph	H	Me	Me—I	>95:5	76%	[68]	A	99%[b]	[62]
21	Me_3	H	Me	Me	Me—I	90:10	87%	[68]	B	79%	[62]
22	Me_3	H	Me	Me	Et—I	88:12	90%	[68]	B	73%	[61]
23	Me_3	H	Me	Me	Bu—I	81:19	77%	[68]	—		
24	Me_3	H	Me	Me	$I(CH_2)_4-I$	84:16	63%	[68]	—		
25	Me_3	H	Me	Me	$CH_2=CHCH_2-Br$	82:18	81%	[68]	B	86%	[62]
26	Me_3	H	Me	Me	$CH_2=CHCH_2-I$	72:28	76%	[68]	—		
27	Me_3	H	Me	Me	$Me_2C=CHCH_2-Br$	75:25	85%	[68]	—		
28	Me_3	H	Me	Me	$PhCH_2-Br$	65:35	81%	[68]	B	84%	[61]
29	Me_3	H	Me	Me	MeS—SMe	11:89	97%	[68]	—		
30	Me_3	Me	Me	Me	$CH_2=CHCH_2-Br$	>95:5	78%	[97]	—		

31	Me_3	Ph	Me	Me	Me—I	92:8	70%	[68]	—		
32	Me_3	$—(CH_2)_3—$		H	$CH_2{=}CHCH_2—Br$	>97:3	70%	[61]	A	96%[b]	[61]
33	Me_3	$—(CH_2)_4—$		H	Me—I	>97:3	87%	[68]	A (F)	84%	
34	Me_3	$—(CH_2)_4—$		H	$CH_2{=}CHCH_2—Br$	>97:3	73%	[68]	—	(97%)[b]	[62]
35	Me_3	$—(CH_2)_4—$		H	$PhCH_2—I$	>97:3	81%	[68]	A	89%[b]	[62]
36	Me_3	$—CHMe(CH_2)_3—$		H	Me—I	2 isomers	81%	[68]	—		
37	Me_3	$—(CH_2)_3CHMe—$		H	Me—I	>95:5	70%	[68]	—		
38	Me_3	$—(CH_2)_3CHMe—$		H	$MeO_2C—Cl$	—	—	—	—	87%	[61]
39	$tBuMe_2$	Ph	H	H	Me—I	>95:5	87%	[79]	—		
40	$tBuMe_2$	Ph	H	Me	Me—I	90:10	54%	[79]	—		
41	$tBuMe_2$	H	Me	Me	Me—I	96:4	74%	[68]	—		
42	$tBuMe_2$	H	Me	Me	$MeO_2C—Cl$	—	84%	[97]	—		

[a] For Methods A–F see Table 2; [b] Mixture of 2 diastereomers

4.2.2. Explanation of the Diastereoselectivity

Although of no importance for most of the subsequent ring opening reactions, the diastereoselectivity of the enolate alkylations deserves attention for mechanistic reasons. Regardless of the *cis/trans*-ratio of the starting materials in tetrahydrofuran as solvent the deprotonation/alkylation sequence provides *trans*-configurated products exclusively or with high preference (Table 3). This incorporation of the electrophile *cis* to the siloxy group predominates even in cases where the sterically more hindered side of the enolate has to be attacked, thus leading to *contrasterical* alkylation (e.g. entries 19, 21–28, 41).

Further experiments have shown that this effect is due to the oxygen-function at C-2, as similar alkoxy cyclopropanes display the same behaviour [69]. Therefore the term "syn-oxyphily" has been suggested for this unusual phenomenon, the explanation of which is still speculative. Since it operates against pure steric effects, it must be of electronic nature with the β-oxygen as the key element.

Very likely enolates of alkyl cyclopropanecarboxylates are generally not planar but pyramidal to a certain extent, thereby avoiding some of the I-strain caused by a methylenecyclopropane type double bond [71]. It is supposed that in oxycyclopropanes the two possible pyramidal enolates *syn-107* und *anti-107* are caused to be energetically different by the RO-groups. Predominating formation of *trans-108* is explained by the higher population of the more stable *anti-108*.

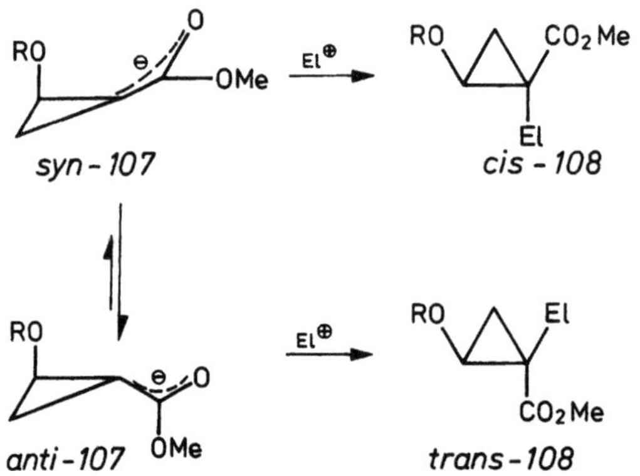

(34)

The higher stability of *anti-107* might simply be attained for electrostatic reasons with the electronegative moieties in *anti* position. However, an anomeric effect with overlap of the occupied enolate π-orbital and the σ*-orbital of the β-C-O-bond can also be operative. Whereas usually the antiperiplanar arrangement of the orbitals involved maximizes the anomeric effect [72], with the rigid cyclopropane geometry this situation is unfavourable. Therefore pyramidalization of the planar enolate to *anti-107* with the larger orbital lobe ("lone pair") at C-1 *syn* to the RO-substituent is energetically more advantageous than the alternative arrangement (see Fig. 1). This might be the first example with evidence of a *syn*-anomeric effect [73]. It should lead to preferred attack of electrophiles at the side with the larger orbital lobe, thus explaining formation of mainly *trans*-products.

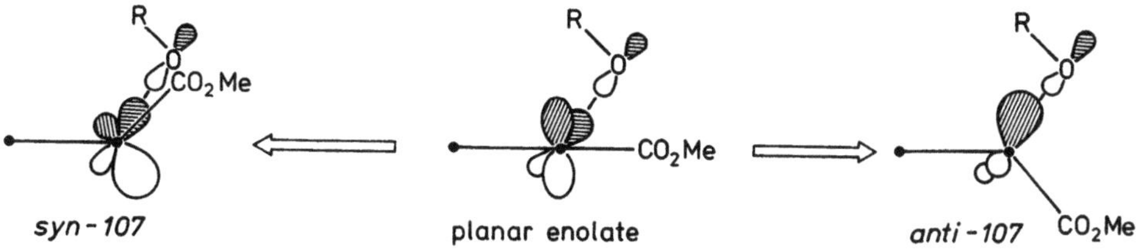

Fig. 1. Pyramidalization of Esterenolate *107* (View along the cyclopropane C-1/C-2 bond, π-type orbital in front, σ*-orbital of the C-2/OR-bond behind)

The stereoelectronic effect of the RO-group is less pronounced, when bulkier electrophiles are employed (Table 3, entries 23, 27, 28), but is increased when the well solvating agent hexamethyl phosphorous amide (HMPA) is used as an additive [61,68]. On the other hand, if one performs the deprotonation/alkylation sequence in the unpolar solvent pentane, a complete reversal of the stereochemical outcome provides the *cis*-product in excess (Eq. 35) [61]. Now a coordination of the lithium cation to the siloxy function might favour structures like *109* (or its oligomers) and cause predominant formation of *cis*-cyclopropanes.

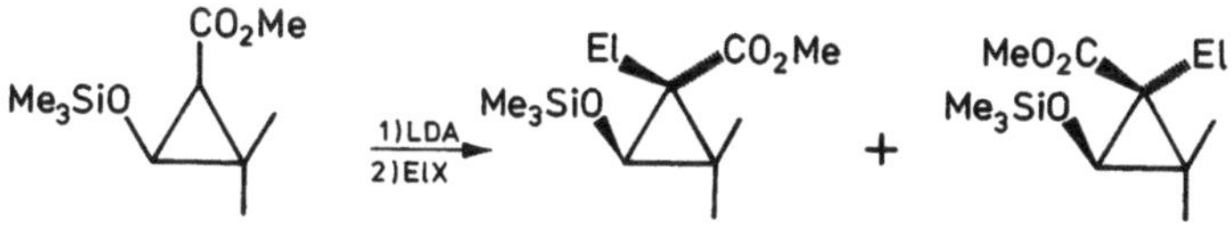

ElX	Solvent	*trans* : *cis*	
MeI	THF	90 : 10	(87%)
MeI	Pentane	7 : 93	(50–80%)
Me_2S_2	THF	11 : 89	(97%)

(35)

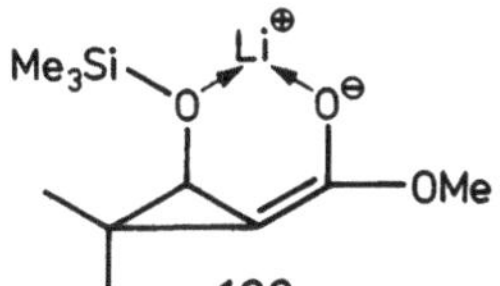

109

With dimethyl disulfide as the electrophile the *cis/trans*-ratio is 90:10 even in tetrahydrofuran. Competition experiments show that this reaction is much faster than the usual alkylations, which afford mainly *trans*-compounds [68]. With the sulfur electrophile a single electron transfer (SET) seems likely generating a cyclopropyl radical as a reactive species, which naturally displays a different selectivity compared to the enolate anion.

4.3 Simple Ring Opening Reactions of Methyl 2-Siloxycyclopropanecarboxylates

4.3.1 One-Pot-Reactions to Products Derived from γ-Oxoesters

In the preceeding sections it has been demonstrated that a great variety of γ-oxoesters can be synthesized by fluoride induced ring cleavage of methyl 2-siloxycyclopropanecarboxylates. As a trimethylsiloxy function is easily converted to a hydroxyl group under acidic or strongly basic conditions [55], this quality can be explored for one-pot-procedures with certain reagents which are able to transform the siloxycyclopropanes to γ-oxoesters and to trap these intermediates giving new types of products.

Equations 36 and 37 illustrate that γ-oxoacids or acetals of γ-oxoesters are available under appropriate conditions [62,74]. Heating with concentrated hydrochloric acid transforms *110* to the α,β-unsaturated lactone *111* (Eq. 38)[62], whereas compounds displaying the suitable substituent pattern can be cyclized to 1,3-cyclopentane diones with bases (Eq. 39) [61].

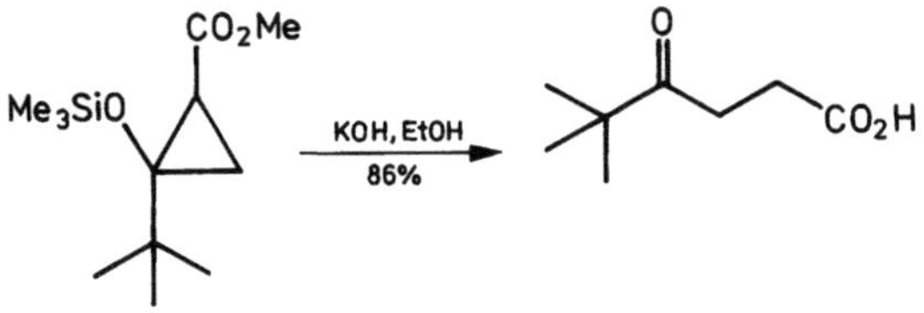

(36)

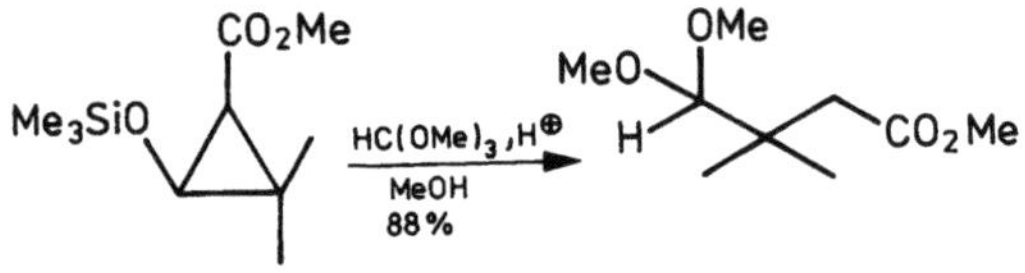

(37)

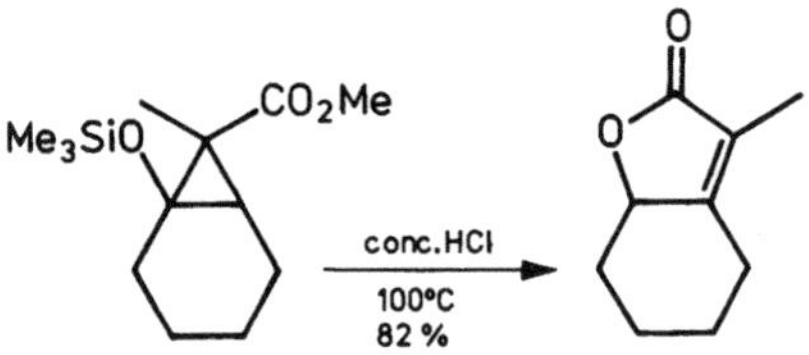

110 *111*

(38)

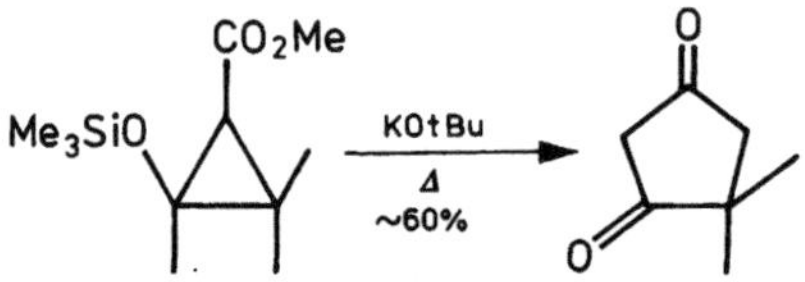

(39)

Nitrogen functions will be incorporated by treatment with phenylhydrazine or hydroxylamine providing the corresponding hydrazones [75] or oximes [61] (Eq. 40). When siloxycyclopropanes containing a masked aldehyde function ($R^1 = H$) are heated with hydroxylamine and formic acid under reflux, dehydration of the intermediate oxime generates β-cyanoester *112* in moderate to good yield (Eq. 41) [61].

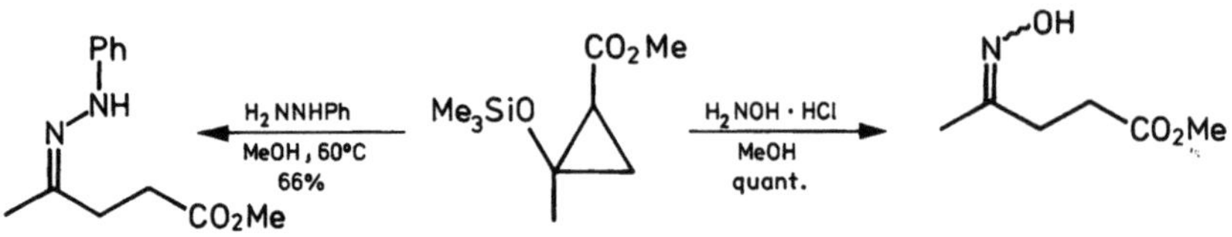

(40)

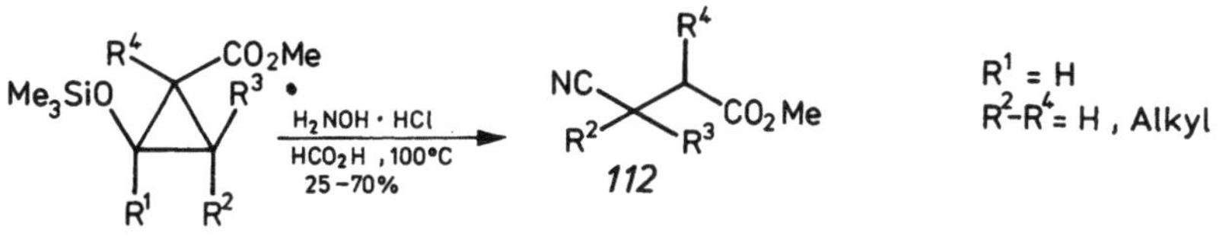

(41)

4,5-Dihydro-2H-3-pyridazinones *113* are obtained in excellent yields from several methyl 2-siloxycyclopropanecarboxylates with hydrazine hydrate as a reagent [75].

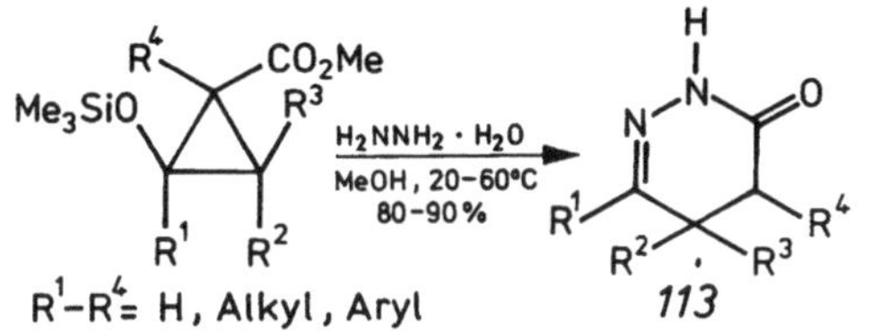

(42)

Although in most cases reaction conditions for these one-pot-transformations are rather harsh, it is conceivable that optimization might lead to milder methods in singular cases. Nevertheless, these straightforward transformations make available a variety of interesting products by very simple and cheap procedures.

4.3.2 Methods Leading to Reduced Products

Treatment of methyl 2-siloxycyclopropanecarboxylates with potassium borohydride in methanol and acidic work-up afford γ-butyrolactones *114* by a very simple one-pot procedure in excellent yields and with high purity (Eq. 43) [76].

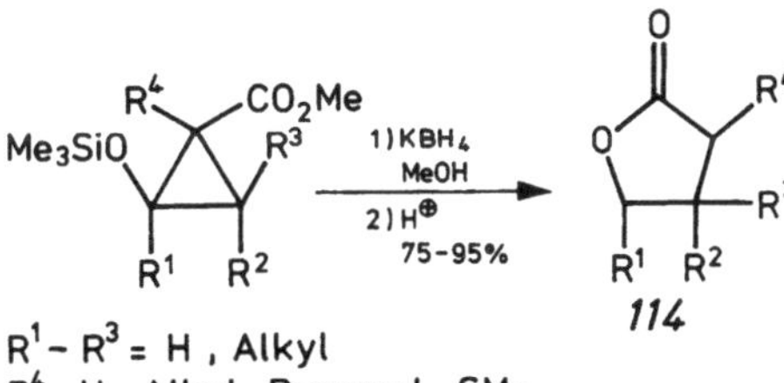

(43)

The multistep reaction very likely starts with desilylation by potassium methylate (generated *in situ*) and ring opening to the γ-oxoester. This is immediately reduced to a γ-hydroxyester, which undergoes lactonization to form the final product *114*. Corresponding to the regioselective synthesis of the starting cyclopropanes, isomeric γ-butyrolactones can easily be constructed. When the reaction is performed in CD_3OD, α-deuterated γ-butyrolactones can be prepared.

The electrochemical reduction of γ-oxoesters like *115*, synthesized via the cyclopropane route, results in cyclization with participation of the olefinic unit [77]. This reaction provides interesting cyclopentanol derivatives, which can be transformed to the corresponding cyclopentenes. Alternatively a fragmentation to medium sized ketones like *116* occurs after saponification and anodic oxidation [77].

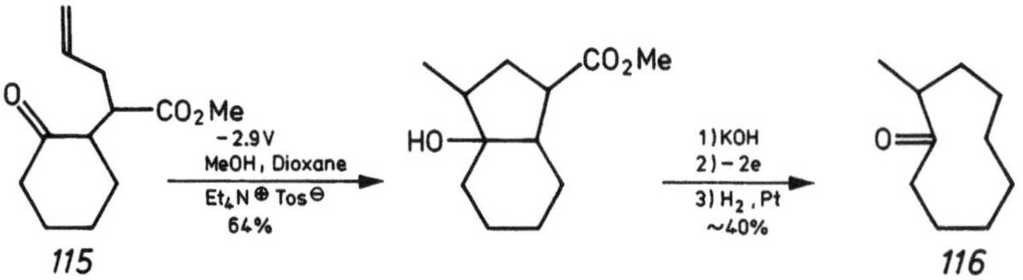

115 116 (44)

The reduction of methyl 2-siloxycyclopropanecarboxylates can also be started at the ester function when lithium aluminum hydride in ether is the reagent. The resulting alcohols undergo the wellknown cyclopropylcarbinyl/homoallyl rearrangement upon treatment with acid to provide β,γ-unsaturated carbonyl compounds *117*. These are synthesized isomerically pure and in good yields in a number of cases, if the two-phase-system 2N hydrochloric acid/pentane is employed[78]. Otherwise the very easy isomerization to the conjugated α,β-unsaturated compounds *118* occurs to some extend, which can intentionally be completed by base catalysis.

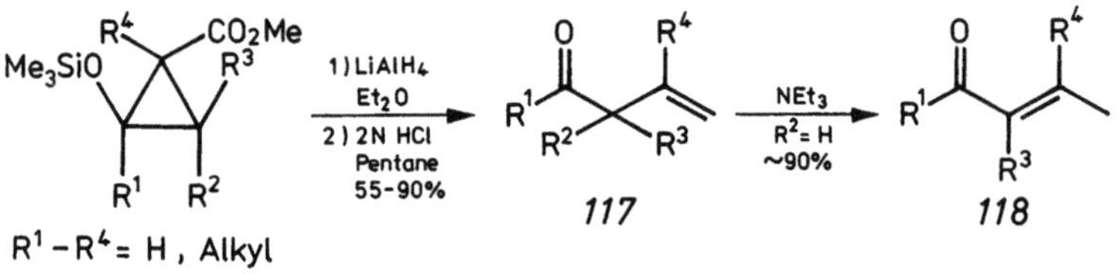

117 118 (45)

Methyllithium as a nucleophile finally provides compounds with a trisubstituted double bond (Eq. 46). Ring cleavage of the intermediate cyclopropylcarbinols under strongly basic conditions gives γ-hydroxy carbonyl compounds like *119*[78]. The regioselectivity of this ring opening is remarkable, since it seems to be governed by the hydroxymethyl group passing through a secondary carbanion as an intermediate.

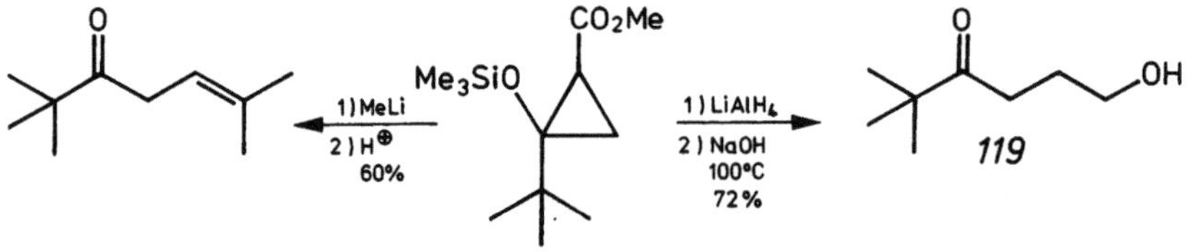

119 (46)

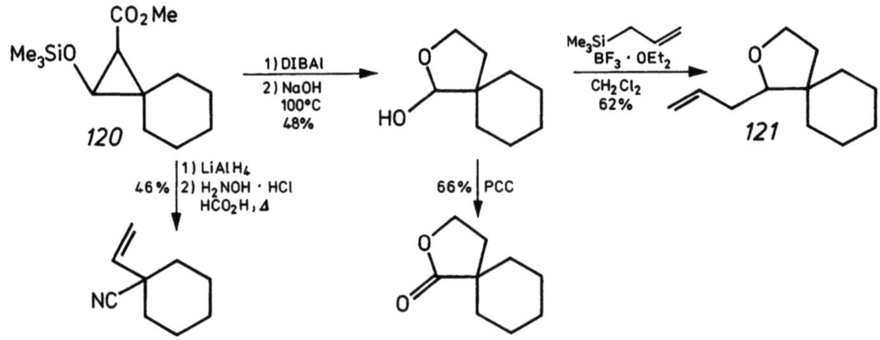

120 121 122 123 (47)

These reductions can be combined with other transformations as exemplified by cyclopropane *120*, which has been converted to the nitrile *122*, the spiro lactone *123*, and the functionalized spiro tetrahydrofuran *121* [61]. All products are eventually derived from cyclohexane carbaldehyde.

Direct reductive cleavage by catalytic hydrogenolysis of a cyclopropane C-C-bond is only possible in exceptional cases, where the three membered ring is further activated by a phenyl or a vinyl group. With unpoisoned catalyst a subsequent reductive desiloxylation occurs to afford esters like *124* (Eq. 48), whereas addition of small amounts of triethylamine allows isolation of the desired siloxy compounds (e.g. *125*, Eq. 49). Interestingly, both reactions demonstrate that the cleavage of the cyclopropane bond proceeds non-stereoselectively with inversion and retention at C-1 and C-2, respectively [79].

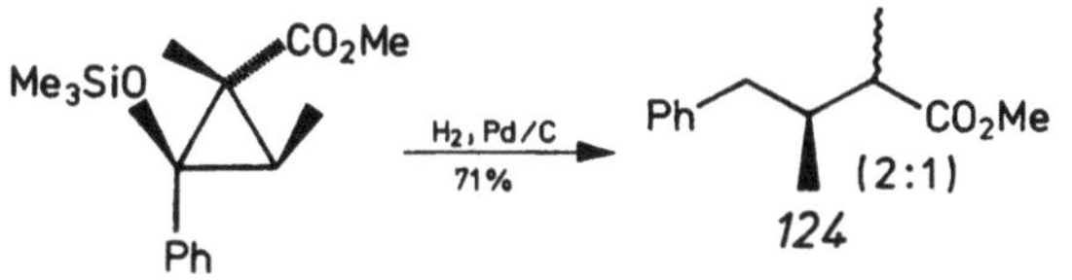

(48)

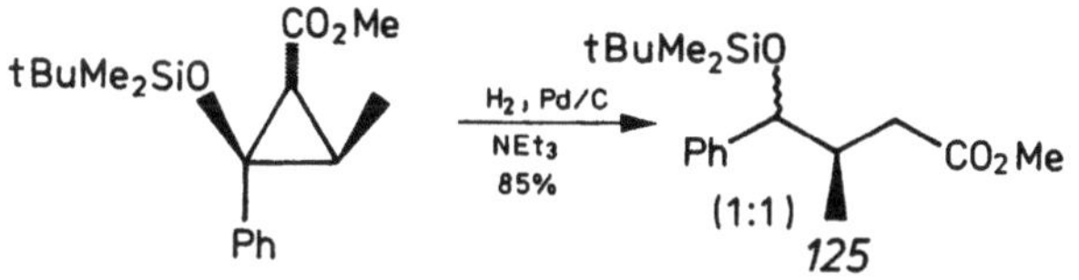

(49)

4.3.3 Methods Leading to Oxidized Products

By addition of bromine at low temperature methyl 2-siloxycyclopropanecarboxylates are regioselectively cleaved to give α-bromo-γ-oxoesters *126* quantitatively. These can be isolated in singular cases, but usually they are directly transformed to α,β-unsaturated γ-oxoesters *127* by subsequent elimination of hydrogen bromide with triethylamine [80]. Thereby *E*-alkenes are obtained in good overall yields, which are interesting acceptor building blocks for further synthetic operations.

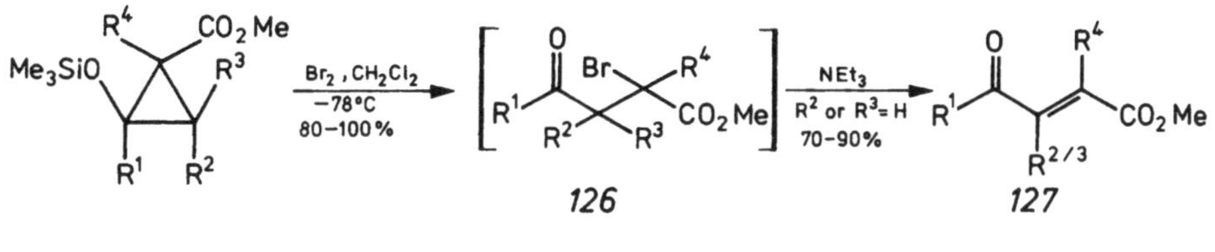

(50)

Due to the great flexibity in the preparation of the cyclopropanes — especially with regard to R^4 — a variety of other alkenes *127* should be available in a regio- and stereoselective manner by this overall methoxycarbonylmethylenation method of a given carbonyl compound.

Reaction with an excess of bromine can be performed under warm-up conditions delivering the α,β-dibromo-γ-oxoester *128* as a mixture of diastereomeres. Interestingly, the ratio of these stereoisomers is strongly influenced by the reaction conditions chosen. Both dibromo compounds are converted to the corresponding acetylene *129* by triethylamine treatment [61]. Thus, using very simple and cheap reagents the siloxycyclopropanes can serve as precursors for γ-oxoesters with a single, a double or a triple bond between C-α and C-β.

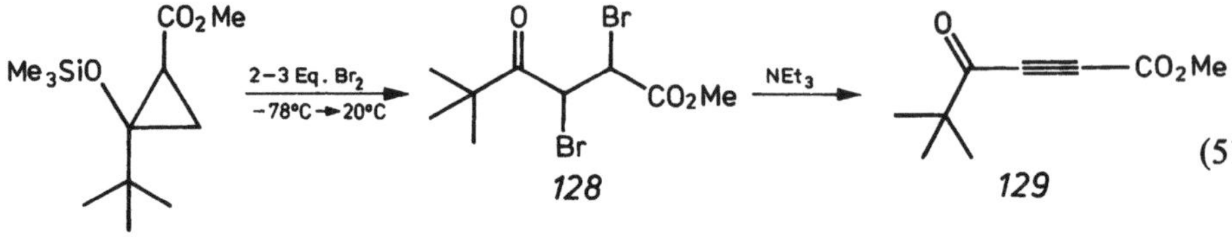

128 *129* (51)

Whereas chlorine or iodine as electrophiles do not provide clean addition products [61], phenylselenenyl chloride opens siloxycyclopropanes forming α-selenenylated γ-oxoesters [81]. Usually, however, this process is better conducted in the presence of Lewis acids at low temperature (vide infra, section 4.5.1).

4.4 Cleavage Reactions of Methyl 2-Siloxycyclopropanecarboxylates Combined with Subsequent C-C-Bond Forming Processes

Although methyl 2-siloxycyclopropanecarboxylates are cleaved by certain electrophiles, only tetracyanoethylene (TCNE) as a carbon electrophile could directly be added to phenyl or vinyl activated cyclopropanes providing cyclopentane derivatives [61].

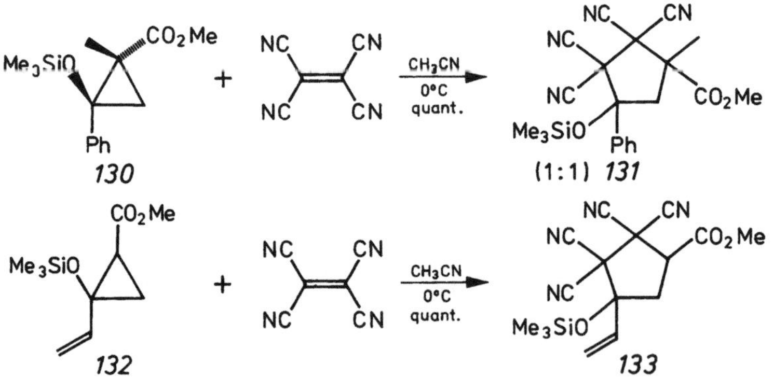

130 (1:1) *131* (52)

132 *133* (53)

Eq. 52 and 53 demonstrate remarkable characteristics of this [3 + 2]-cycloaddition: starting with a pure diastereomer *130*, two stereoisomeric cyclopentanes *131* are obtained. This stereorandom outcome is most simply rationalized assuming a stepwise mechanism with a 1,5-zwitterion as an intermediate in the cycloaddition. The vinylcyclopropane *132* only gives five-membered ring products *133* and no cycloheptene derivative, which would result from a conceivable [5 + 2]-cycloaddition. Less activated olefins or cyclopropanes do not undergo a similar [3 + 2]-cycloaddition. Due to the specific substitution pattern, the cyclopentane formation from these siloxycyclopropanes is of no preparative value.

Synthetically more interesting are ring openings which were performed under conditions allowing *in situ* reactions of the resulting γ-oxoesters. Thus siloxycyclopropane *134* is cleaved to the formylester *135*, which can directly be transformed to the substituted γ-butyrolactone *136* by subsequent treatment with allylbromide/zinc [82] and acid [65].

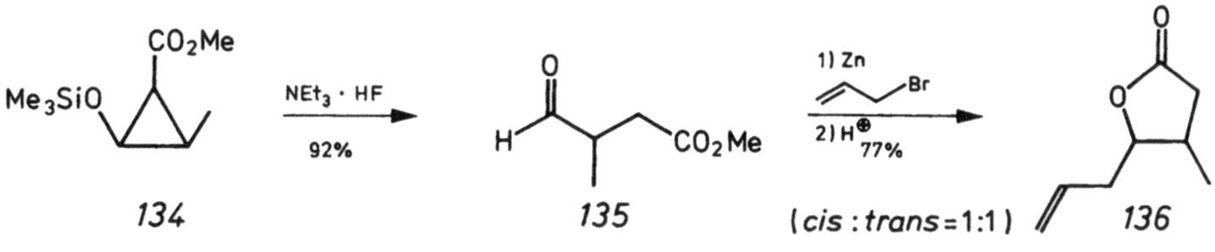

134 *135* (*cis* : *trans* = 1:1) *136*

(54)

Whereas scope and limitations of this procedure are yet unknown, vinyl ketones obtained as intermediates from methyl-2-alkenyl 2-siloxycyclopropanecarboxylates have already demonstrated their extreme versatility for synthetic purposes.

4.4.1 Michael Additions to Vinyl Ketones Generated from Vinylcyclopropanes

Siloxycyclopropane *132* — which is a masked vinyl ketone — has served as test substrate for many *in situ* transformations. It cleanly reacts with several of O-, N-, and S-nucleophiles under mild acidic or basic conditions leading to polyfunctionalized γ-oxoesters *137* (Eq. 55) [83], which for instance can be further converted to γ-butyrolactones, as exemplified by the one-pot-synthesis of *138* [76].

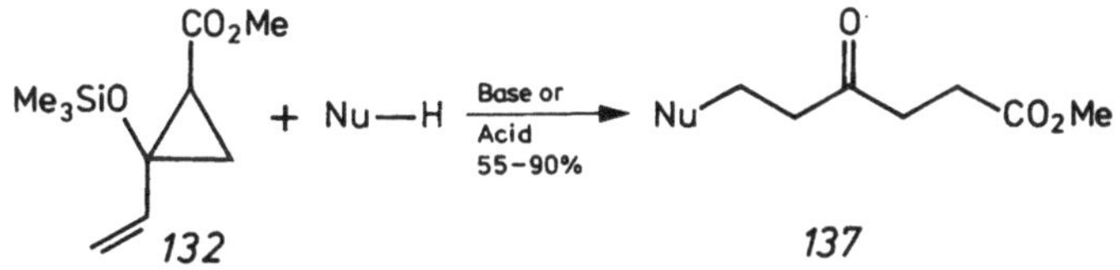

132 *137*

Nu = MeO, PhS, Et_2N, O_2N, $PhSO_2$

(55)

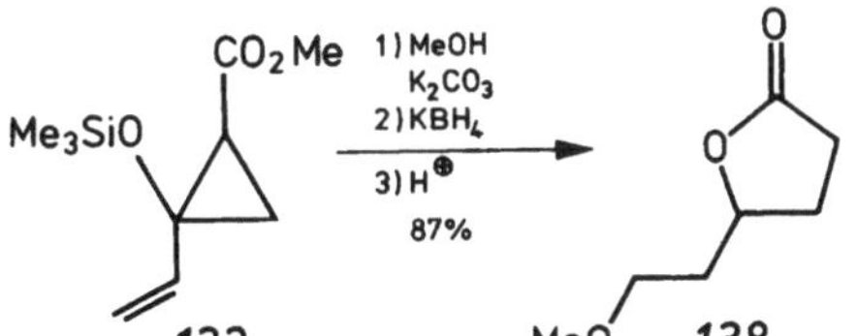

132 *138*

(56)

More interestingly, Michael additions of CH-acidic compounds *139* onto the vinyl ketone — generated from *132* by action of the catalyst Triton B (0.08 equiv. in methanol) — smoothly afford adducts *140* in good yields (Eq. 57) [83]. Under these conditions methoxide acts as the desilylating agent as well as the base to form the corresponding carbanions.

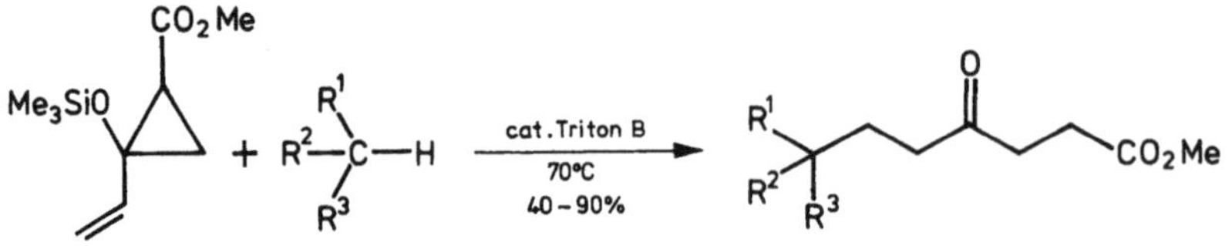

132 139 140

$R^1, R^2 = CO_2Me, COMe, SO_2Ph, NO_2, H$ $R^3 = H, Me$

(57)

Nitroalkanes provide products, which are especially versatile building blocks, since the nitro group introduced can be converted to several other functionalities. The examples shown in Eq. 58–61 include synthesis of regioisomeric 7-nitro-4-oxo-esters, of a cyclic system *141*, and of a propenyl substituted derivative — all proceeding with good overall yield from easily available, inexpensive starting materials [83,84].

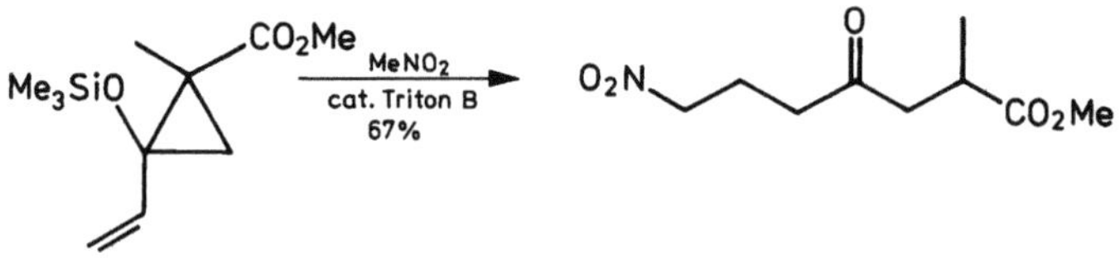

(58)

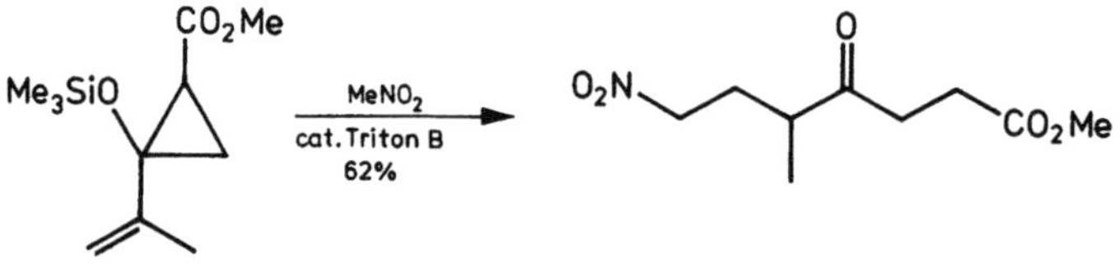

(59)

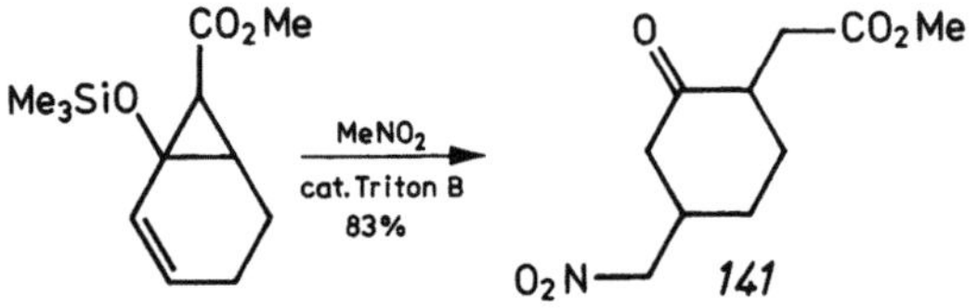

141
(*trans* : *cis* = 87 : 13)

(60)

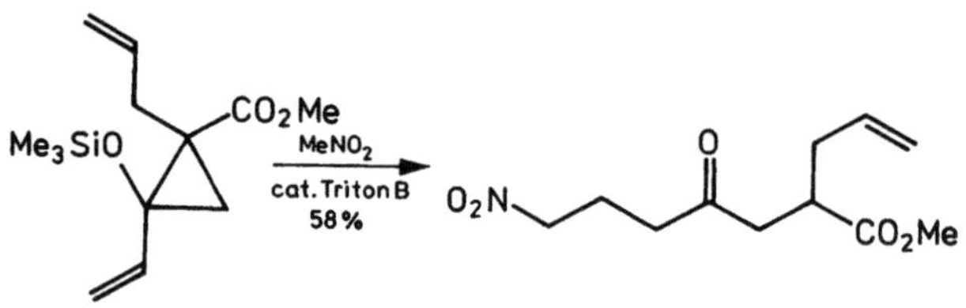

(61)

Transformations of the nitro function in *142* to a carbonyl group employing variants of the Nef reaction or its reduction to a cyclic nitrone *143*, which is capable to undergo 1,3-dipolar cycloadditions, underscore the high synthetic potential of these nitroalkane adducts (Eq. 62) [84].

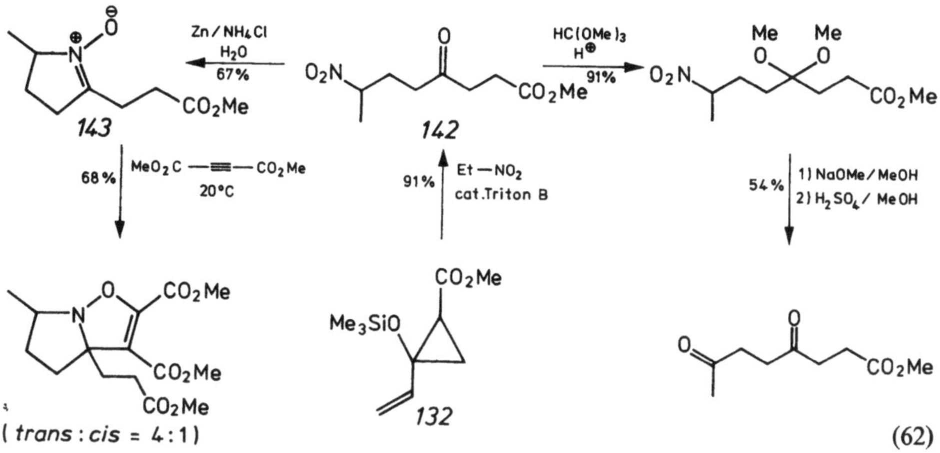

(62)

Free radical additions of phenylthio or stannyl radicals to 2-alkenyl 2-siloxycyclopropanes afford similar products although a completely different mechanism is operative [84]. This direct generation of protected γ-oxoesters *144* and *145* is of interest since the silyl enol ether function might be usable for regioselective C-C-bond formation and the allyl stannane moiety in *145* could be activated for subsequent transformations. Yet further examples have to demonstrate utility and scope of this mode of ring opening.

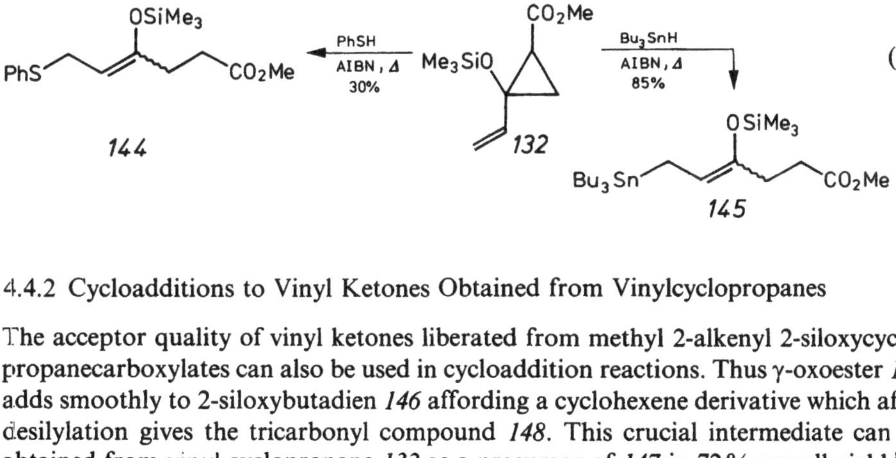

(63)

4.4.2 Cycloadditions to Vinyl Ketones Obtained from Vinylcyclopropanes

The acceptor quality of vinyl ketones liberated from methyl 2-alkenyl 2-siloxycyclopropanecarboxylates can also be used in cycloaddition reactions. Thus γ-oxoester *147* adds smoothly to 2-siloxybutadien *146* affording a cyclohexene derivative which after desilylation gives the tricarbonyl compound *148*. This crucial intermediate can be obtained from vinyl cyclopropane *132* as a precursor of *147* in 72% overall yield [85]. Its chemoselective methylation, lactonization, and dehydration make norbisabolid available — a constituent of the root bark of *atalantia monophylla*.

More fascinating, however, are intramolecular modes of the Diels-Alder reaction. The vinyl cyclopropanes in discussion open a very short and flexible entry to substrates suitable to undergo this cycloaddition. Cyclopropane *149* — easily obtained by alkylation of the unsubstituted compound *132* (see Table 3, entry 12) — after ring cleavage with fluoride gives a trienone *150*, for instance, which has ideal electronic and steric properties for an intramolecular [4 + 2]-cycloaddition. After reaction at

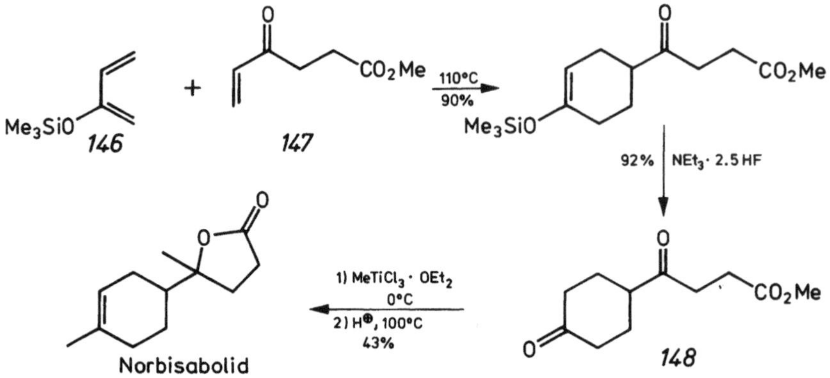

(64)

room temperature a 6:1 mixture of two *cis*-octalones *151*/*152* is identified, from which the major isomer *151* can be isolated by crystallization [86]. This compound is probably formed *via* a boat-like *endo*-transition state TS with the ester function in a sterically favourable position.

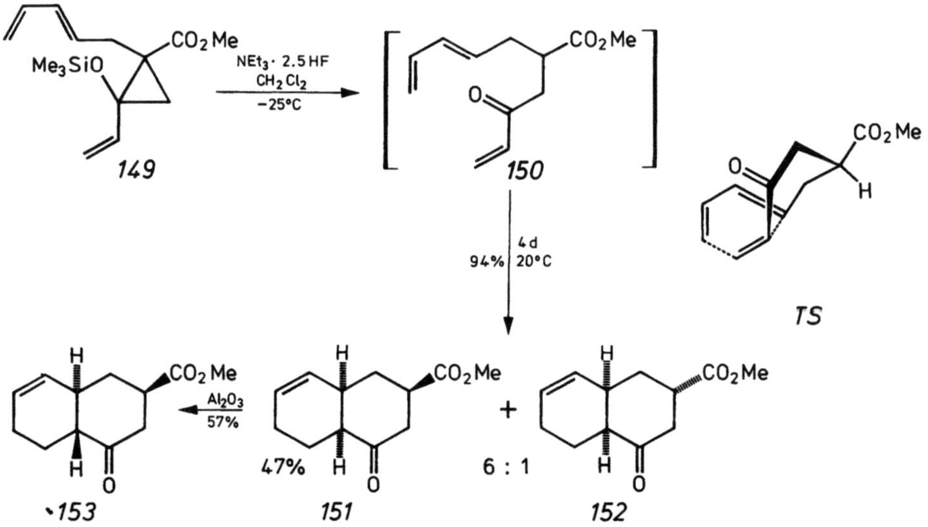

When the crude reaction mixture is chromatographed on Al_2O_3 the *trans*-octalone *153* is obtained in 57% overall yield as a result of epimerization. The potential of this route to prepare bicyclic or polycyclic carbon skeletons under mild conditions and in a stereocontrolled manner is evident.

4.4.3 Fluoride Triggered Ring Opening/Addition Reactions of Alkyl 2-Siloxycyclopropanecarboxylates

If the cleavage of siloxycyclopropanes with fluoride anions could be performed under strictly anhydrous conditions, the ester enolate generated by ring opening should be interceptable by electrophiles different from protons. This process could be realized in the methylation of cyclopropane *154* (Eq. 66), however the degree of alkylation does not exceed 60%. The intramolecular variant also gives a mixture of alkylated and protonated product (Eq. 67)[62].

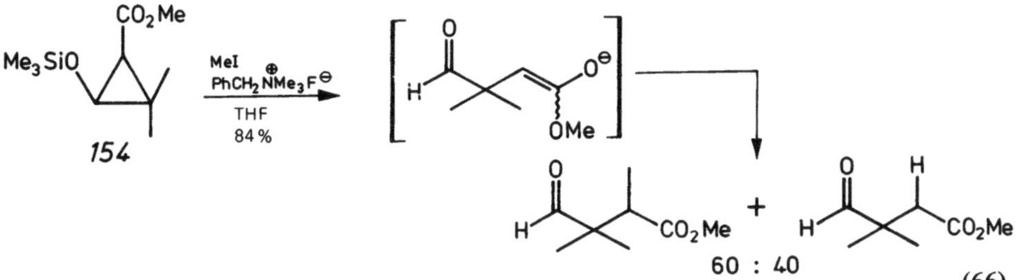

(66)

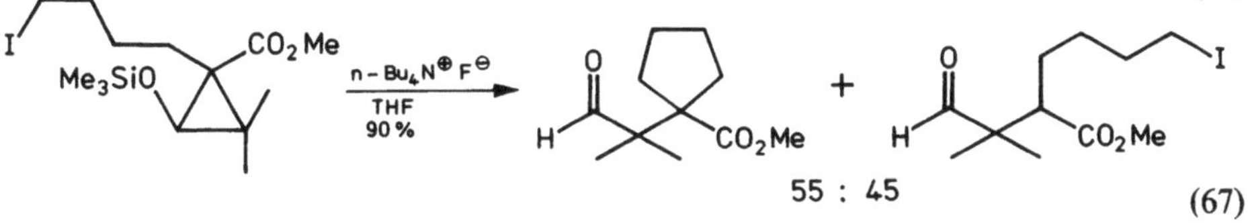

(67)

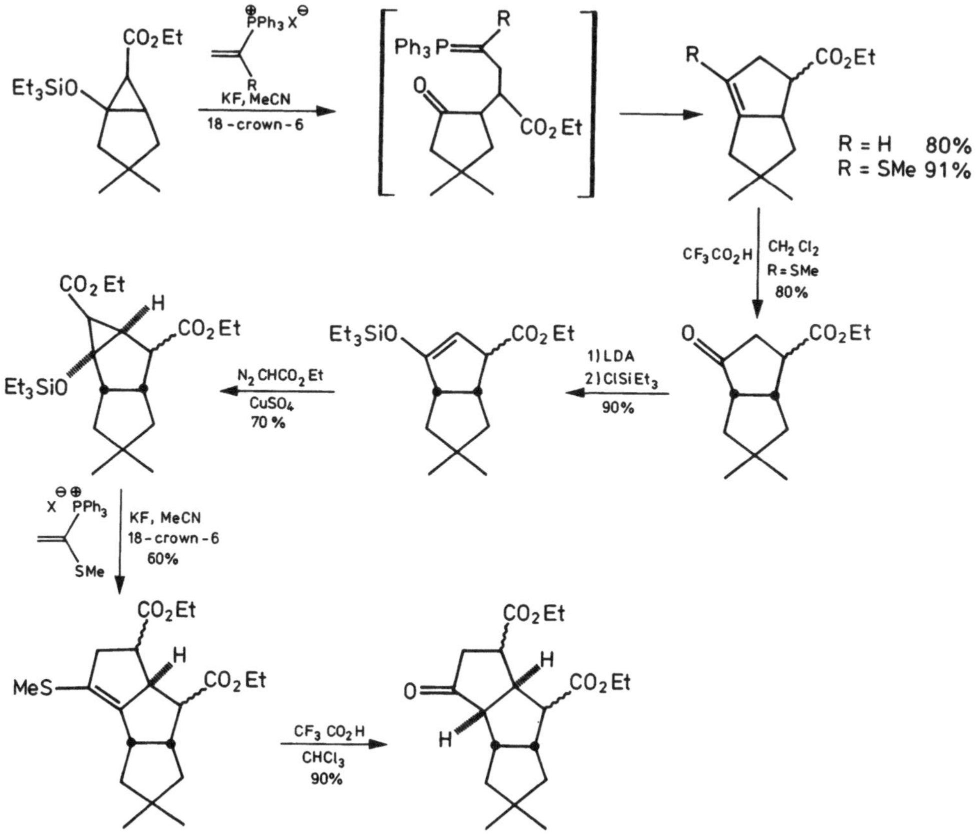

Scheme 6. Fluoride induced repetitive [3 + 2]-annulation using ethyl 2-siloxycyclopropanecarboxylates

Very likely the ammonium fluorides are the proton sources and therefore the reason for incomplete conversions, since potassium fluoride in acetonitrile gives high yields in a very elegant [3 + 2]-annulation process [87]. It combines a Michael addition to a vinyl phosphonium salt with an intramolecular Wittig reaction and proceeds only in the presence of 18-crown-6 with satisfying yield. This cyclopentene synthesis has been executed in a repetitive manner to prepare linear triquinanes as illustrated in Scheme 6. Unfortunately, the sequence is non-stereoselective with regard to the ethoxycarbonyl functions.

4.5 Lewis Acid Induced Reactions of Methyl 2-Siloxycyclopropanecarboxylates

4.5.1 Catalytic Processes

Whereas methyl 2-siloxycyclopropanecarboxylates are thermally stable up to temperatures as high as 170 °C, they readily rearrange at low temperatures under the influence of appropriate Lewis acids. Catalytic amounts (0.05–0.4 equiv.) of iodotrimethylsilane within minutes to days promote a quantitative ring opening of cyclopropanes *155* to the corresponding silyl enol ethers *156* (Eq. 68, Table 4) [88].

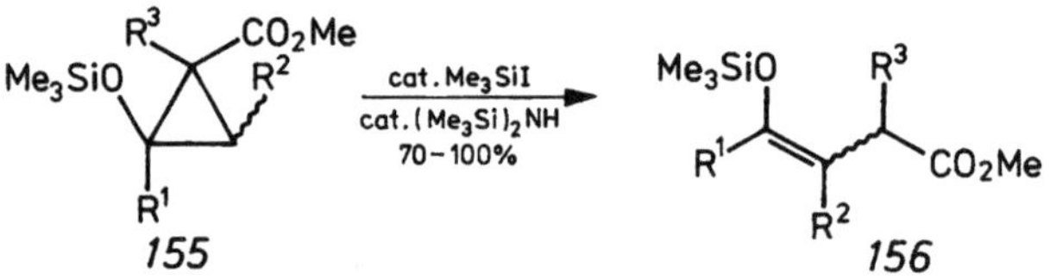

(68)

Although the presence of the base hexamethyl disilazane facilitates the isolation of pure olefins *156*, it is not essential for the occurrence of the rearrangement. ^{1}H-NMR spectroscopy reveals that the "simple" ring opening under proton transfer *155* → *156* is actually a cascade of silylgroup and proton shifts [61], which finally establishes the thermodynamic equilibrium. Therefore, with small substituents R^1 or R^2 mixtures

Table 4. Isomerization of Cyclopropanes *155* to Silyl Enol Ethers *156* According to Eq. 68

Entry	R^1	R^2	R^3	Reaction-time	*Z/E*	Yield
1	CMe_3	H	H	1 d	100/0	97 %
2	CMe_3	H	Me	1 d	100/0	100 %
3	CMe_3	H	SMe	4 d	100/0	100 %
4	CMe_3	H	$CH_2CH{=}CH_2$	2 d	100/0	71 %
5	Ph	H	H	5 h	90/10	77 %
6	$CH{=}CH_2$	H	H	30 min	65/35	49 %
7	H	H	H	20 min	80/20	100 %
8	H	Me	H	30 min	50/50	95 %
9	Me	H	H	20 min	70/20/10[a]	89 %
10	$-(CH_2)_4-$		H	20 min	84/0/16[a]	72 %
11	$CHMe_2$	H	H	6 h	25/0/75[a]	83 %

[a] Formation of the possible regioisomeric silyl enol ether

of *E*/*Z*-isomers are obtained (entries 5–9) and compounds capable to enolize α and α′ to the keto function give the possible regioisomeric silyl enol ethers (entries 9–11). It is evident that cyclopropane *157* can only rearrange to the γ,δ-unsaturated ester *158* (Eq. 69).

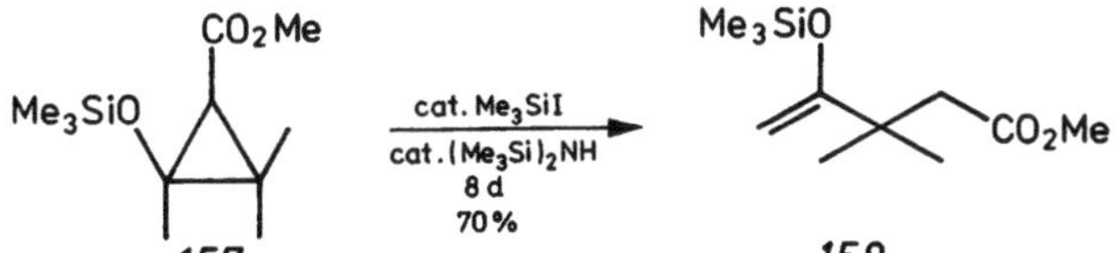

157 *158* (69)

The functionalized silyl enol ethers *156* are useful synthetic intermediates since electrophiles can now be introduced either directly in the β-position by known methodology [55] or in the α-position after deprotonation with LDA to an allyl anion (Eq. 70) [61]. Both pathways should enormously widen the scope of specifically substituted γ-oxoesters and their derivatives obtained via siloxycyclopropanes.

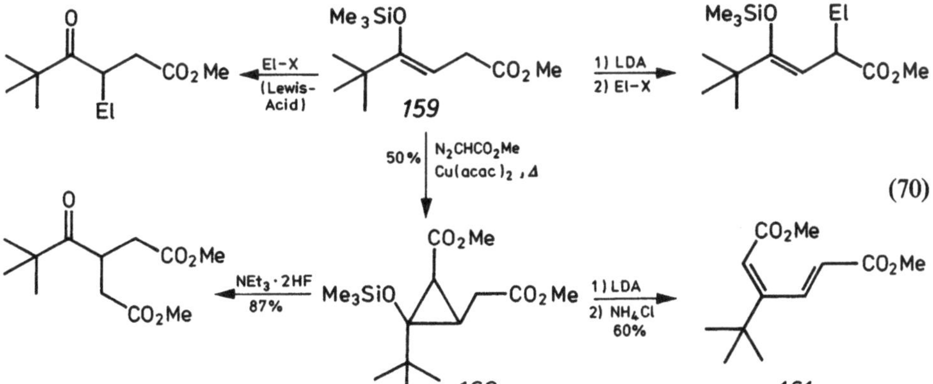

159 *160* *161* (70)

Cyclopropanation of the enol ether *159* and normal ring cleavage introduces a second methoxycarbonylmethyl group, but deprotonation of the intermediate cyclopropane *160* now occurs at the more acidic side chain CH_2 and not at the cyclopropane core. Therefore a ring opening followed by elimination yields the electrondeficient diene *161* (Eq. 70) [61].

Treatment of siloxycyclopropane *162* with equimolar amounts of iodotrimethylsilane and triethylamine generates the electronrich diene *163* in high yield [88]. Many further applications of compounds like *163* — cycloadditions, for instance, or reactions with electrophiles — should be manageable.

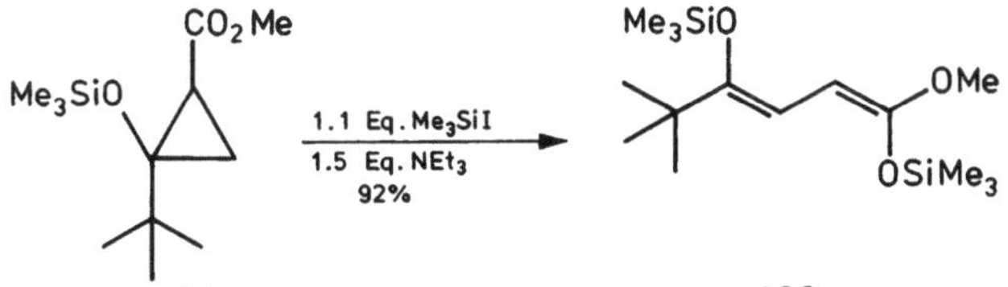

162 *163* (71)

A remarkable double cyclopropanation with the Simmons-Smith reagent and ring cleavage under basic conditions provides two carboxylic acids *164* and *165* in almost equal amounts. None of the expected β-benzoyl propionic acid could be isolated. Although formation of the crucial intermediate *168* has been explained with a slightly different mechanism by the authors [89], according to the Me_3SiI-catalyzed rearrangement described above the pathway suggested in Eq. 72 via *167* is more likely. Either iodotrimethylsilane is formed *in situ* or ZnI_2 is the catalyst transforming *166* to *167*. Surprisingly, the ring opening of the cyclopropanolate generated from *168* under basic conditions occurs with negligible regioselectivity.

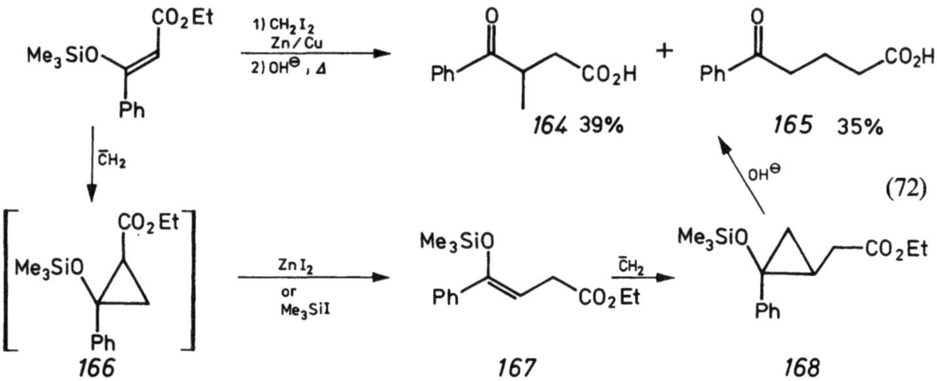

(72)

The soft "counterion" iodide is essential for the Me_3SiI-induced ring cleavage *155* → *156*. Employment of hard Lewis acids like titanium tetrachloride as catalysts leads to a smooth *cis/trans* equilibration of methyl 2-siloxycyclopropanecarboxylates even at very low temperatures [90]. The equilibrium seems to be governed by steric effects mainly, although not all aspects of this process are understood. A straightforward mechanistic interpretation suggests a coordination of the Lewis acid to the carbonyl oxygen, which strongly accelerates heterolytic cleavage of the donor-acceptor activated cyclopropane *169*. In the zwitterionic intermediate *170* rotation around single bonds and reclosure to *169* is possible, thus the thermodynamic equilibrium at the cyclopropane stage can be established.

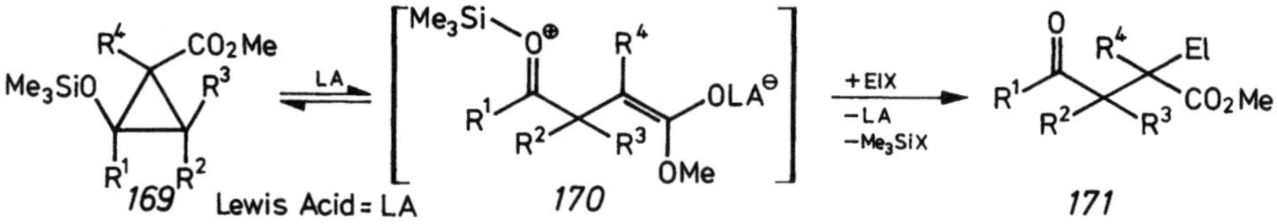

(73)

This mechanistic picture is supported by the fact that the ketene acetal moiety in *170* can be trapped by suitable electrophiles providing adducts *171*. Ring cleaving selenenylation and sulfenylation is therefore possible at —78 °C in the presence of a few drops of $TiCl_4$ (Table 5, entries 1—5) [91]. Without Lewis acid no reaction with the sulfur electrophile could be observed at room temperature and incorporation of the phenylselenenyl group occurs only rather slowly.

Table 5. Lewis Acid Promoted Addition Reactions of Electrophiles to Cyclopropanes *169* According to Eq. 73 and 74

Entry	R^1	R^2	R^3	R^4	El-X	LA	Yield
1	H	Me	Me	Me	PhSe—Cl	$TiCl_4$	69%
2	Me	Me	Me	H	PhSe—Cl	$TiCl_4$	88%
3	CMe_3	H	H	H	PhSe—Cl	$TiCl_4$	85%
4	CMe_3	H	H	Me	PhSe—Cl	$TiCl_4$	82%
5	CMe_3	H	H	H	ArS—Cl[a]	$TiCl_4$	87%
6	Me	Me	Me	H	$CH_2{=}NMe_2Cl$	$TiCl_4$	55%
7	Me	Me	Me	H	$CH_2{=}NMe_2OTf$	—	74%
8	CMe_3	H	H	H	$CH_2{=}NMe_2OTf$	—	89%
9	CMe_3	H	H	Me	$CH_2{=}NMe_2OTf$	—	80%
10	Ph	H	H	H	$CH_2{=}NMe_2OTf$	—	56%
11	$-(CH_2)_4-$		H	H	$CH_2{=}NMe_2OTf$	—	65%

[a] Ar = o-$NO_2C_6H_4$

4.5.2 Reactions with Equimolar Amounts of Lewis Acids

In contrast to transformations described in the preceeding section additions of N,N-dimethylmethaniminium chloride or carbonyl compounds, respectively, to siloxycyclopropanes *169* require equimolar quantities of Lewis acid. Probably the promotor is deactivated by the resulting Lewis basic products. Aminomethylation can be performed with the iminium chloride and $TiCl_4$ (Table 5, entry 6), but it proceeds more easily and reproducibly with the corresponding iminium triflate — generated *in situ* from the iminium chloride by anion exchange with trimethylsilyl trifluoromethanesulfonate [91]. Very likely a rearrangement induced by traces of trimethylsilyl triflate present forms an O-methyl O-silyl ketene acetal *172* (detectable by ^{1}H-NMR) [61], which should be the true reacting nucleophilic species (Eq. 74).

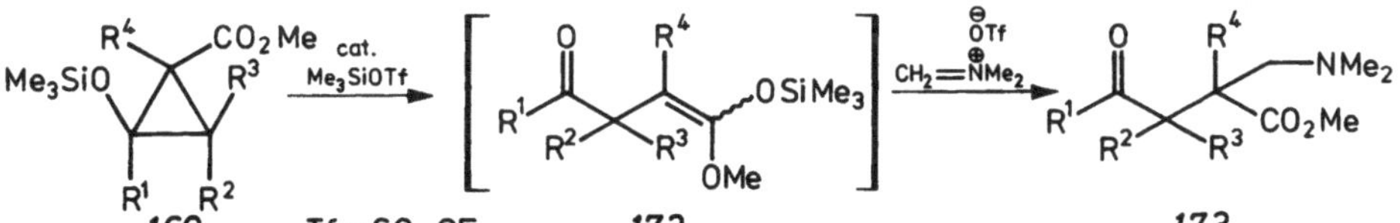

(74)

Several α-aminomethylated γ-oxoesters *173* can be prepared in good yield. However, for reasons not understood so far compounds *169* with R^1 = H do not undergo this C-C-bond forming cyclopropane cleavage. Products *173* can serve as precursors for syntheses of acrylate derivatives or α-methylene γ-butyrolactones (Eq. 75) [91].

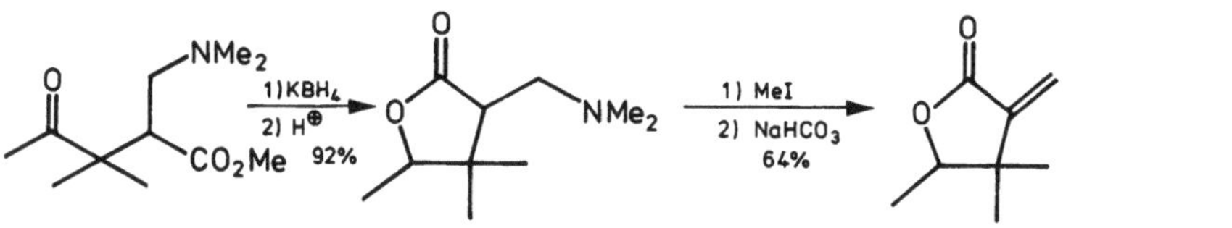

(75)

Synthetically even more versatile trifunctional intermediates result from the addition of carbonyl compounds onto methyl 2-siloxycyclopropanecarboxylates [92]. Benzophenone, titanium tetrachloride, and *162*, for instance, provide an excellent yield of the α-hydroxyalkylated γ-oxoester *174*, which predominates in the equilibrium with its cyclic hemiacetal *176* (γ-lactol). It can undergo elimination to the unsaturated ester *175*, but as Scheme 7 illustrates, *174*/*176* can also serve as the starting material to several highly substituted furan(one) derivatives.

Whereas most transformations are selfexplanatory, the conversion to the tetrahydrofuran-3-carboxylate *178* should be regarded with special attention. This highly stereoselective reaction proceeds with an oxacarbenium ion as intermediate and could also be extended to certain silylated C-nucleophiles (*vide infra*) [93].

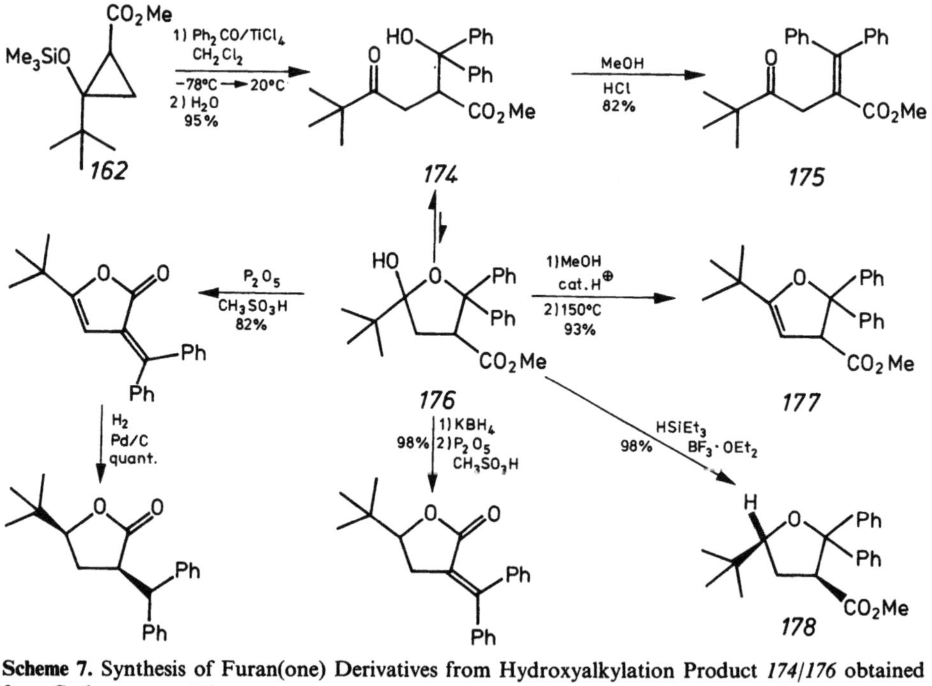

Scheme 7. Synthesis of Furan(one) Derivatives from Hydroxyalkylation Product *174*/*176* obtained from Cyclopropane *162*

The synthesis of dihydrofuran derivatives such as *177* has been performed to explore scope and limitations of the Lewis acid promoted hydroxyalkylation of siloxycyclopropanes. Table 6 shows that aromatic as well as aliphatic ketones can efficiently be incorporated. Enolization of ketones does not occur and a 1-methyl group at the cyclopropane is no obstacle for the reaction, which now binds the carbonyl compound to a quarternary center with surprisingly high efficiency (entry 5). Albeit there are some restrictions with regard to the substitution pattern of the cyclopropanes, bicyclic siloxycyclopropanes also give good yields (e.g. entry 6 and Eq. 76). Further examples of the tetrahydrofuran synthesis from intermediate γ-lactols with

Table 6. Synthesis of Dihydrofuran Derivatives Analogous to *177* (Scheme 7)

Entry	Starting Material	Carbonyl Compound	Method[a]	Product	Yield [%]
1	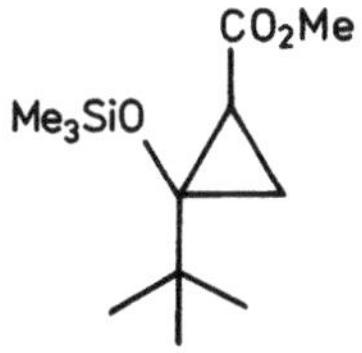	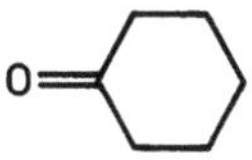	B	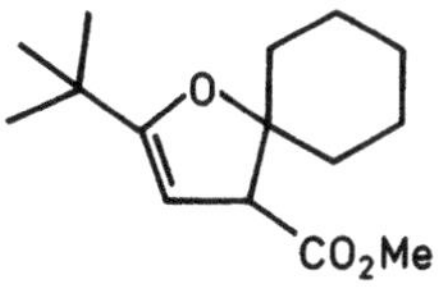	88
2	//	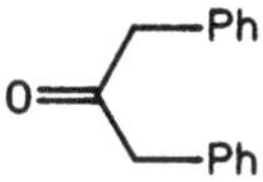	C	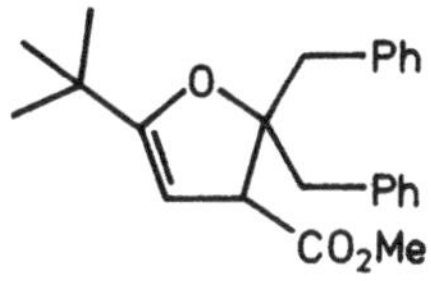	58
3	//	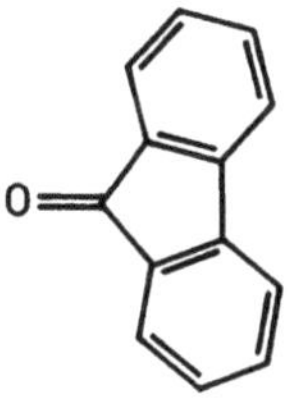	C	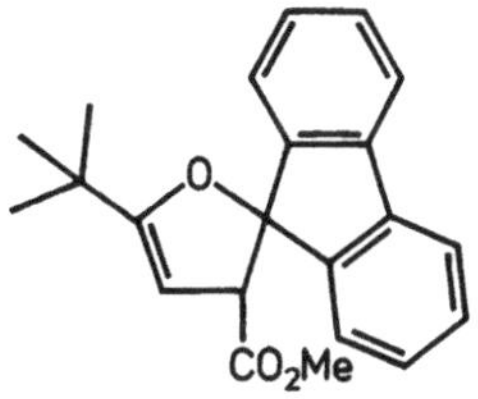	61
4	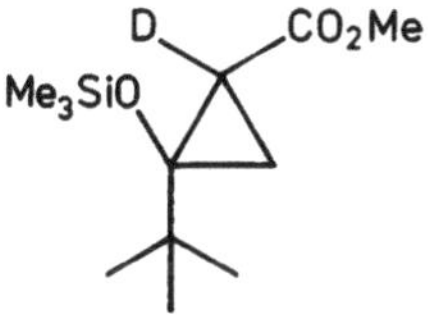	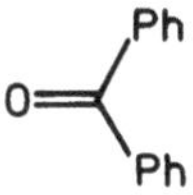	A	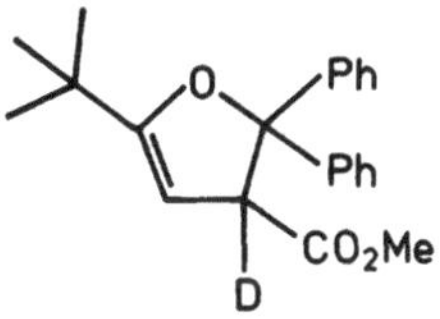	75
5	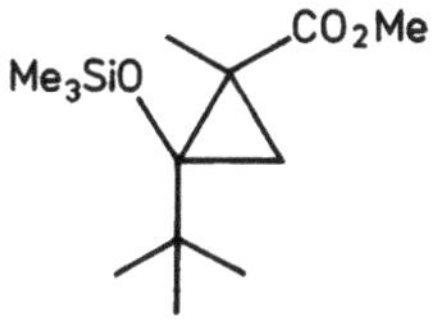	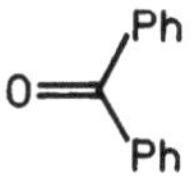	C	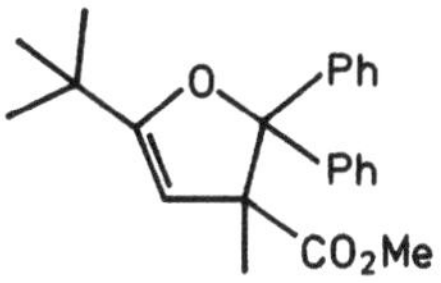	88
6	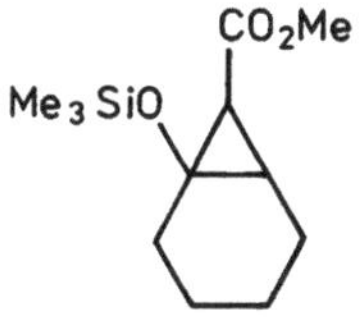	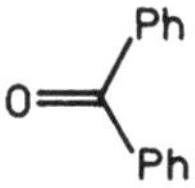	C	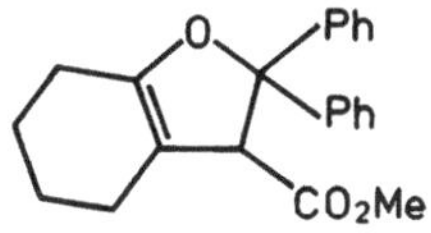	52

[a] Method A : see Scheme 7
Method B : crude acetal is thermolized
Method C : crude γ-lactol is thermolized

triethylsilane/BF_3 are displayed in Eq. 76 and 77. Here bicyclic products with spiro or linear annulated rings result [92].

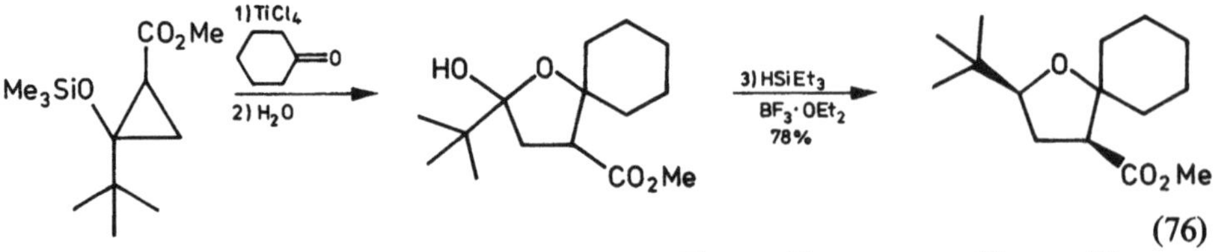

(76)

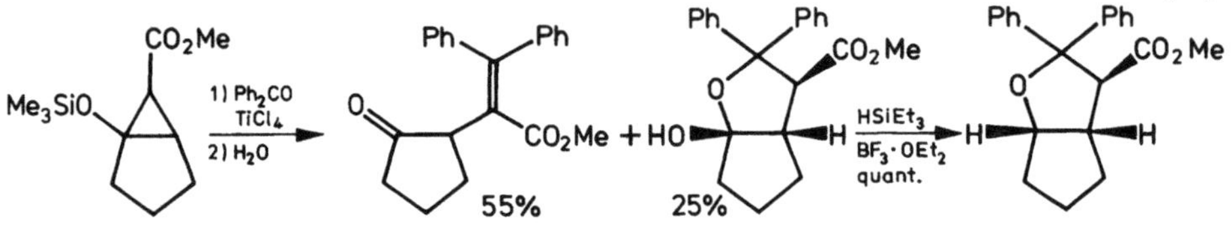

(77)

Intriguing mechanistic details of the $TiCl_4$-promoted C-C-bond forming process were discovered during attempts to execute the addition of *162* to benzaldehyde [61]. The usual conditions with premixing of the carbonyl component and $TiCl_4$ followed by warm up together with *162* leads to two diastereomeric chlorinated products *179*, whose origin from the primary adduct(s) via a S_N1-type substitution is obvious (Eq. 78). However, treatment of cyclopropane *162* with the Lewis acid and warm up *without* aldehyde generates a new species as indicated by a colour change of the resulting solution from wine red to yellow. This intermediate is reactive enough to add cleanly to benzaldehyde at —78 °C providing *one* diastereomer after aqueous work up.

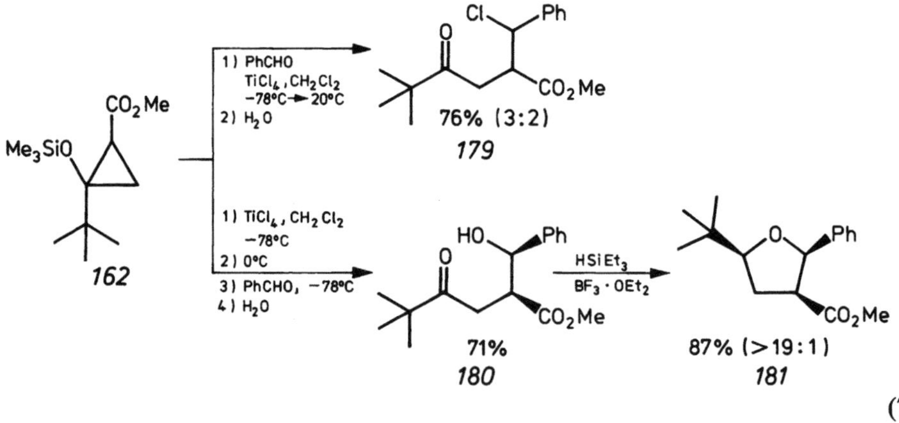

(78)

The *anti*-stereochemistry of the adduct *180* is confirmed by its smooth transformation to *cis*-tetrahydrofuran derivative *181*. Therefore, combination of two highly stereoselective processes brings about the efficient preparation of the specifically substituted heterocycle *181* [61].

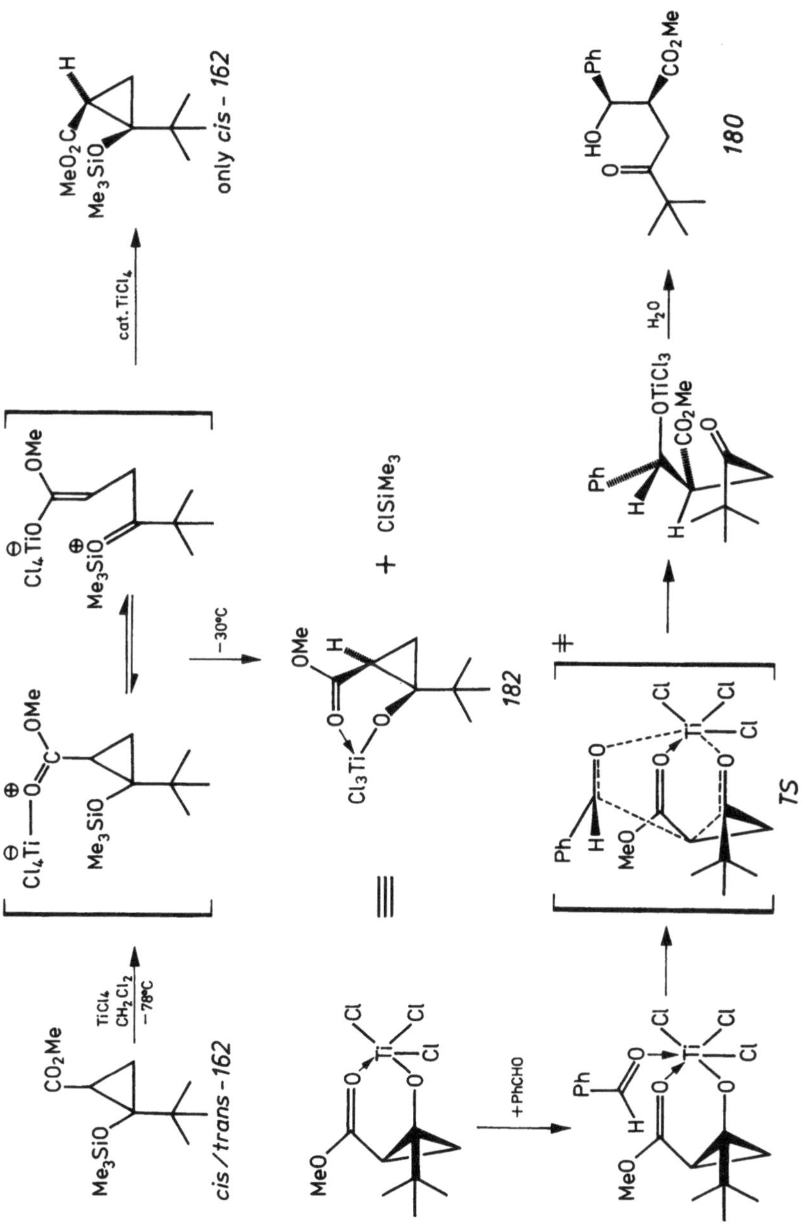

Scheme 8. Generation and Stereoselective Reaction of Titanoxycyclopropane *182*

^{13}C-NMR spectroscopy proves that titanoxycyclopropane *182* is the crucial species, which is formed at ca. —30 °C and reacts with the aldehyde at lower temperature without chlorination. Scheme 8 illustrates generation of *182* and its stereoselective addition to the aldehyde with a transition state *TS* locating the phenyl group in the sterically favourable *anti*-position with respect to the large *tert*-butyl group.

This mechanism is therefore completely different from that established for the formally related reaction of 1-ethoxy 1-trimethylsiloxycyclopropane *183*, which gives γ-butyrolactones with aldehydes [94]. In this homoaldol addition the β-titanated ester *184* is the reactive intermediate.

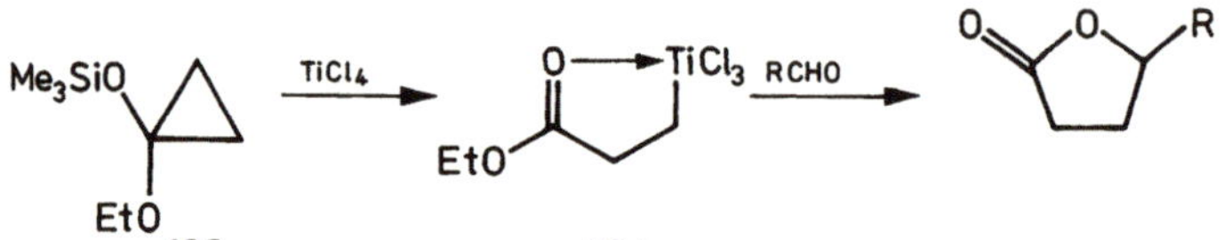

183 *184* (79)

However, in the usual procedure only aldehydes can be employed as electrophiles for *183*, whereas donor-acceptor-substituted cyclopropanes as *162* are active even towards very sluggish ketones like benzophenone. The mechanism depicted in Scheme 8 illustrates that it is the ester function, which serves as a suitable handle for the Lewis acid to start the reaction and to facilitate cleavage of the cyclopropane bond with concomitant formation of a new C-C-bond.

The generality of this stereoselective process has yet to be demonstrated but synthesis of *185* (Eq. 80) from *162* and isobutyroaldehyde in satisfying yield shows that aliphatic aldehydes can also be used as electrophiles [61].

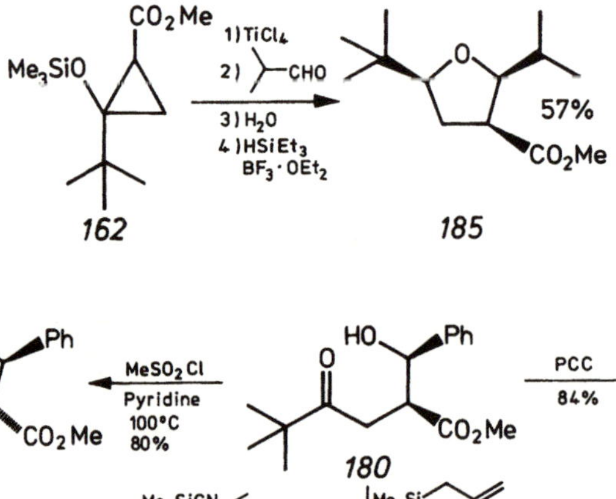

162 *185* (80)

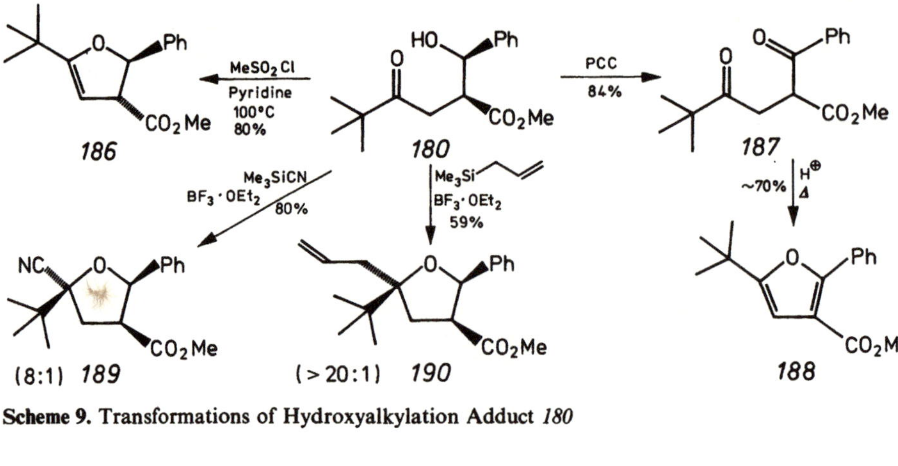

Scheme 9. Transformations of Hydroxyalkylation Adduct *180*

Adduct *180* can be transformed to various furan derivatives (Scheme 9) [61]. Dehydration is accompanied by *cis/trans* isomerization giving *trans*-dihydrofuran *186*, and oxidation to ketone *187* followed by condensation provides the trisubstituted furan *188*. Cyano and allyl trimethylsilane, respectively, lead to tetrahydrofuran derivatives *189* and *190* with high stereoselection, which carry new substituents suitable for further manipulation. These reactions work equally well for the ketone adduct *176* and a bicyclic γ-lactol affording *191–193* [93].

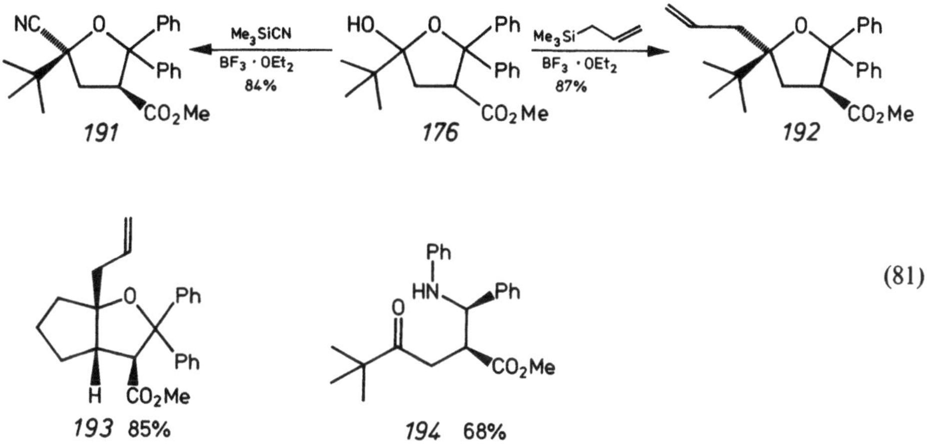

(81)

Other C-electrophiles besides iminium salts and carbonyl compounds have not yet extensively been tested for C-C-forming cyclopropane ring cleavage. However, whereas acyl or tertiary alkyl halides do not give addition products, benzaldimine at least provides the expected secondary amine *194* with good yield and diasteroselectivity [61].

The reactions portrayed so far in this paragraph involve hard Lewis acids, which coordinate at the cyclopropane carbonyl oxygen and thereby assist different modes of ring cleavage. The soft Lewis acid mercury(II) acetate might directly attack the cyclopropane C-1 of *162*. The mercury compound *195* is formed in high yield but, unfortunately, allows only low conversions to the desired addition product *196* [95]. It results from free radical reaction of a reductively generated radical, which is obviously deactivated by the ester function. The main product therefore is the γ-oxoester *197* formed by capture of hydrogen.

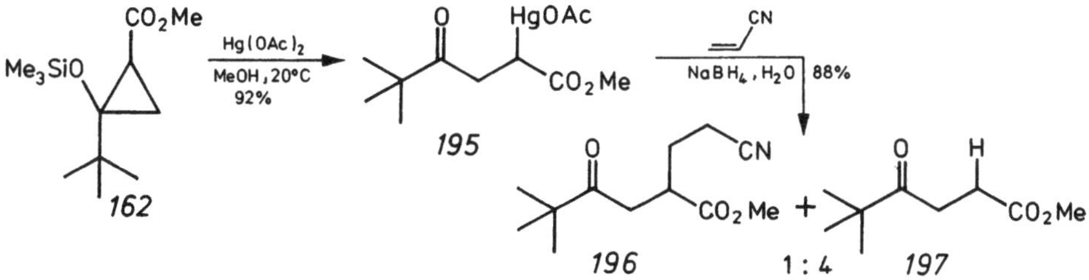

(82)

4.6 Synthesis of Heterocycles via Enolates of Methyl 2-Siloxycyclopropanecarboxylates

4.6.1 Furan Derivatives by Addition to Carbonyl Compounds

As mentioned in the last paragraph, the methodology employing Lewis acids for activation of siloxycyclopropanes has certain restrictions with regard to the substitution pattern. Therefore an alternative path to prepare hydroxyalkylated adducts would be most valuable. It can actually be achieved via enolates of methyl 2-siloxycyclopropanecarboxylates, which add cleanly to many carbonyl compounds (Eq. 83)[96].

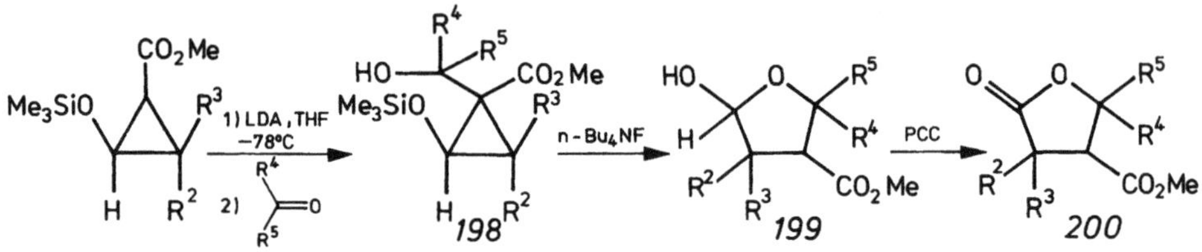

(83)

However, due to the free hydroxy function primary adducts *198* with a trimethylsiloxy group are rather labile in most cases and can usually not be isolated in good yield [97]. Therefore direct treatment of the crude reaction product with n-Bu_4NF is highly advantageous to complete ring cleavage to the corresponding γ-lactols *199*. Of course, this two-step-procedure to synthesize hydroxyalkylation products *199* is not stereoselective. Thus for $R^4 \neq R^5$ four diastereomeric γ-lactols are formed, which can be oxidized with pyridinium chlorochromate to provide paraconic esters *200* in satisfying overall yield [96].

Table 7. Synthesis of Paraconic Esters *189* According to Eq. 83

Entry	R^2	R^3	R^4	R^5	Yield	*cis*:*trans*
1	Me	Me	Ph	Ph	70%	—
2	Me	Me	Me	Me	67%	—
3	Me	Me	$-(CH_2)_5-$		54%	—
4	$-(CH_2)_5-$		Me	Me	43%	—
5	$-(CH_2)_5-$		$-(CH_2)_5-$		51%	—
6	$-(CH_2)_5-$		Ph	Ph	27%	—
7	Me	Me	Ph	H	57%	2:3
8	Me	Me	Me	H	52%	1:2
9	Me	Me	CHMePh	H	52%	1:3[a]
10	Me	Me	CH=CHMe	H	38%	1:3
11	$-(CH_2)_5-$		Me	H	51%	1:1

[a] Isomers due to the exocyclic center of chirality

Table 7 demonstrates that aromatic, aliphatic, and α,β-unsaturated ketones or aldehydes can be used as electrophiles. With regard to the cyclopropane substituents, R^1 has to be hydrogen for the synthesis of γ-butyrolactones *200*, but other examples (*vide infra*) show that the crucial C-C-bond forming step also proceeds with $R^1 \neq$ H.

As siloxycyclopropanes with $R^1 = H$ were not suitable for the Lewis acid promoted hydroxyalkylation, the two principle methods are supplementary.

In two cases the direct oxidation with PCC of the primary adduct *198* — presumably already contaminated with the corresponding γ-lactol — has been tried and surprisingly the β-hydroxylated γ-butyrolactones *201* are formed (Eq. 84) [96b].

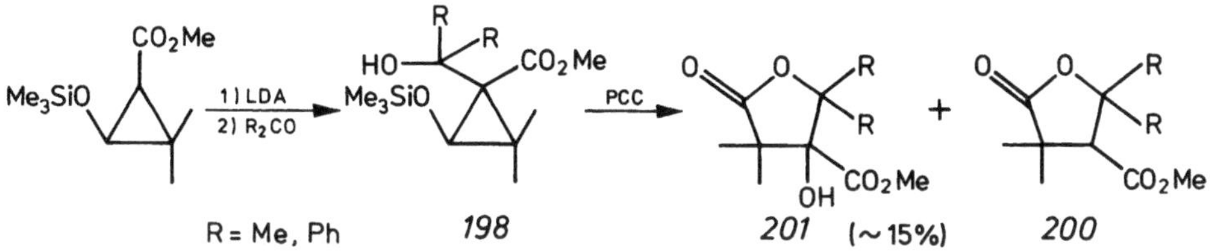

(84)

A second method to approach synthetically valuable and easily isolable compounds, consists of the Lewis acid promoted addition of silylated nucleophiles to γ-lactols. Again the reagent combination $HSiEt_3/BF_3$ excellently serves for reductive removal of the anomeric hydroxy group in *202* providing methyl tetrahydrofuran-3-carboxylates *203* with satisfying overall yield from *97* (Eq. 85, Table 8) [93b]. As to be expected for systems with $R^4 \neq R^5$ a *cis/trans*-mixture of *203* is obtained, reflecting the missing stereoselectivity in the step forming the γ-lactol.

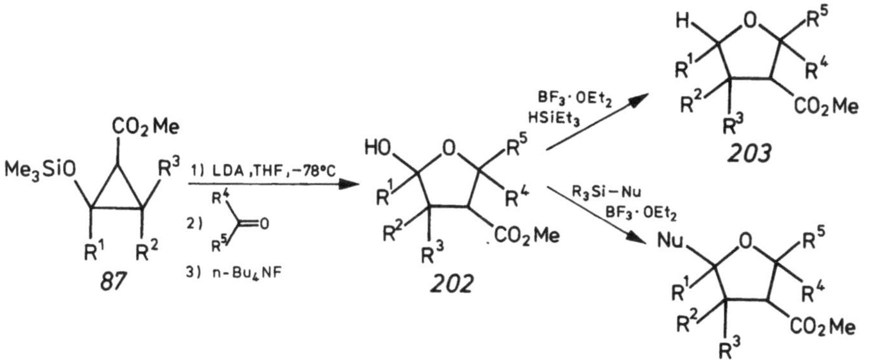

(85)

Table 8. Synthesis of Tetrahydrofuran-3-carboxylates *203* According to Eq. 85

Entry	R^1	R^2	R^3	R^4	R^5	Yield	*cis*:*trans*
1	H	Me	Me	Ph	Ph	47%	—
2	H	Me	Me	Me	Me	71%	—
3	H	$-(CH_2)_5-$		Ph	Ph	20%	—
4	H	$-(CH_2)_5-$		$-(CH_2)_5-$		48%	—
5	H	Me	Me	Me	H	51%	1:3
6	H	Me	Me	Ph	H	79%	2:3
7	H	Me	Me	CHMePh	H	71%	1:2[a]
8	H	$-(CH_2)_5-$		Me	H	54%	1:1

[a] 4 Isomers due to the exocyclic center of chirality

As already discussed above several C-nucleophiles also add smoothly to γ-lactols (Eq. 85, Table 9). BF_3 is the promotor of choice allowing efficient and highly diastereoselective reactions with allyl and cyanotrimethylsilane or with bis(trimethylsilyl)acetylene which introduce the corresponding substituents at C-5 of the methyl tetrahydrofuran-3-carboxylates *204*. In the case of propargyltrimethylsilane an allenyl group is transferred to the heterocyclic [93].

Table 9. Synthesis of 5-substituted Tetrahydrofuran-3-carboxylates *204* According to Eq. 85

Entry	R^1	R^2	R^3	R^4	R^5	$R_3Si{-}Nu$	Yield	*cis*:*trans*
1	H	Me	Me	Me	Me	$Me_3Si{-}CH_2{-}CH{=}CH_2$	56%	1:6
2	H	$-(CH_2)_5-$		Me	Me	$Me_3Si{-}CH_2{-}CH{=}CH_2$	67%[a]	1:4
3	H	Me	Me	Me	Me	$Me_3Si{-}CN$	48%	1:3
4	H	$-(CH_2)_5-$		Me	Me	$Me_3Si{-}CN$	92%[a]	4:5
5	H	Me	Me	Me	Me	$Me_3Si{-}C{\equiv}C{-}SiMe_3$	77%	<1:20
6	H	$-(CH_2)_5-$		Me	Me	$Me_3Si{-}C{\equiv}C{-}SiMe_3$	62%[a]	<1:20
7	H	Me	Me	Me	Me	$Me_3Si{-}CH_2{-}C{\equiv}CH$	92%[a]	2:3[b]

[a] Yield based on isolated γ-lactol *202*; [b] In the product Nu = $C{=}C{=}CH_2$

The stereoselectivity of this reaction rises when more bulky nucleophiles are employed (compare entries 7, 3, 1, and 5). This is most impressively demonstrated by comparison of the γ-lactol reduction with its allylation leading to *205* or *206*, respectively (Scheme 10). Formation of tetrahydrofuran derivative *208*, dihydrofuran *209*, or unsaturated α-methylen-γ-butyrolactone *207* illustrate that various modes of straightforward work-up procedures provide two different five membered heterocycles [93b, 96]. A second example without the geminal dialkyl substitution at C-3 of the siloxycyclopropane depicted in Eq. 86 making available the annulated tetrahydrofuran-3-carboxylate *210* underlines the generality of the C-C-bond forming hydroxyalkylation reaction via ester enolates.

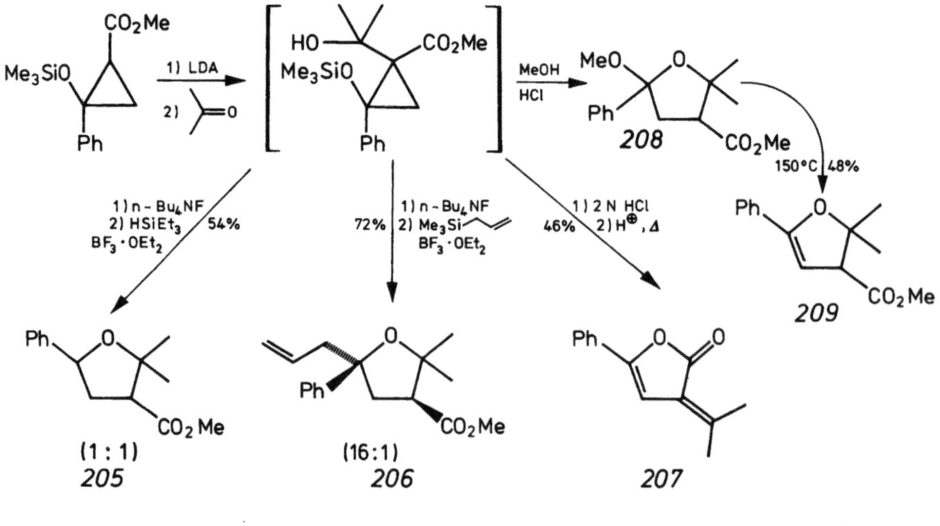

Scheme 10. Synthesis of Different Furan(one) Derivatives from a Siloxycyclopropane and Acetone

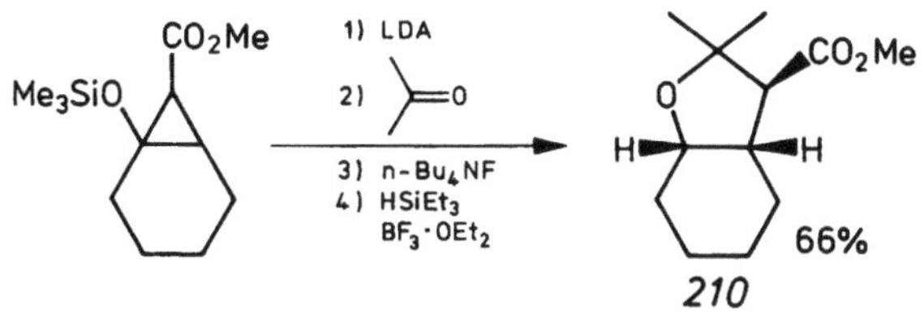

(86)

As mentioned earlier, the primary adducts *198* are isolable with good yields in exceptional cases only. To determine the stereoselectivity of the hydroxyalkylation step, the enolate of the more stable *tert*-butyldimethylsiloxy derivative *211* has been combined with acetone. The two diastereomeric adducts *212* and *213* could be isolated and separated by chromatography [96 b)]. Interestingly the product *213* formed by the "contrasterical" approach of the electrophile predominates although the effect is much less pronounced than with alkyl halides (see above).

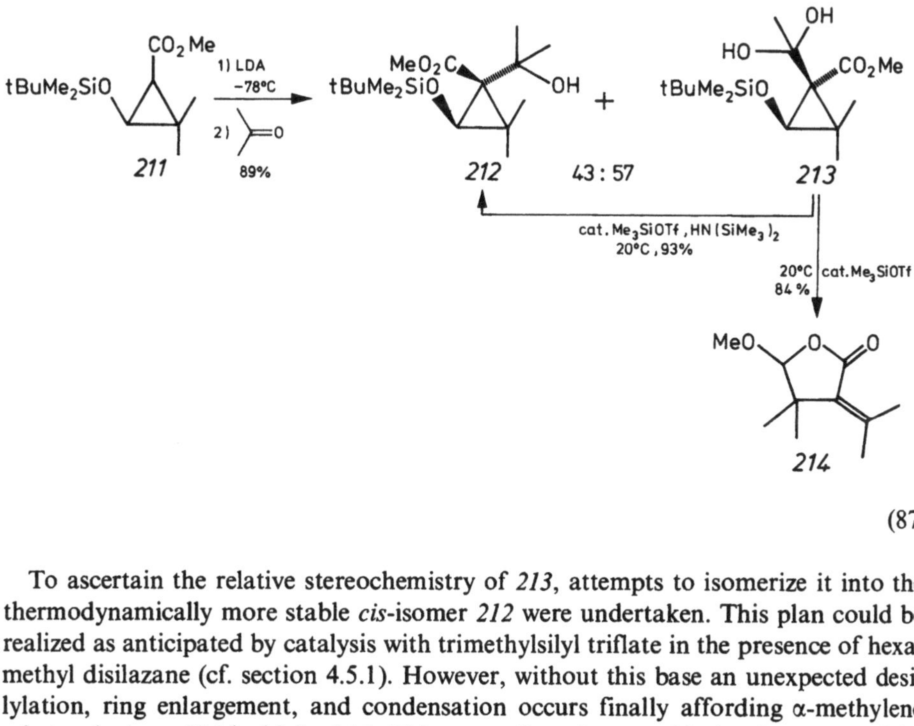

(87)

To ascertain the relative stereochemistry of *213*, attempts to isomerize it into the thermodynamically more stable *cis*-isomer *212* were undertaken. This plan could be realized as anticipated by catalysis with trimethylsilyl triflate in the presence of hexamethyl disilazane (cf. section 4.5.1). However, without this base an unexpected desilylation, ring enlargement, and condensation occurs finally affording α-methylene γ-butyrolactone *214* in high yield. This type of product could also be prepared by use of trimethylsiloxycyclopropanes *215* and *218*, respectively (Eq. 88, 89). In the first case, the intermediate unsaturated ester *216* has been isolated and further converted to lactone *217* by acid treatment [96 b)].

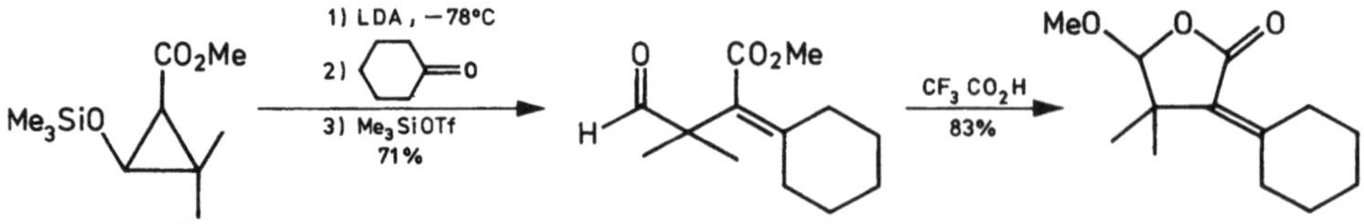

215 216 217

(88)

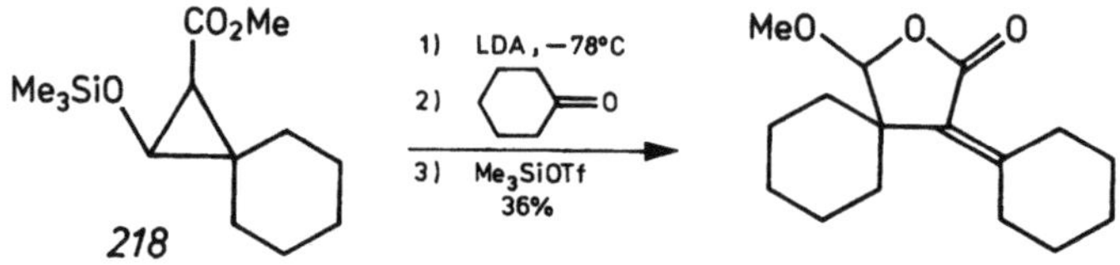

218

(89)

All examples shown in this section demonstrate that with methyl 2-siloxycyclopropanecarboxylates as key building blocks a great variety of furan(one) derivatives can be gained in an extremely flexible manner. Eq. 90 depicts the origin of the parts in such tetrahydrofuran-3-carboxylates.

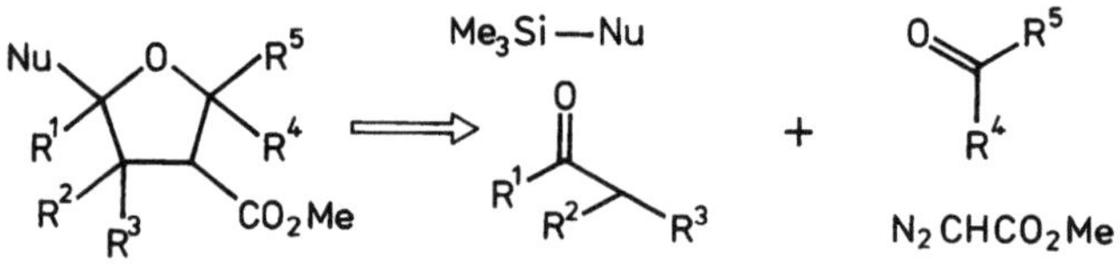

(90)

4.6.2 Thiophene and Pyrrole Derivatives by Addition to Carbon Disulfide and Phenylthioisocyanate

An attempted synthesis of dithioesters *220* by stepwise treatment of enolates from cyclopropanes *97* with carbon disulfide and methyl iodide lead to an unexpected new product without a cyclopropane ring. By spectroscopic means and study of the reactions dihydrothiophene structure *222* has been established [98].

Control experiments show that the ring enlargement *219* → *221* occurs at the anionic stage even at −78 °C. The driving force for this 1,3-sigmatropic shift — or, more specifically, heterovinyl-cyclopropane/heterocyclopentene rearrangement — should be the loss of ring strain and the better stabilization of the negative charge in *221*. The primary adducts *222* or their desilylated derivatives are usually not isolated, but directly converted to thiophenes. The aromatization could be achieved either by acid promoted silanol elimination providing *223* (R^3 = H) or by treatment with Lewis acid which induces a Wagner-Meerwein alkyl shift to afford *224* (R^1 = H).

The overall yields are moderate at best (Table 10), since purification of *223* or *224* is tedious and detrimental due to unknown side products formed by CS_2. Nevertheless several functionalized thiophenes are available by this unique route which might be optimized in singular cases and can possibly lead to other sulfur containing heterocycles. For instance, if the primary adduct *222* obtained from cyclopropane *215* is oxidized with pyridinium dichromate the thiolactone *225* can be prepared [98].

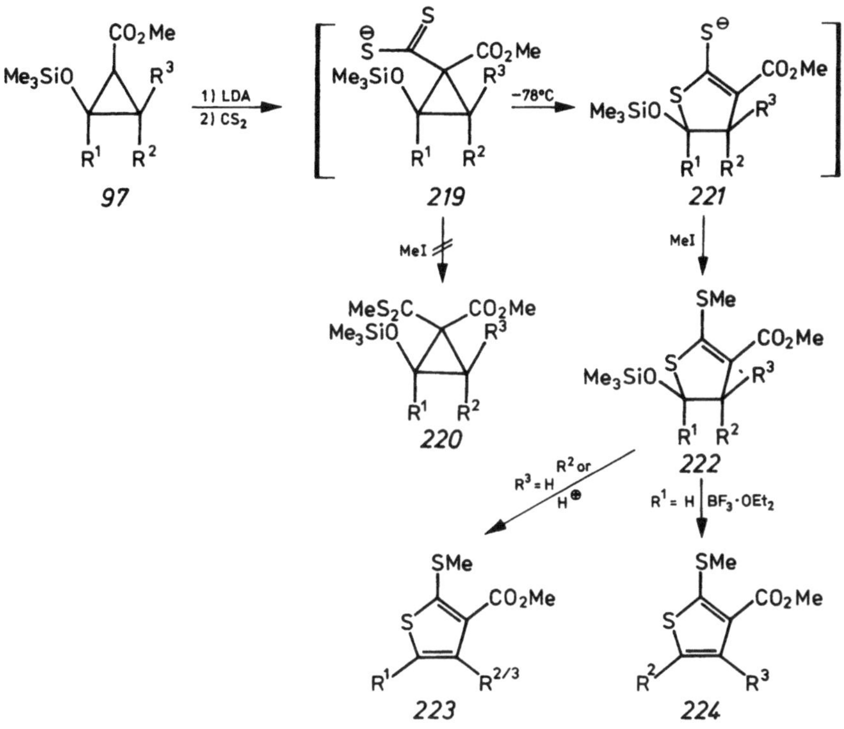

(91)

Table 10. Synthesis of Thiophenes *223* or *224* According to Eq. 91

Entry	R^1	R^2	R^3	Product	Method	Yield
1	H	Me	Me	*224*	$BF_3 \cdot OEt_2$	35%
2	H	$-(CH_2)_5-$		*224*	$BF_3 \cdot OEt_2$	42%
3	Ph	H	H	*223*	CF_3CO_2H	20%
4	CMe_3	H	H	*223*	CF_3CO_2H	41%
5	Me	H	H	*223*	CF_3CO_2H	7%
6	$-(CH_2)_4-$		H	*223*	CF_3CO_2H	16%
7	$-(CH_2)_5-$		H	*223*	CF_3CO_2H	29%

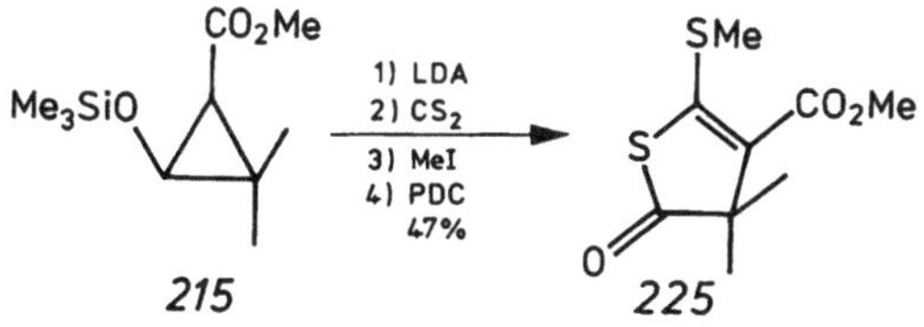

(92)

It is evident that this addition/ring enlargement sequence has to be tested with other heterocumulenes too. So far only phenylthioisocyanate has been used, but completely analogous to CS_2 dihydropyrrole derivatives *227* are now generated, which are either isolated or directly aromatized to pyrrole-3-carboxylates *228* after acid treatment (Eq. 93)[97].

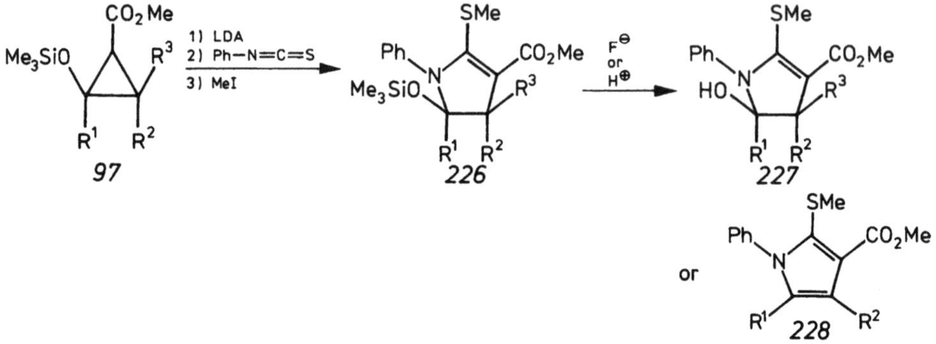

(93)

Yields are scattering very likely due to purification problems in some cases (Table 11). Similar to the thiophene series, oxidation of the hemiaminal *229* leads to the unsaturated lactam *230*.

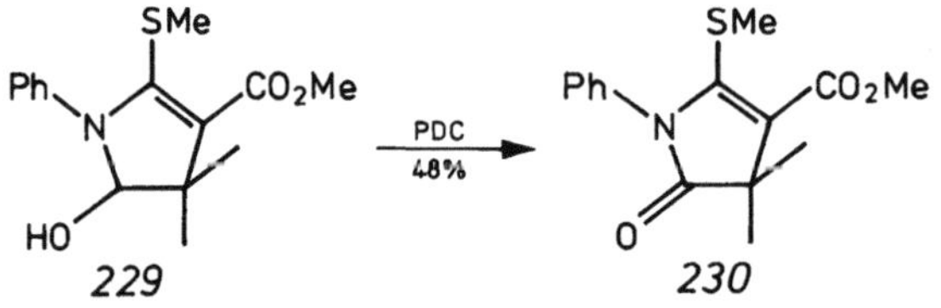

(94)

Table 11. Synthesis of Pyrrole Derivatives *227* or *228* According to Eq. 93

Entry	R^1	R^2	R^3	Method	Product	Yield
1	H	Me	Me	F^-	*227*	57%
2	H	$-(CH_2)_5-$		F^-	*227*	56%
3	Ph	H	H	H^+	*227*	12%
4	CMe_3	H	H	H^+	*228*	16%
5	$-(CH_2)_5-$		H	H^+	*228*	59%

As mentioned before, this route to heterocycles by ring enlargement of three-membered carbocycles should be extendable to other electrophiles having cumulated double bonds X=Y=Z. Addition of ketenes or allenes, on the other hand, might provide interesting carbocyclic systems.

5 Sulfur-, Nitrogen-, and Carbonfunctions as Donorsubstituents

5.1 Activation of Cyclopropanes by Alkylthio and Arylthio Groups

Cyclopropanes carrying alkyl- or arylthio substituents have found many applications in organic synthesis. Important developments are due to Trost's efforts, who recently reviewed the topic in this series [5]. There the subsequent reactions of key compound *232* have been described. The donor-acceptor substituted cyclopropane *231* serves as starting material for *232* as outlined in Eq. 95 [99].

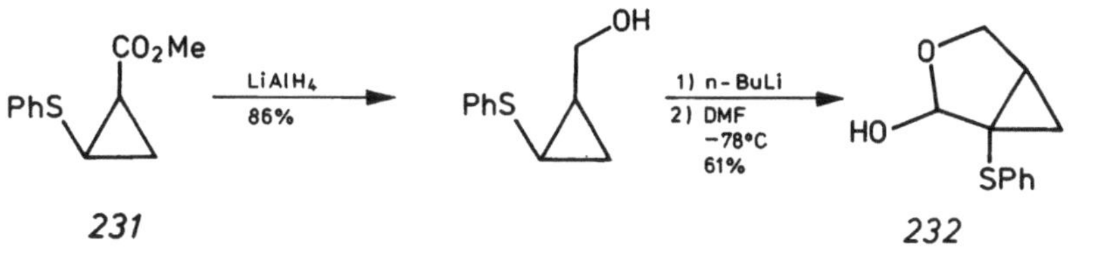

(95)

Cyclopropanes with this combination of substituents did not receive too much attention elsewhere. Ring cleavage of the ester *233* — obtained by the addition elimination path(f) (cf. Scheme 1) — could be achieved by the Julia method providing diene *234* [100].

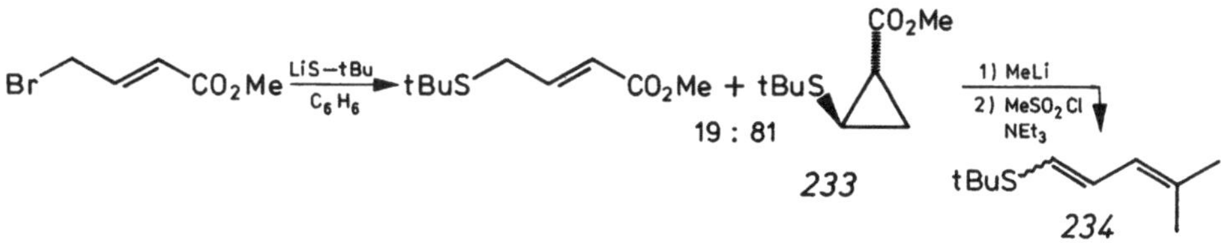

(96)

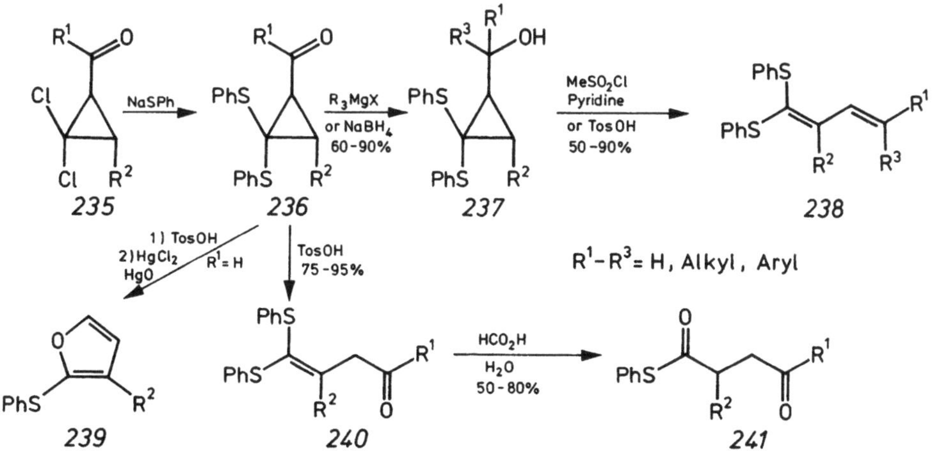

Scheme 11. Transformations of the Cyclopropyl Ketones *236*

Similarly, carbinols *237* are converted to dienes *238* having two benzenethio substituents (Scheme 11)[101]. Here again starting materials are dichloro-cyclopropyl ketones *235* (cf. Scheme 5), which can be transformed to the desired dithioacetals *236* by treatment with sodium benzene thiolate. Cyclopropanes *236* are ring opened to β,γ-unsaturated ketones *240*, γ-oxothioesters *241*, or for R^1 = H expanded to furan derivative *239* [102, 103].

5.1 Activation of Cyclopropanes by Amino Groups

The dichlorocyclopropyl ketones *242* are also precursors for aminosubstituted cyclopropanes *243* as illustrated in Eq. 97, which yield N,N-dialkyl γ-oxoalkanamides *244* [104].

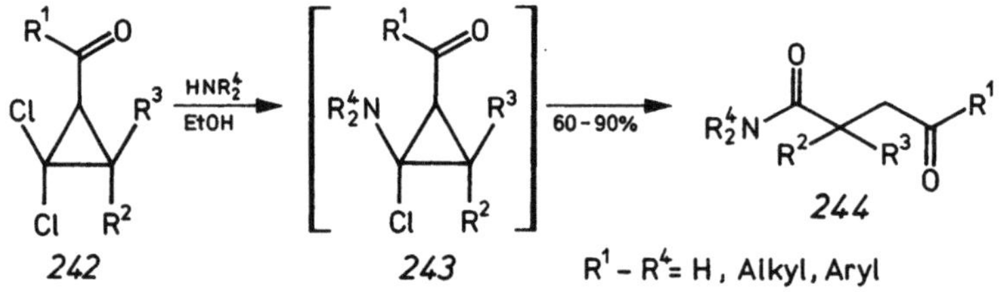

(97)

Addition of electrophilic carbenes to enamines usually does not proceed with good efficiency, very likely because of the disturbance by the Lewis basic nitrogen [15]. If however the less basic enamide derivatives are used as olefins, high conversions to donor-acceptor cyclopropanes are possible. Thus cyclic carbamate *245*, which itself originates from an oxycyclopropane, gives the bicyclic compound *246* almost quantitatively. Its cleavage with aqueous base provides lactone *247* that could be coupled with tryptophyl bromide to afford *248*, a direct precursor of the alkaloid eburnamonine [105].

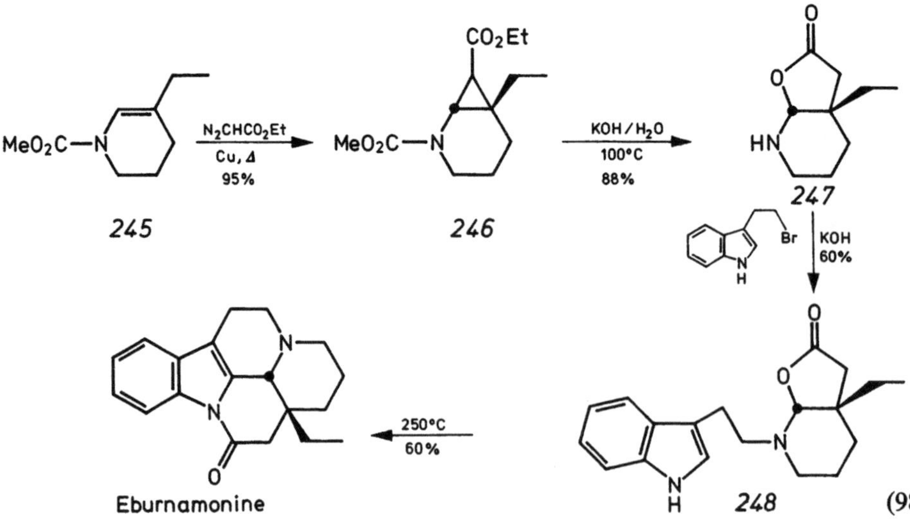

(98)

Homopyrrole derivative *250* is available in rather low yield by cyclopropanation of N-methoxycarbonyl pyrrole *249* with ethyl diazoacetate. Heating of *250* in the presence of CuBr brings about "normal" ring cleavage under aromatization, whereas gas phase thermolysis gives dihydropyridine derivative *252*. Here the intermediacy of the azatriene *251* and its electrocyclic closure to *252* are to be assumed (cf. ring openings of homofuran compounds, section 3.2)[106].

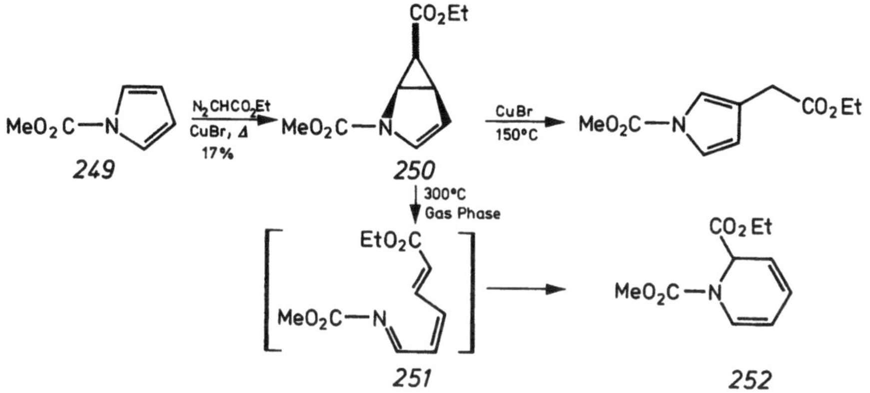

(99)

Methylenation of some β-amino α,β-unsaturated esters or ketones under Simmons-Smith conditions is hampered by low activity of the olefins and formation of side products[107].

The indirect introduction of an acceptor function is more efficient. The example in Eq. 100 shows that arylthiosubstituted aminocyclopropanes — prepared from an enamine and an arylthio carbene — can be oxidized to a sulfoxide or a sulfone, respectively, which allows ring cleavage under protic conditions affording γ-oxosulfoxides or γ-oxosulfones as products[108, 109].

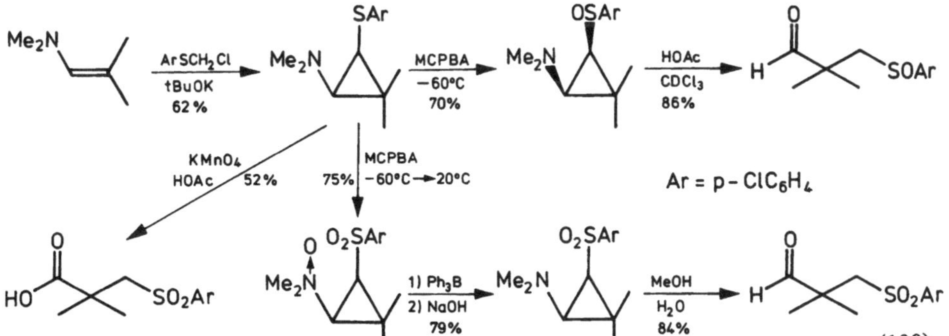

(100)

5.3 Activation of Cyclopropanes by a Trimethylsilylmethyl Group

The weakest donor function treated in this article will be the Me_3SiCH_2- unit which is able to stabilize a positive charge by the well known β-effect[55]. On the other hand, very strong donor quality is attained from this substituent by treatment with fluoride that generates a carbonionic centre. Using $n\text{-}Bu_4NF$ this type of initiation converts cyclopropane *253* into methyl 4-pentenecarboxylate *254*. The starting material is easily prepared from allyl trimethylsilane [110].

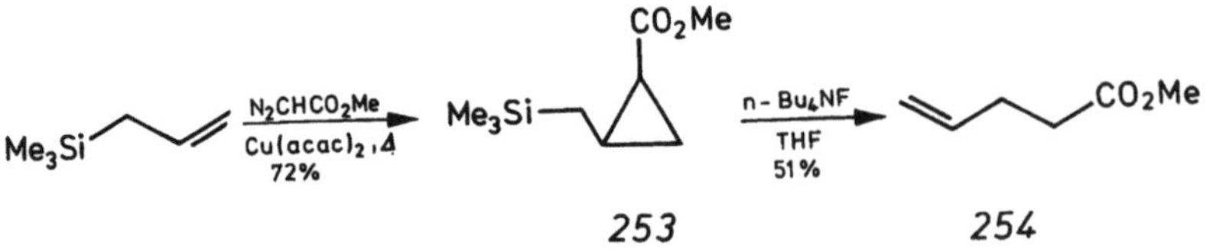

(101)

The ethyl ester *258* (Eq. 103) has been recovered unchanged after treatment with the boron trifluoride acetic acid complex [111], whereas cyclopropane *255* with an additional 2-methyl group opens under these conditions to provide γ-butyrolactone *257* [112]. Apparently the intermediate tertiary carbenium ion *256* is sufficiently stabilized by the trimethylsilylmethyl and the methyl group to be generated from *255*.

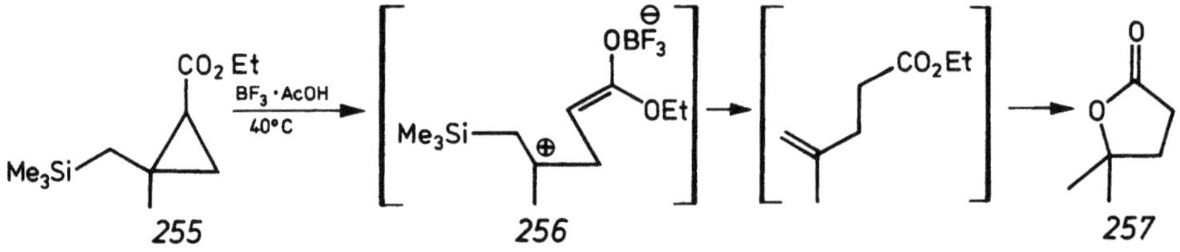

(102)

Enhancement of the acceptor quality can be gained by switching from the ester group to a keto function. This has been realized by treating *258* with sodium or lithium salts of sulfones. In the resulting β-ketosulfones *259* the cyclopropane ring is

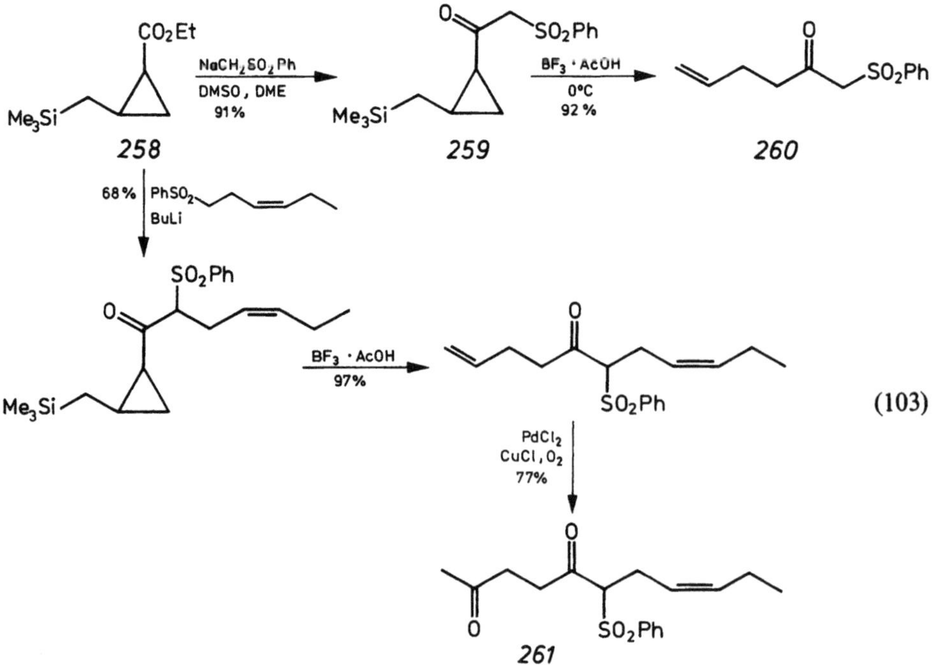

(103)

more strongly activated by the acceptor group — as compared to *258* —, since a secondary carbenium ion as intermediate preceeds *260*. This rather flexible synthesis of unsaturated β-ketosulfones could be used for an approach to diketone *261*, a known precursor of *cis*-jasmone [112].

Not surprisingly, two trimethylsilylmethyl groups further increase reactivity of cyclopropanes. Therefore *262* opens to *263* even at −78 °C yielding a new functionalized allylsilane as the product. An attempt to transform *262* into the corresponding β-ketosulfone *264* leads to the unusual cleaved compound *265*, which carries an allyl and vinylsilane unit as well as the β-ketosulfone moiety. It is evident that this mode of cyclopropane opening is initiated under the basic reaction conditions by deprotonating the intermediate *264* in the side chain α to one of the trimethylsilyl groups [113].

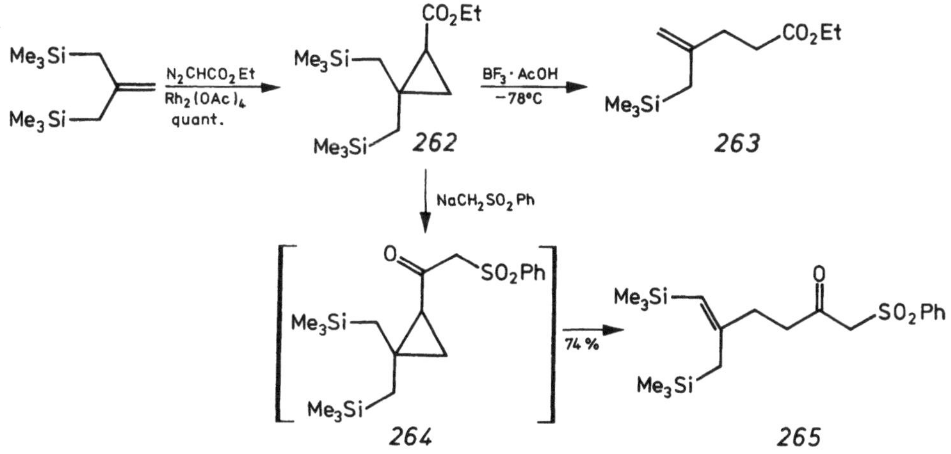

(104)

Assisted by Lewis acid the acyl imidazole *266* can be cleaved to provide the expected carboxylic acid *267*. With fluoride in the presence of a Michael acceptor the intermediate enolate can be trapped. A subsequent intramolecular Claisen condensation

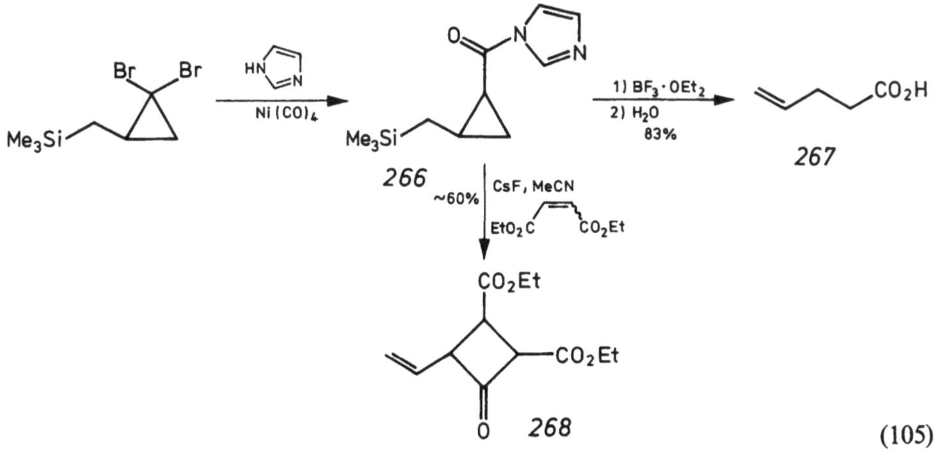

(105)

forms the cyclobutanone *268* in reasonable yield [114]. The starting material *266* results from nickel tetracarbonyl-induced reductive carbonylation of the corresponding dibromocyclopropane *265* [115].

6 Conclusions and Further Perspectives

Syntheses of donor-acceptor-substituted cyclopropanes are possible by a variety of methods which are often very efficient and flexible. Although these cyclopropanes are usually stable enough to be isolated and handled without problems the vicinal location of activating substituents allows relief of the ring strain under rather specific conditions. These ring opening reactions cause donor-acceptor-substituted cyclopropanes to be a versatile synthetic tool which can be used for transformations often not easily achievable by other means.

Many combinations of substituents at the cyclopropane ring have been realized, but not all are of synthetic value. In this respect alkyl 2-siloxycyclopropanecarboxylates are of particular versatility. They allow many modes of ring cleavages which can be combined with change of functionality or with C-C-bond forming reactions providing a manifold of polyfunctional 1,4-dicarbonyl compounds, carbocycles, and heterocyclic systems as products.

The fundamental chemistry of donor-acceptor-substituted cyclopropanes is now well understood. This solid platform should allow many applications of known processes and exploration of new reaction types. A future challenge will be asymmetric syntheses which should be achievable, for instance, using Lewis acids containing enantiomerically pure ligands. Even more attractive might be cyclopropane formation under the influence of a suitable optically active catalyst. This intriguing approach could lead to enantioselective syntheses of many compounds in a most economical way. Finally it can be expected in the near future that transition metal induced reactions will also play an important role in this area of small ring chemistry.

7 Acknowledgement

The author would like to thank his coworkers I. Reichelt, H. Holzinger, E. Kunkel, W. Bretsch, C. Brückner, E. L. Grimm, R. Zschiesche, T. Kunz, and A. Wienand for their engaged and skilful contributions to the field reviewed. Their work has been performed at the Institut für Organische Chemie der Universität Würzburg. The chance to start this research program at this institute as well as many fruitful discussions with the staff members are gratefully acknowledged. Very valuable support has been obtained from the Deutsche Forschungsgemeinschaft, the Fonds der Chemischen Industrie, the Universitätsbund Würzburg, and the Karl-Winnacker-Stiftung (Hoechst AG).

8 References

1. For a recent review see Reissig, H.-U.: Organic Synthesis Via Cyclopropanes: Principles and Applications, in: The Chemistry of the Cyclopropyl Group. Chapter 11 (ed. Rappoport, Z.) John Wiley and Sons, Chichester, 1987
2. Greenberg, A., Stevenson, T. A.: Structure and Energies of Substituted Strained Organic Molecules, in: Molecular Structure and Energetics, Vol. 3 (eds. Liebman, J. F., Greenberg, A.), VCH, Weinheim 1986, pp 193–266
3. De Meijere, A.: Angew. Chem. *91*, 867 (1979); Angew. Chem. Int. Ed. Engl. *18*, 809 (1979)
4. Seebach, D.: Angew. Chem. *91*, 259 (1979); Angew. Chem. Int. Ed. Engl. *18*, 239 (1979)
5. Trost, B. M.: Topics in Current Chemistry *133*, 3 (1986)
6. Krief, A.: ibid. *135*, 2 (1987)
7. Salaüun, J.: ibid. *144*, 1 (preceding review)
8. We are very grateful to Prof. Dr. E.-U. Würthwein (Universität Münster) for performing these calculations. Also see ref. 79
9. Rambaud, R.: Bull. Soc. Chim. Fr. 1552 (1938); Rambaud, R., Brini-Fritz, N., Durif, S.: Bull. Soc. Chim. Fr. 681 (1957). Also seen D'yakanov, I. A., Lugavsova, N. A.: Zh. Obshch. Khim. *21*, 839 (1951)
10. Julia, M., Le Thullier, G.: Bull. Soc. Chim. Fr. 717 (1966); Julia, M., Baillarge, M.: Bull. Soc. Chim. Fr. 734, 743 (1966)
11. Wenkert, E.: Acc. Chem. Res. *13*, 27 (1980); Wenkert, E.: Heterocycles *14*, 1703 (1980)
12. Wenkert, E., Mueller, R. A., Reardon, Jr., E. J., Sathe, S. S., Scharf, D. J., Tosi, G.: J. Am. Chem. Soc. *92*, 7428 (1970)
13. Wenkert, E., Buckwalter, B. L., Craveiro, A. A., Sanchez, E. L., Sathe, S. S.: ibid. *100*, 1267 (1978)
14. McMurry, J. E. C., Glass, T. E.: Tetrahedron Lett. 2575 (1971)
15. Wenkert, E., McPherson, S. A., Sanchez, E. L., Webb, R. L.: Synth. Commun. *3*, 255 (1973)
16. Kunz, H., Lindig, M.: Chem. Ber. *116*, 220 (1983)
17. Wenkert, E., Buckwalter, B. L., Sathe, S. S.: Synth. Commun. *3*, 261 (1973); also see: Kulinkovich, O. G., Tischenko, I. G., Sorokin, V. L.: Zh. Org. Khim. *21*, 1663 (1985)
18. Evans, D. A., Sims, C. L., Andrews, G. C.: J. Am. Chem. Soc. *99*, 5453 (1977)
19. Wenkert, E., Halls, T. D. J., Kwart, L. D., Magnusson, G., Showalter, H. D. H.: Tetrahedron *37*, 4017 (1981)
20. Wenkert, E., Alonso, M. E., Buckwalter, B. L., Chou, K. J.: J. Am. Chem. Soc. *99*, 4778 (1977)
21. Menicagli, R., Malanga, C., Lardicci, L.: J. Chem. Res. (S) 20 (1985)
22. a) Adams, J., Belley, M.: Tetrahedron Lett. *27*, 2075 (1986); b) Adams, J., Belley, M.: J. Org. Chem. *51*, 3878 (1986)
23. Wenkert, E., Goodwin, T. E., Ranu, B. C.: J. Org. Chem. *42*, 2137 (1977); Tetrahedron Lett. *27*, 2075 (1986)
24. Schenck, G. O., Steinmetz, R.: Liebigs Ann. Chem. *668*, 19 (1963)
25. Rokach, J., Girad, Y., Guindon, Y., Atkinson, J. G., Larue, M., Young, R. N., Masson, P., Holme, G.: Tetrahedron Lett. *21*, 1485 (1980)
26. Wenkert, E., Bazukis, M. L. F., Buckwalter, B. L., Woodgate, P. D.: Synth. Commun. *11*, 533 (1981)
27. a) Rokach, J., Adams, J.: Acc. Chem. Res. *18*, 87 (1985); b) Rokach, J., Adams, J., Perry, R.: Tetrahedron Lett. *24*, 5185 (1983); c) Adams, J., Rokach, J.: Tetrahedron Lett. *25*, 35 (1984); d) Leblanc, Y., Fitzsimmons, B. J., Adams, J., Perez, F., Rokach, J.: J. Org. Chem. *51*, 789 (1986)
28. Padwa, A., Wisnieff, T. J., Walsh, E. J.: J. Org. Chem. *51*, 5036 (1986)
29. a) Kato, T., Katagiri, N.: Chem. Pharm. Bull. *21*, 729 (1973); b) Kato, T., Katagiri, N., Sato, R.: J. Chem. Soc., Perkin Trans 1, 525 (1979); c) Kato, T., Katagiri, N., Sato, R.: Chem. Pharm. Bull. *27*, 1176 (1979); d) Kato, T., Katagiri, N., Sato, R.: J. Org. Chem. *45*, 2587 (1980)
30. Takano, S., Sugahara, T., Ishiguro, M., Ogasawara, K.: Heterocycles *6*, 1141 (1977)
31. Wenkert, E., Greenberg, R. S., Raju, M. S.: J. Org. Chem. *50*, 4681 (1985)
32. Scarpati, R., Cioffi, M., Scherillo, G., Nicolaus, R. A.: Gazz. Chim. Ital. *96*, 1164 (1966)
33. a) Graziano, M. L., Scarpati, R.: J. Chem. Soc. Perkin Trans 1, 289 (1985); b) Pelletier, O., Jankowski, K.: Can. J. Chem. *60*, 2383 (1982)

34. Graziano, M. L., Iesce, M. R.: Synthesis 762 (1985)
35. Graziano, M. L., Iesce, M. R., Scarpati, R.: J. Heterocyc. Chem. *23*, 553 (1986)
36. Dowd, P., Kaufman, C., Paik, Y. H.: Tetrahedron Lett. *26*, 2283 (1985)
37. Abdallah, H., Grée, R., Carrié, R.: Tetrahedron *41*, 4339 (1985)
38. Wenkert, E., Alonso, M. E., Buckwalter, B. L., Sanchez, E. L.: J. Am. Chem. Soc. *105*, 2021 (1983)
39. a) Fischer, E. O., Dötz, K. H.: Chem. Ber. *103*, 1273 (1970); b) Dötz, K. H., Fischer, E. O.: Chem. Ber. *105*, 1356 (1972)
40. Wienand, A., Reissig, H.-U.: unpublished results
41. Baldwin, S. W., Blomquist, Jr., H. R.: Tetrahedron Lett. *23*, 3883 (1982)
42. Caplin, G. A., Ollis, W. D., Sutherland, I. O.: J. Chem. Soc. (C) 2302 (1968)
43. a) Tishchenko, I. D., Kulinkovich, O. G., Masalov, N. V.: Zh. Org. Khim. *16*, 1203 (1980); b) Kulinkovich, O. G., Tishchenko, I. D., Romashin, L. N.: Zh. Org. Khim. *21*, 90 (1985)
44. Kulinkovich, O. G., Tishchenko, I. D., Masalov, N. V.: Khim. Geterotsikl. Soedin. *12*, 1603 (1984)
45. Kulinkovich, O. G., Tishchenko, I. D., Romashin, L. N., Savitskaya, L. N.: Synthesis 378 (1986)
46. Kulinkovich, O. G., Tishchenko, I. D., Masalov, N. V.: ibid. 886 (1984)
47. Kulinkovich, O. G., Tishchenko, I. D., Sorokin, V. L.: Zh. Org. Khim. *20*, 2548 (1984)
48. Banwell, M. G.: J. Chem. Soc., Chem. Commun. 1453 (1983)
49. Kobayashi, Y., Taguchi, T., Morikawa, T., Takase, T., Takanashi, H.: J. Org. Chem. *47*, 3232 (1982)
50. Parham, W. E., McKown, W. D., Nelson, V., Kajigaeshi, S., Ishikawa, N.: ibid. *38*, 1361 (1973)
51. a) Kulinkovich, O. G., Tishchenko, I. D., Sorokin, V. L.: Synthesis 1058 (1985); b) Kulinkovich, O. G., Tishchenko, I. D., Sorokin, V. L.: Zh. Org. Khim. *21*, 1658 (1985); c) Kulinkovich, O. G., Tishchenko, I. D., Sorokin, V. L.: Zh. Org. Khim. *21*, 1663 (1985)
52. Pohmakotr, M., Pisutjaroenpong, S.: Tetrahedron Lett. *26*, 3613 (1985)
53. Padwa, A., Wannamaker, M. W.: ibid. *27*, 2555 (1986)
54. a) Tochtermann, W., Rösner, P.: Chem. Ber. *114*, 3725 (1981); b) Wenkert, E., Craveiro, A. A., Sanchez, E. L.: Synth. Commun. *7*, 85 (1977)
55. a) Colvin, E.: Silicon in Organic Synthesis, Butterworth, London 1981; b) Weber, W. P.: Silicon Reagents in Organic Synthesis, Springer, Berlin 1983
56. LeGoaller, R., Pierre, J.-L.: C. R. Acad. Sci. Paris, Série C *276*, 193 (1973); also see: LeGoaller, R., Pierre, J.-L.: Can. J. Chem. *55*, 757 (1977)
57. Kunkel, E., Reichelt, I., Reissig, H.-U.: Liebigs Ann. Chem. 512 (1984)
58. Marino, J. P., de la Pradilla, R. F., Laborde, E.: J. Org. Chem. *49*, 5279 (1984)
59. Ollivier, J., Salaün, J.: J. Chem. Soc., Chem. Commun. 1269 (1985)
60. Coates, R. M., Sandefur, L. O., Smillie, R. D.: J. Am. Chem. Soc. *97*, 1619 (1975)
61. Reissig, H.-U.: unpublished results
62. Kunkel, E., Reichelt, I., Reissig, H.-U.: Liebigs Ann. Chem. 802 (1984)
63. a) Hünig, S., Wehner, G.: Synthesis 180 (1975); b) Franz, R.: J. Fluorine Chem. *15*, 423 (1980); c) $NEt_3 \cdot 3HF$ is commercially available from Riedel-deHaen AG, $NEt_3 \cdot HF$ is available from Aldrich Chemical Co.
64. For recent use of optically active catalysts for cyclopropanation see: Fritschi, H., Leutenegger, U., Pfaltz, A.: Angew. Chem. *98*, 1028 (1986); Angew. Chem. Int. Ed. Engl. *25*, 1005 (1986) and references cited therein
65. Kunz, T., Reissig, H.-U.: unpublished results
66. Saigo, K., Kurihara, H., Miura, H., Hongo, A., Kubota, N., Nohira, H.: Synth. Commun. *14*, 787 (1984)
67. a) Brook, A. G., Kucera, H. W., Pearce, R.: Can. J. Chem. *49*, 1618 (1971); b) Dalton, J. C., Bourque, R. A.: J. Am. Chem. Soc. *103*, 699 (1981)
68. Reichelt, I., Reissig, H.-U.: Liebigs Ann. Chem. 531 (1984)
69. Reichelt, I., Reissig, H.-U.: Chem. Ber. *116*, 3895 (1983)
70. a) Pinnick, H. W., Chang, Y.-H., Foster, S. C., Govindan, M.: J. Org. Chem. *45*, 4505 (1980); b) Kai, Y., Knochel, P., Kwiatkowski, S., Dunitz, J. D., Oth, J. F. M., Seebach, D., Kalinowski, H.-O.: Helv. Chim. Acta *65*, 137 (1982)
71. Compare: Wagner, H.-U., Boche, G.: Z. Naturforsch., Teil B *37*, 1339 (1982) and references cited therein

72. Deslongchamps, P.: Stereoelectronic Effects in Organic Chemistry, Pergamon Press, Oxford 1983
73. The involvement of an anomeric effect was mentioned in ref. 68 (footnote 25); also compare ref. 53
74. Saigo, K., Okagawa, S., Nohira, H.: Bull. Chem. Soc. Jpn. *54*, 3603 (1981)
75. Reichelt, I., Reissig, H.-U.: Synthesis 786 (1984)
76. Grimm, E. L., Reissig, H.-U.: J. Org. Chem. *50*, 242 (1985)
77. Schäfer, H. J., Bitenc, M.: unpublished results; Bitenc, M.: Doctoral Thesis, Univ. Münster 1987
78. Bretsch, W., Reissig, H.-U.: Liebigs Ann. Chem., 175 (1987)
79. Brückner, C., Reissig, H.-U.: Chem. Ber. *120*, 617 (1987)
80. Reichelt, I., Reissig, H.-U.: Liebigs Ann. Chem. 820 (1984)
81. Reissig, H.-U., Reichelt, I.: Tetrahedron Lett. *25*, 5879 (1984)
82. Pétrier, C., Luche, J.-L.: J. Org. Chem. *50*, 910 (1985)
83. Grimm, E. L., Zschiesche, R., Reissig, H.-U.: ibid. *50*, 5543 (1985)
84. Zschiesche, R., Reissig, H.-U.: unpublished results
85. Zschiesche, R., Reissig, H.-U.: Liebigs Ann. Chem., 387 (1987)
86. Zschiesche, R., Grimm, E. L., Reissig, H.-U.: Angew. Chem. *98*, 1104 (1986); Angew. Chem. Int. Ed. Engl. *25*, 1086 (1986)
87. Marino, J. P., Laborde, E.: J. Org. Chem. *52*, 1 (1987)
88. a) Reissig, H.-U.: Tetrahedron Lett. *26*, 3943 (1985); b) For a related rearrangement by transition-metal-catalysis see: Doyle, M. P., van Leusen, D.: J. Org. Chem. *47*, 5326 (1982)
89. Saigo, K., Yamashita, T., Hongu, A., Hasegawa, M.: Synth. Commun. *15*, 715 (1985)
90. Reissig, H.-U., Böhm, I.: Tetrahedron Lett. *24*, 715 (1983)
91. Reissig, H.-U., Lorey, H.: Liebigs Ann. Chem. 1914 (1986)
92. Reissig, H.-U., Reichelt, I., Lorey, H.: ibid. 1924 (1986)
93. a) Brückner, C., Lorey, H., Reissig, H.-U.: Angew. Chem. *98*, 559 (1986); Angew. Chem. Int. Ed. Engl. *25*, 556 (1986); b) Brückner, C., Holzinger, H., Reissig, H.-U.: J. Org. Chem. submitted for publication
94. a) Nakamura, E., Kuwaijma, I.: J. Am. Chem. Soc. *99*, 7360 (1977); b) Nakamura, E., Oshino, H., Kuwaijma, I.: J. Am. Chem. Soc. *108*, 3745 (1986)
95. Giese, B., Horler, H.: unpublished results; Horler, H.: Doctoral Thesis, Techn. Hochschule Darmstadt 1985
96. a) Brückner, C., Reissig, H.-U.: J. Chem. Soc., Chem. Commun. 1512 (1985); b) Brückner, C., Reissig, H.-U.: J. Org. Chem. submitted for publication
97. Brückner, C.: Doctoral Thesis, Univ. Würzburg 1986
98. Brückner, C., Reissig, H.-U.: Angew. Chem. *97*, 578 (1985); Angew. Chem. Int. Ed. Engl. *24*, 588 (1985)
99. Trost, B. M., Ornstein, P. L.: J. Org. Chem. *47*, 748 (1982)
100. Little, R. D., Dawson, J. R.: J. Am. Chem. Soc. *100*, 4607 (1978)
101. Kulinkovich, O. G., Tishchenko, I. G., Roslik, N. A., Reznikov, I. V.: Synthesis 383 (1983)
102. Kulinkovich, O. G., Tishchenko, I. G., Roslik, N. A.: ibid. 931 (1982)
103. Kulinkovich, O. G., Tishchenko, I. G., Roslik, N. A.: Khim. Geterotsikl. Soedin. 132 (1984)
104. Tishchenko, I. G., Kulinkovich, O. G., Masalov, N. V.: Synthesis 268 (1982)
105. a) Wenkert, E., Hudlicky, T., Showalter, H. D. H.: J. Am. Chem. Soc. *100*, 4893 (1978); b) also see ref. 19
106. Tanny, S. R., Grossman, J., Fowler, F. W.: ibid. *94*, 6495 (1972)
107. Bieräugel, H., Akkerman, J. M., Lapierre Armande, J. C., Pandit, U. K.: Recl. Trav. Chim. Pays-Bas *95*, 266 (1976)
108. Rynbrandt, R. H., Dutton, F. E., Chidester, C. G.: J. Am. Chem. Soc. *98*, 4882 (1976)
109. Rynbrandt, R. H., Dutton, F. E.: J. Org. Chem. *40*, 2282 (1975)
110. Reichelt, I., Reissig, H.-U.: Liebigs Ann. Chem. 828 (1984)
111. Ochiai, M., Sumi, K., Fujita, E.: Chem. Lett. 79 (1982)
112. Ochiai, M., Sumi, K., Fujita, E.: Chem. Pharm. Bull. *31*, 3931 (1983)
113. Ochiai, M., Sumi, K., Fujita, E.: Tetrahedron Lett. *23*, 5419 (1982)
114. Hirao, T., Misu, D., Agawa, T.: J. Chem. Soc., Chem. Commun. 26 (1986)
115. Hirao, T., Harano, Y., Yamana, Y., Oshiro, Y., Agawa, T.: Tetrahedron Lett. *24*, 1255 (1983)

Functionalised Cyclopropenes as Synthetic Intermediates

Mark S. Baird

School of Chemistry, Bedson Building, The University Newcastle Upon Tyne NE1 7RU, Great Britain

Table of Contents

Although the use of cyclopropanes in synthesis has been an area of very considerable interest for many years, much less attention has been drawn to the potential and actual uses of cyclopropenes. This may perhaps be a reflection of the perceived difficulty in handling such species, but probably also results from the rather limited routes to cyclopropenes, and in particular those bearing other functional groups, compared to the variety available which lead to the saturated analogues. However, the development of more flexible routes to cyclopropenes makes available not only the exploitation of those reactions which are peculiar to the unsaturated system but also, through stereocontrolled addition to the double bond, provides more efficient routes to cyclopropanes of particular stereochemistry. This review will attempt to identify the progress which has been made in the synthesis of cyclopropenes, to show the rich variety of their synthetic transformations, and to point to areas where future progress may occur. It will not cover purely structural or spectroscopic aspects of cyclopropene chemistry, which were dealt with in most excellent fashion by Closs [1], and in several later reviews [2]. Nor will the photochemical or metal induced reactions be dealt with in great detail, the former area having been covered in reviews by Padwa [3] and the latter in an article in the present series by Binger and Buch [4].

1 Developments in the Preparation of Cyclopropenes

The routes available for generating a cyclopropene ring include those involving the formation of the 1,3- (or 2,3-) bond, with or without allylic rearrangement, of one or both of the 1,2-bonds, of both 1,3- and 2,3- bonds, and of both 1,3- and one of the 1,2-bonds in (1).

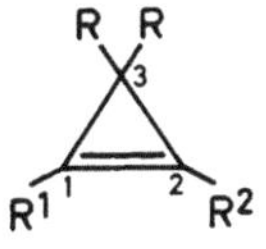

1

1.1 Base Induced Elimination in Allylic Systems

One of the earliest routes to cyclopropenes suitable for large scale use was the dehydrohalogenation of an allylic chloride using strong base. This method is very effective for cyclopropene itself [5]. It is also effective for simple alkylcyclopropenes, though the base used is critical as in some cases the cyclopropene reacts further to give a methylenecyclopropane; thus, reaction of 3-chloro-2-methylprop-1-ene with sodium amide in tetrahydrofuran at 65 °C can provide large quantities of 1-methylcyclopropene, but with potassium amide the product is methylenecyclopropane [6,7]. In general, however, lithium amide appears to be the base of choice [8,9]:

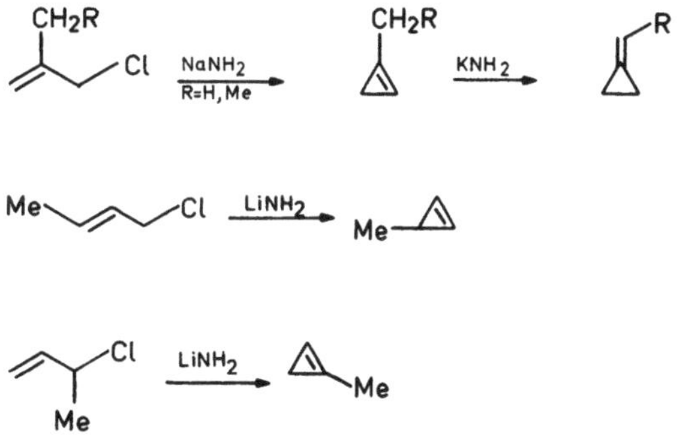

The position of the alkyl substituent in the product indicates that cyclisation occurs with rearrangement of the double bond, ie., by 1,1-elimination and formal formation and cyclisation of a vinylcarbene. Although the overall yields are not always good, the reagents are readily available and large quantities of the simple alkylcyclopropenes can be produced. 1,2-Dimethylcyclopropene has been prepared in a similar process by treatment of methallyl chloride with two equivalents of phenyl lithium, followed by quenching with methyl iodide; presumably, the initial reaction leads to 1-methylcyclopropene which is converted in situ to the 2-lithio-species [10]. The elimination of HBr from brominated alkylidenemalonates also leads to cyclopropenes, though in low yield [11]:

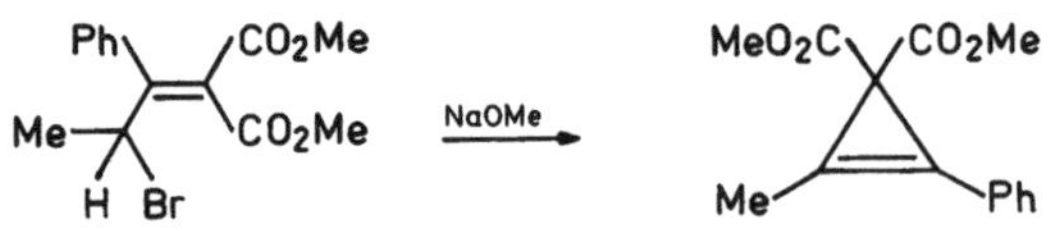

In a recently developed method, the single bond of a cyclopropene is obtained by 1,3-dehalogenation in an allylic system; thus reaction of the iodochlorides (2) with an alkyl lithium at –78 °C leads to a range of cyclopropenes [12]:

The required iodochlorides can be prepared from propargylic alcohols by carbometallation followed by iodinolysis:

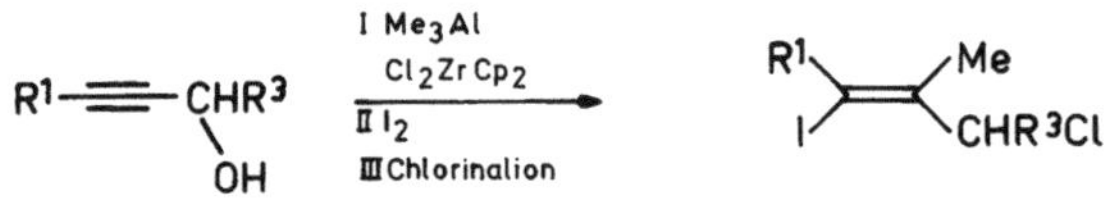

A variation leads to cyclopropanes:

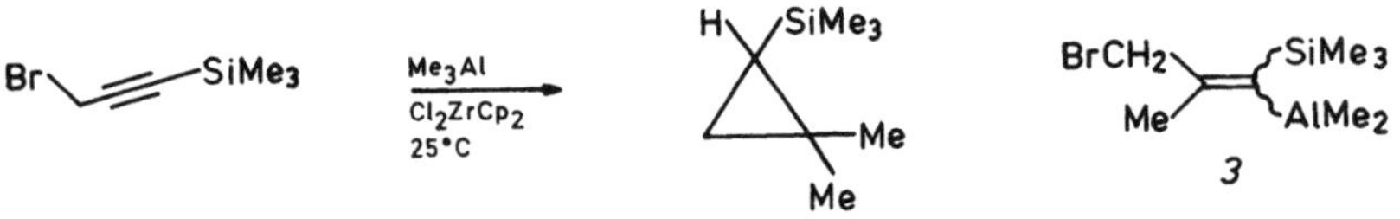

It is thought that addition of trimethylaluminium across the alkyne leads to (3), which undergoes 1,3-elimination to produce 1-methyl-2-trimethylsilylcyclopropene; addition of trimethylaluminium to the double bond follows, with the expected regiochemistry [13].

1.2 Formation of Both Components of the 1,2-Bond: Elimination in 1,3-Halogenated Propanes

Although this review does not in general cover cyclopropenones, it is important to note the elegant preparation of cyclopropenone dimethylacetal by reaction of 2,3-dichloropropene with N-bromosuccinimide in methanol to produce (4), followed by treatment with potassium amide in liquid ammonia [14]:

4

The double dehalogenation of 1,1,3,3-tetrachlorides provides an attractive route to some cyclopropenes:

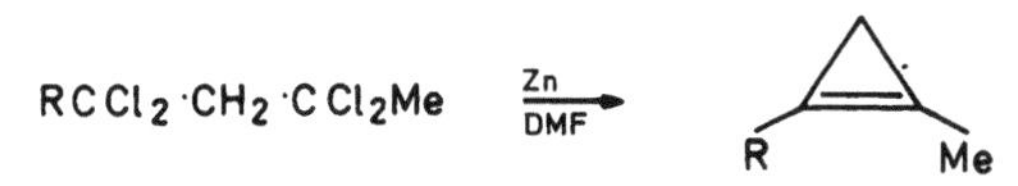

The starting polyhalides are obtained in reasonable yield by addition of, eg., 1,1,1-trichloropropane to 2-chloropropene, initiated by $Fe(CO)_5$ — HMPA. Treatment with zinc in DMF leads to good yields of 1,2-dialkylcyclopropenes, but the reaction is much less satisfactory when one of the substituents is hydrogen or chlorine [15], and it does not seem to have been tested with more heavily functionalised systems.

The double dehydrobromination of α,α′-dibromoalkanedicarboxylates (5) is a good route to cycloalkene diesters when n = 2–4, but unfortunately is not successful with n = 1; when (5, n = 1) is treated with one equivalent of potassium t-butoxide, a reasonable yield of the E- and Z-bromides (6, X = Br) is obtained, but a second equivalent of base leads to E-(6, X = Bu^{t}O), presumably by addition of t-butylalcohol, to an intermediate cyclopropene [16].

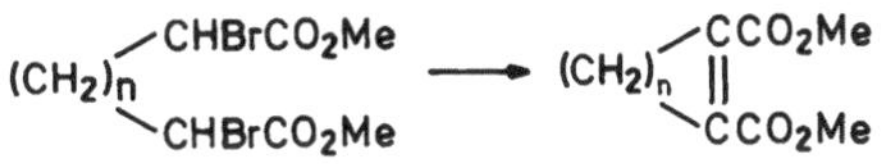

5

6

An unusual alternative involves the reaction of the dibromide (7) with base:

7

This formally occurs by two dehydrohalogenations, 1,3- and 1,2-, though the sequence of these is not certain [17].

1.3 1,2-Elimination in Cyclopropanes

Early work on 1,2-elimination in cyclopropanes centred on the thermal elimination of amines from cyclopropyltrialkylammonium salts [1]. An elegant example is a route to optically active cyclopropenes involving the resolution of the acid (8, R = CO_2H), conversion to the quaternary ammonium salt (8, R = $^+NMe_3$) and elimination over platinised asbestos at 360 °C [18]:

8

However, the ready availability of halocyclopropanes has led to extensive studies of their 1,2-dehydrochlorination, and amines are now rarely used as cyclopropene precursors. Although the reaction of 1,1-dichlorocyclopropanes with strong base does in certain situations lead to cyclopropenes, it is frequently the case that the initially formed 1-halocyclopropene does not survive under the reaction conditions, undergoing either addition of a nucleophile to the alkene bond or prototropic shifts followed by further dehydrohalogenation. Two main variations on this method are available which proceed under conditions where further reaction does not, in general, occur, that is 1,2-dehalogenation and 1,2-dehalosilylation. Each of these three alternatives will be considered in turn.

1.3.1 1,2-Dehydrohalogenation

Because 1,1-dihalocyclopropanes are so readily available by carbene addition to alkenes, their dehydrohalogenation to 1-halocyclopropenes provides, in principle, one of the most attractive routes to functionalised cyclopropenes. However, most early studies of the reaction did not lead to the cyclopropenes themselves, but to products of their further reaction. The main problems arise when the 1-halocyclopropene (9) can undergo prototropic shifts by removal of a proton from C_2' or C_3, or when the base used is also a good nucleophile and addition to the cyclopropene can occur:

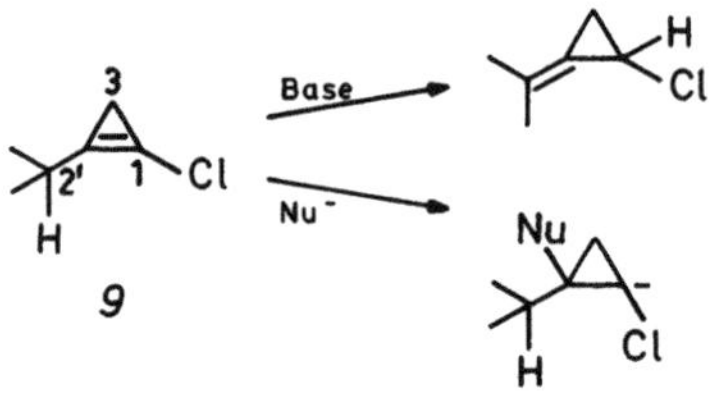

9

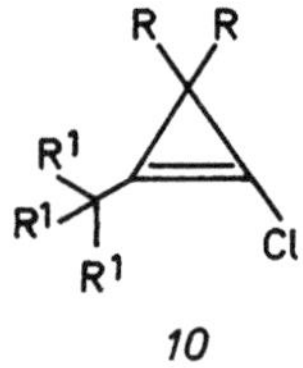

10

The former problem is absent in systems such as (10), and the second problem is sometimes lessened when a mono-halocyclopropane is used in place of a dihalide, the rate of addition to the derived cyclopropene being decreased due to the lack of halogen on the double bond. The following discussion is divided into two sections to cover the frequently met situations:

Dehydrohalogenation of Monohalocyclopropanes

Problems with subsequent reaction of a 1-halocyclopropene may be avoided by reduction of the dihalocyclopropane to a monohalocyclopropane prior to dehydrohalogenation. Thus 3,3-dimethylcyclopropene may be obtained in multi-gram quantities by treatment of 1-bromo-3,3-dimethylcyclopropane with potassium t-butoxide in DMSO at −78 °C [19]. Dehydrohalogenation of a series of related 3,3-dialkyl substituted monochloro- or monobromo-cyclopropanes leads to moderate yields of cyclopropenes (11, R = alkyl, alkenyl, aryl, CN, Cl, R^1 = H, Ph, t-Bu, R^2 = H) [20]; Indeed, dehydrobromination of (12) leads to either mono- or di-cyclopropenes [21]. Reaction of a dihalocyclopropane with an alkyl lithium at low temperature followed by carboxylation of the derived 1-lithio-1-halocyclopropane provides a convenient source of 1-halocyclopropane carboxylates; dehydrohalogenation leads to cyclopropene esters (11, $R^2 = CO_2R$)[22]. In the same way the tricyano-system (13) eliminates on reaction with triethylamine even at −78 °C[23].

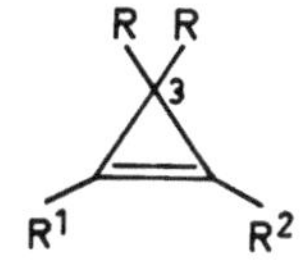

11

12

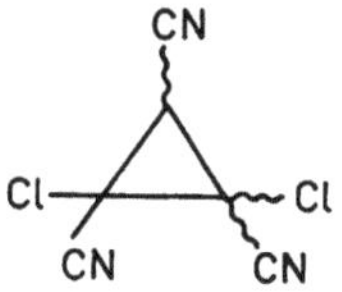

13

A new adaptation of this method involves passing the halocyclopropane at low pressure over potassium t-butoxide on a solid support, condensing the products at low temperature. In this way, bromocyclopropane is converted to cyclopropene, although the yields in preparative reactions are low[24]. Using $KOBu^t$ supported on chromosorb-W, 1-chloro-2-methylenecyclopropane is converted into methylenecyclopropene, which is unstable above −75 °C[25], and 1-chloro-2-vinylcyclopropane is converted into 1-vinylcyclopropene[26]. In the case of the dichloride (14), however, a simple double dehalogenation does not occur, and 1-vinylcyclopropene again results, in a process involving an overall reduction. The reaction has been explained in terms of initial formation of the monochloride (15) followed by radical cleavage of the cyclopropene bond and elimination of a chlorine atom to produce (16), and then hydrogen atom abstraction; some support for the final step is provided by the isolation of acetone, presumably derived from t-butylalcohol by hydrogen abstraction[26].

14

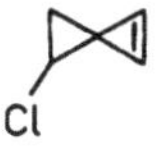

15

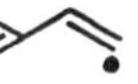

16

When the dehydrochlorination of (14) is carried out in solution, the reaction follows a rather different course, and (17) and (18) are isolated. These products may both be explained by a reaction sequence involving addition of t-butoxide to the monochloride (15)[26].

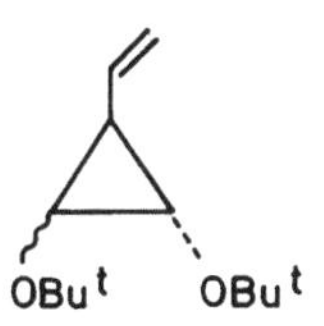

17

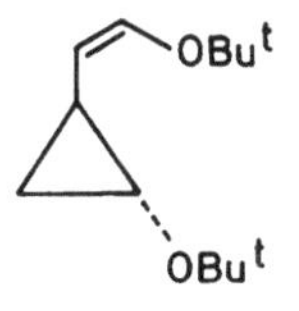

18

Dehydrohalogenation of 1,1-Dihalocyclopropanes

The dehydrohalogenation of cyclopropanes bearing three alkyl or aryl substituents as in (19, R^1 = t-Bu, Ph) leads to halocyclopropenes which cannot undergo prototropic shifts by removal of an allylic hydrogen. When R^1 = Ph and R^2 = H, alkyl or aryl, good yields of chlorocyclopropenes may be obtained[27–29]. In other cases ring opening is also observed:

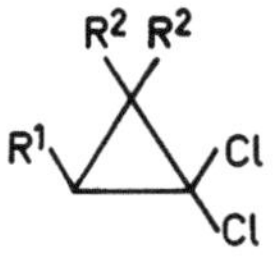

19

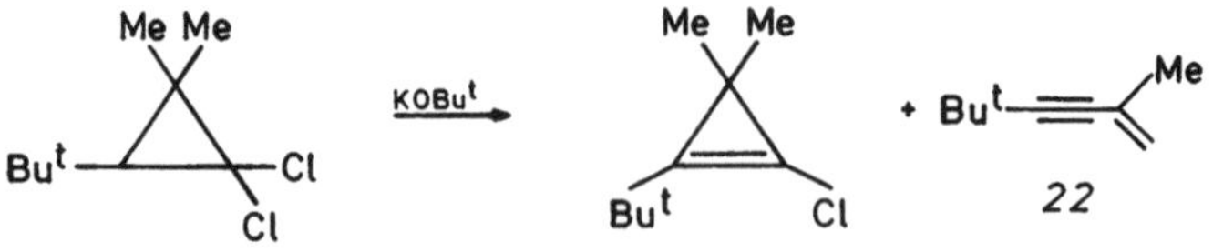

20 21 22

The origin of the alkyne in this reaction was not clear [30], but this will be discussed again below. In the same way the methylene cyclopropenes (24) can be obtained from the corresponding 1,1-dibromocyclopropanes, although the products are extremely sensitive to water, undergoing rapid hydrolysis at ambient temperature to produce (25) and (26), presumably by initial hydration of either of the two double bonds [31].

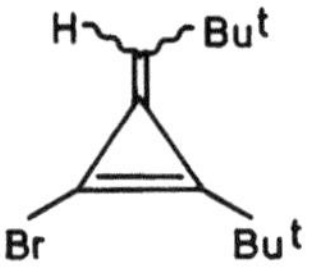

24

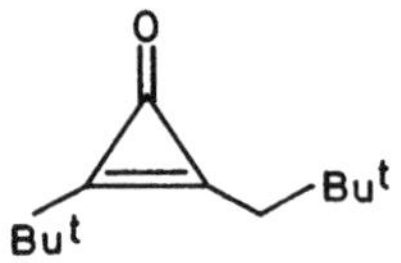

25

$Bu^t-CH_2-\overset{O}{\overset{\|}{C}}-C\equiv C-Bu^t$

26

The reaction of the dimethyl-derivative (27) with butoxide ion might be expected to produce the chlorocyclopropene (28); however, in practice two eliminations occur to produce (31) and the carbene (30), which can be trapped by an added alkene. Both products may be derived from (28), by a 1,4- or a formal 1,2-elimination respectively; a study using a ^{14}C-label at C-1 of (27) showed that the carbene (30) was formed with the label exclusively at C-1, suggesting elimination via (29) [32]. However, in a related study, the isolated cyclopropene (28) labelled with ^{12}C at C-1 has been shown to react with methyl lithium to produce the carbene (30) labelled only at C-2; this suggests either that the reaction of (28) with butoxide follows a completely different course to that with methyl lithium, or that (28) is not involved in the reaction of (27) with base [33]. In a similar reaction the dichloride (32) has been shown to react with t-butoxide in DMSO to produce the allene (33); the product may be explained in terms of initial elimination to produce (34), followed either by rearrangement to the alkyne (35) and then elimination or by direct 1,4-elimination as in (36), followed in either case by a prototropic shift. Whatever the mechanism, a ^{12}C-label at Ca in (32) is found at Ca in (33) [33].

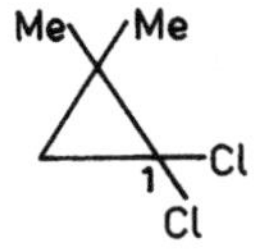

27

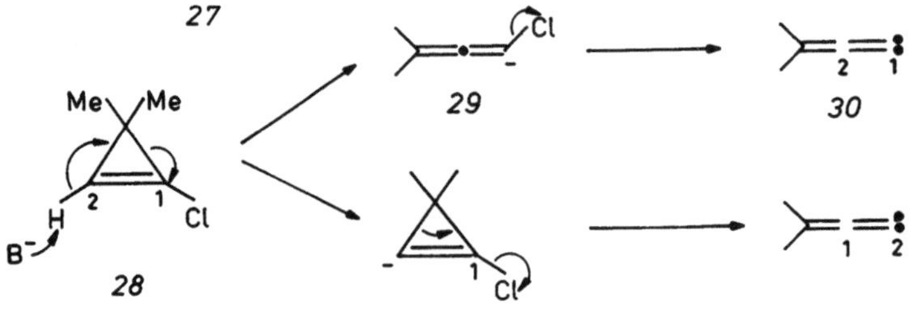

28 29 30

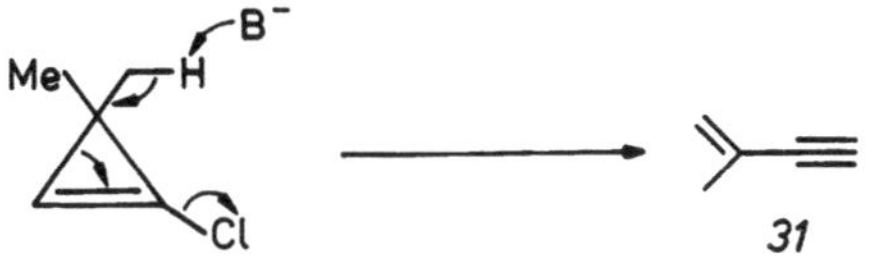

31

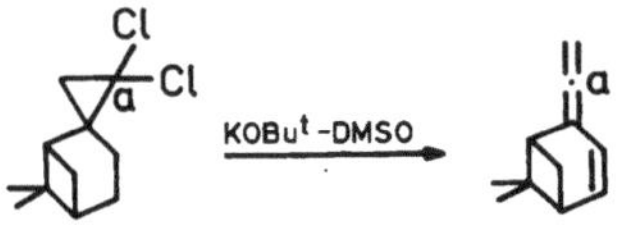

32 33

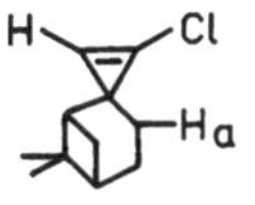

34

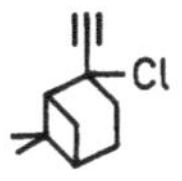

35

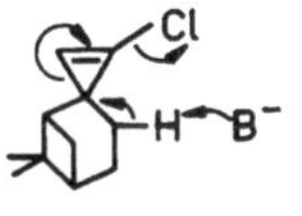

36

An exception to the general rule that halocyclopropenes cannot be isolated when prototropic shifts are available is seen in the reactions of (37, R = $Me(CH_2)_7$ or H) with potassium hydroxide in ethanol at reflux, which are reported to lead to 1,2-elimination to bromocyclopropenes [34].

A variation on the above elimination is also possible in certain trihalocyclopropanes. Thus (38) undergoes clean base-induced elimination of HBr to provide a route to 1,1-difluorocyclopropa[a]naphthalene [35].

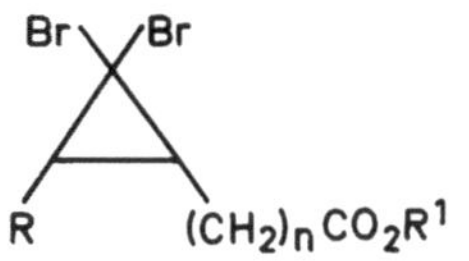

37

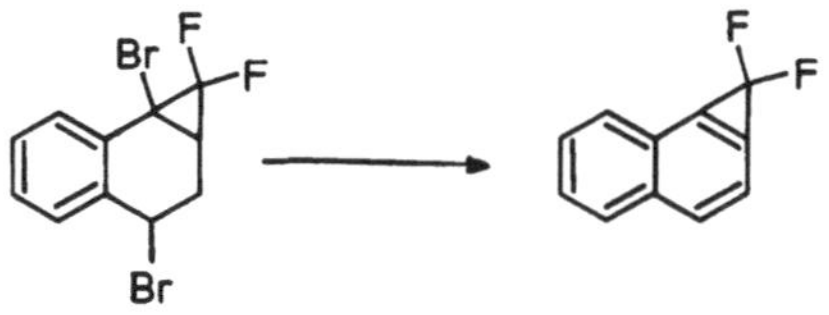

38

1.3.2 1,2-Dehalogenation

Dehalogenation of perhalocyclopropanes using zinc provides a very simple route to tetrahalocyclopropenes [36]. Moreover, a number of the problems of further reaction inherent in the 1,2-dehydrohalogenation of halocyclopropanes (see above) may be removed by use of dehalogenation of a 1,2-dihalocyclopropane with an alkyl lithium. In this way, reaction of (39, R = R^1 = Me) with butyl lithium at −78 °C leads to 1,2-dimethylcyclopropene, while (39, R, R^1 = $(CH_2)_3$ or $(CH_2)_4$) leads to the highly strained bicyclic species (40, n = 3,4) which may be trapped as Diels-Alder adducts [37]. The 1,2-dihalocyclopropanes can be prepared from the corresponding dicarboxylic acids by the Hunsdiecker reaction [37], but otherwise are not widely reported, though they can be obtained by reduction of tri- or tetrahalocyclopropanes (or by addition of halogen to a cyclopropene!). In general, it is simpler to dehalogenate 1,1,2-trihalocyclopropanes or 1,1,2,2-tetrahalocyclopropanes themselves; these are readily available by dihalocarbene addition to halogenated alkenes. The presence of the additional halogens has the added advantage of making the initial lithium-halogen exchange more facile. Thus treatment of (41) with methyl lithium at —90 °C leads to (42) in a reaction which can be explained by lithium-bromine exchange at C^1 followed by 1,2-loss of lithium halide [38, 39].

39

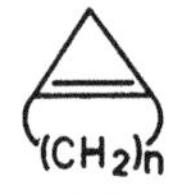

40

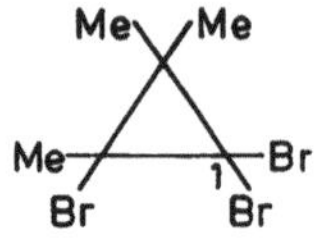

41

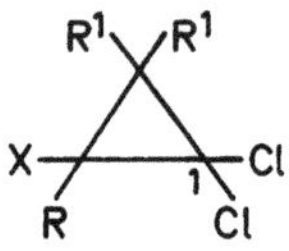

43

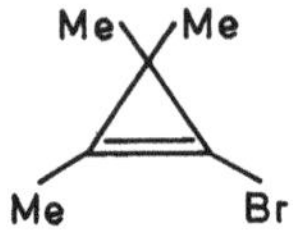

42

The reaction is successful with a variety of alkyl substituents at 2- or 3-positions. It is also successful with the geminal-dichlorides (43), though somewhat more vigorous conditions are required. When X = Cl and R = alkyl, the reaction occurs in ca 10 min. at 20 °C, presumably initiated by lithium-halogen exchange at C^1; however, when R = H, the reaction follows a different course, and lithium-hydrogen exchange followed by 1,2-elimination of LiCl is observed [39]. When X = Br, the reaction occurs at somewhat lower temperature and may be initiated by lithium-bromine exchange; indeed with R = H or alkyl, overall 1,2-elimination of BrCl is observed [38, 39]. This dehalogenation procedure can be carried out on a multi-gram scale and in many cases the 1-halo-cyclopropenes are stable for long periods at ambient temperature. Because the solvent is ether and work-up is simple, it can be applied to relatively unstable products such as (28) [39]. An added advantage is that lithium-halogen exchange in the product 1-halocyclopropene occurs readily on addition of a second equivalent of alkyl lithium (see Section 2), enabling a wide range of substituents to be introduced at C^1 of the cyclopropene by further reaction with electrophiles.

1.3.3 1,2-Dehalosilylation

Dehalosilylation of 1-halo-2-trialkylsilylcyclopropanes occurs under mild conditions and in general without further rearrangement, when brought about by fluoride ion. Thus treatment of (45, X = Cl, Br) with caesium fluoride leads to 1-chloro- or 1-bromo-cyclopropenes in good yield, while (46) is converted to the highly strained cyclopropene (47) which can be trapped as a Diels-Alder adduct with a furan [40]. In the same way reaction of (48, X = Cl) with tetra-n-butyl ammonium fluoride at —20 °C leads to 1-bromo-2-chlorocyclopropene (49, X = Cl) which can be trapped in 80–90% yield by cyclopentadiene [41]. Moreover the cyclopropene, which may be stored for several days at that temperature [42], undergoes Diels-Alder reactions with 1,2-bis-methylene-cyclohexanes and therefore acts as an efficient synthon in several routes to cycloproparomatics [42, 43]. Reaction of the tribromide (48, X = Br) with fluoride ion provides a convenient route to (49, X = Br), while treatment with butyl lithium leads to (49, X = $SiMe_3$) [44].

1.3.4 Other Eliminations

An alternative to the above eliminations, which in principle offers a great many advantages, is the metal induced elimination of the elements of ROBr from a 2-halocyclopropyl ether. Thus the sequence of addition of dibromocarbene to an enol ether, reduction to a monobromide and reaction with magnesium in tetrahydrofuran provides a route to cyclopropene itself[45].

The oxidation of a phenylselenylcyclopropane with elimination of PhSeOH has also been reported, but in the particular case examined the product, (50), was unstable and could only be detected indirectly by trapping with methanol as the ring opened ester (51)[46]:

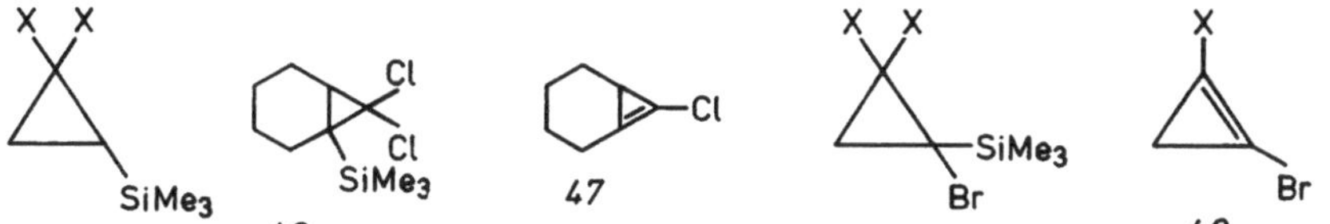

45 46 47 48 49

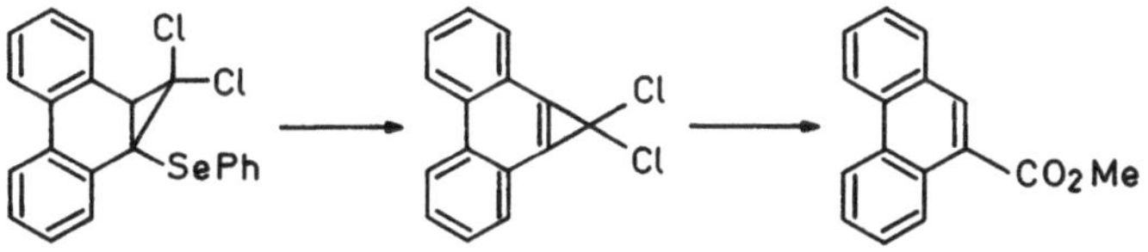

50 51

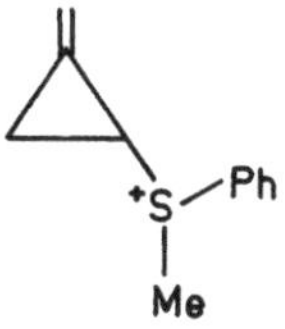

52

Reaction of the sulphonium salt (52) with cyclopentadienide ion leads to methylenecyclopropene, though this can only be trapped in low yield by addition to cyclopentadiene[47].

1.4 From Alkynes

The reaction of a diazo-compound with an acetylene either photolytically or in the presence of a catalyst represented one of the earliest routes to cyclopropenes, and is especially useful for 3-functionalised system[1]. In some examples the diazo-compound adds to the acetylene to produce a pyrazole which on photolysis leads either to a vinylcarbene or to the ring-closed cyclopropene. Thus 2-diazopropane adds to acetylenes carrying electron withdrawing groups such as esters or nitriles to give good yields of pyrazoles; these are converted by photolysis to electrophilic cyclopropenes. In the case of acetylenic ketones some dienic products may also be obtained[48]:

The adduct (53, R = CO_2Me, R^1 = CH = CMe_2) is quantitatively converted to the cyclopropene ester (54, R = CO_2Me, R^1 = CH = CMe_2) on photolysis; (53, R^1 = CO_2Me, R = CH = CMe_2) leads to the same cyclopropene, the intermediate carbene (55) closing regioselectively at the methoxycarbonyl-substituted terminus, presumably due to steric effects in the intermediate[49]. Other examples of this reaction are not always so efficient. Thus, although (53, R = R^1 = Ph) is converted to (54,

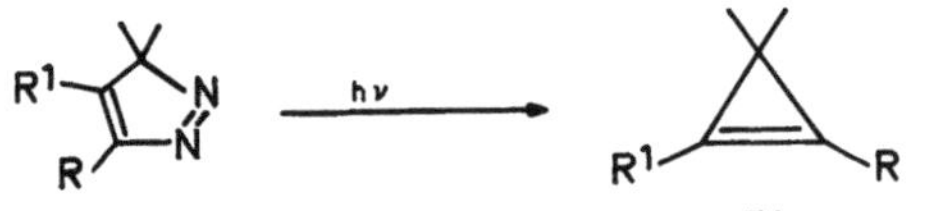

53 54

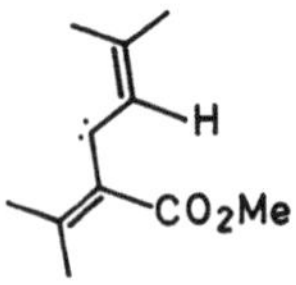

55

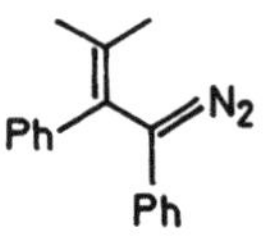

56

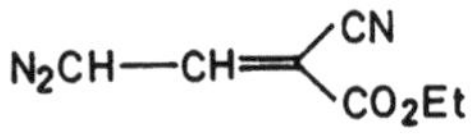

57

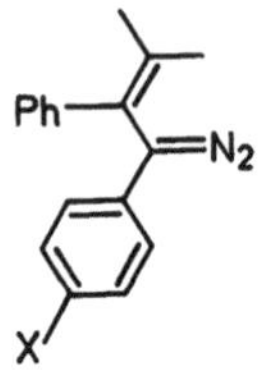

58

$R = R^1 = Ph$) in very high yield by photolysis, the pyrazole is formed in only 3% yield from diphenylethyne. [50] The formation of cyclopropenes from pyrazoles apparently occurs via the corresponding diazo-compound. Thus photolysis of (53, $R = R^1 = Ph$) at between 330 and 410 nm has been shown to lead to the diazo-compound (56) which on further photolysis or pyrolysis is converted to (54) [50]. Indeed, photolysis of vinyldiazoalkanes provides a good route to cyclopropenes; thus photolysis of E- (57) leads to 3-alkoxycarbonyl-3-cyanocyclopropene [51], while (58) leads either to the corresponding cyclopropene or to reversal to the pyrazole [52].

Photolysis of the 3H-pyrazoles (59) leads to good yields of the corresponding cyclopropenes (60) in reactions which have also been shown to proceed through the diazo-compounds. With (59, $R^3 = Ph$) an additional product is the indene (61), apparently derived by ring closure of the carbene (62) followed by a 1,5-hydrogen shift. Moreover the pyrazoles (59, $R^1 = POPh_2$, X = O) lead to allenes (63) as well as cyclopropenes. In this case the cyclopropene is quantitatively converted to allene on further photolysis, presumably by ring opening to the vinyl carbene followed by a 1,2-shift of the diphenylphosphoryl group [53].

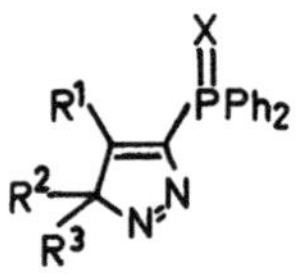

59

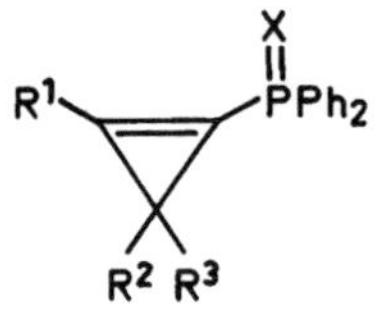

60

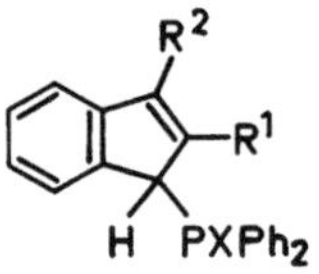

61

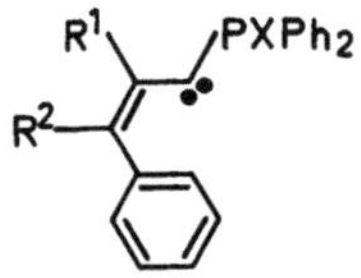

62

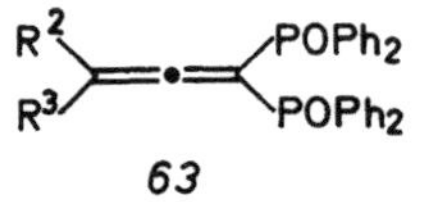

63

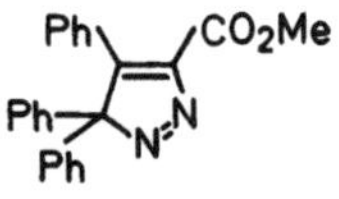

64

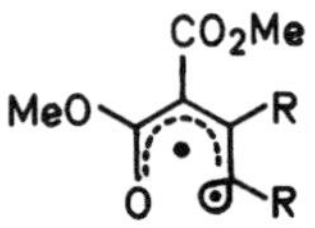

65

Photolysis of pyrazoles such as (64) has also been shown to produce indenes as well as cyclopropenes. The indene is apparently derived from a triplet intermediate, while the cyclopropene is singlet derived [54]. In a related example, direct photolysis of dimethyl diazomalonate in an alkyne leads to moderate yields of cyclopropene, whereas sensitized photolysis leads to furans by ring closure of an intermediate diradical (65) [55].

In other cases cyclopropenes have been obtained by direct reaction of an alkyne with a diazo-compound in the presence of a suitable catalyst. Typical of these is the reaction of ethyl diazoacetate with alkynes in the presence of copper, which is reported to lead to about 40–50% conversion to cyclopropene per equivalent of diazo-compound. This has been applied to the synthesis of the important naturally occurring cyclopropene, sterculic acid, (66) [56]:

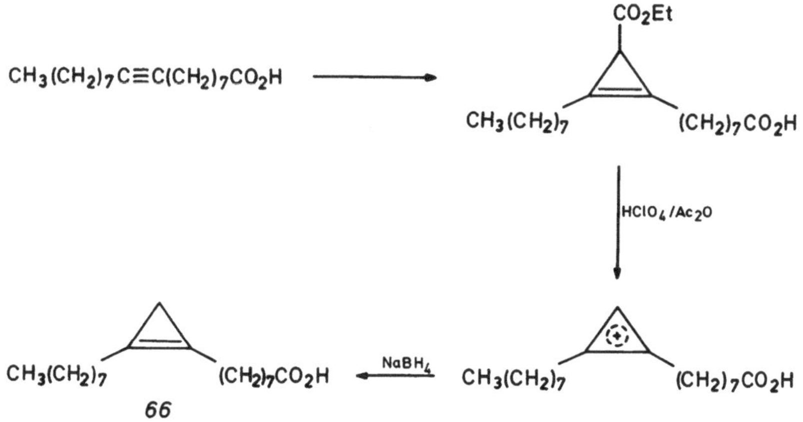

66

The addition of alkoxycarbonylcarbene derived by catalysed decomposition of methyl diazoacetate to several simple, and in particular terminal, alkynes leads to low yields [57], but the reaction with 1-trimethylsilylalkynes proceeds reasonably efficiently; subsequent removal of the silyl-group either by base or fluoride ion provides a route to 1-alkyl-3-cyclopropenecarboxylic acids. In the same way 1,2-bis-trimethylsilyl-ethyne can be converted to cyclopropene-3-carboxylic acid itself [58]. The use of rhodium carboxylates instead of copper catalysts also generally leads to reasonable yields of cyclopropenes, even from terminal alkynes [59].

The addition of dihalocarbenes to alkynes is again a rather inefficient process and usually leads to the isolation of the cyclopropenone rather than the 3,3-dichlorocyclopropene. In a rather unusual example, however, 2-butyne is reported to be converted to (67). This product is apparently derived by addition of dichlorocarbene to the corresponding methylenecyclopropene, derived in turn by elimination of HCl from the primary adduct (68). The cyclopropene (67) does not appear to ring open to a vinylcarbene, but can be trapped in Diels-Alder reactions with cyclopentadiene [60]. A related addition of dichlorocarbene to ethyl 2-butynoate also leads to a low yield of the 3,3-dichlorocyclopropene, which may be hydrolysed to the cyclopropenone [61].

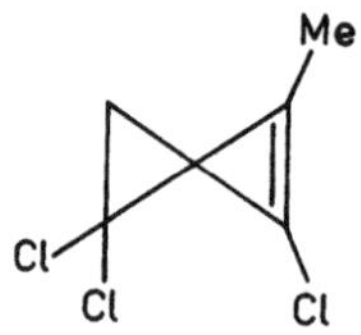

67

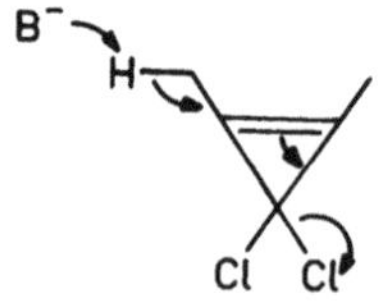

68

Additions of other types of carbene to alkynes are not common, though $Me_2C = C$:, generated from 2-methylpropenyl triflate and base, does add to alkynes to produce transient methylenecyclopropenes which may be trapped as Diels-Alder adducts [62].

1.5 From Cyclopropenium Salts

Reduction of cyclopropenium ions by hydride ion provides a viable route to cyclopropenes [63a]. With disubstituted ions borohydride has been found to attack exclusively at the unsubstituted position, presumably for steric reasons; this has been used in the synthesis of sterculic acid (66) (see above) [56]. Reduction of 1,2-diaryl-3-chlorocyclopropenium ions with dimethylamine-borane in methanol-water is also reported to be efficient [63b]. Cyclopropenium salts are also readily trapped by Grignard reagents; thus 1,2-diphenylcyclopropenium ion is converted to the rather unstable 1,2-diphenyl-3-methylcyclopropene; this may be stored as a charge-transfer complex with 9-cyanomethylene-2,4,7-trinitrofluorene and regenerated quantitatively [64]. A very interesting variation is the trapping of cyclopropenium ions by amide ion to produce the unsaturated analogues of 1-aminocyclopropanecarboxylic acid:

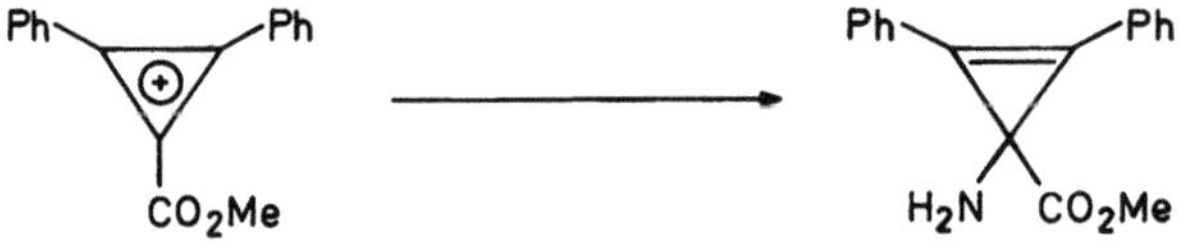

although the reaction has apparently not been extended to less substitued systems, nor has addition to the double bond to produce 1-ACCs themselves been examined [65].

Reaction of the diphenylmethoxycyclopropenium ion with (69) provides a simple route to functionalised methylenecyclopropenes [66]:

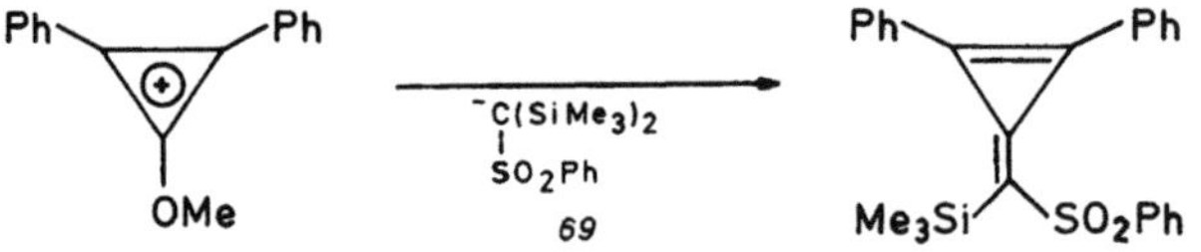

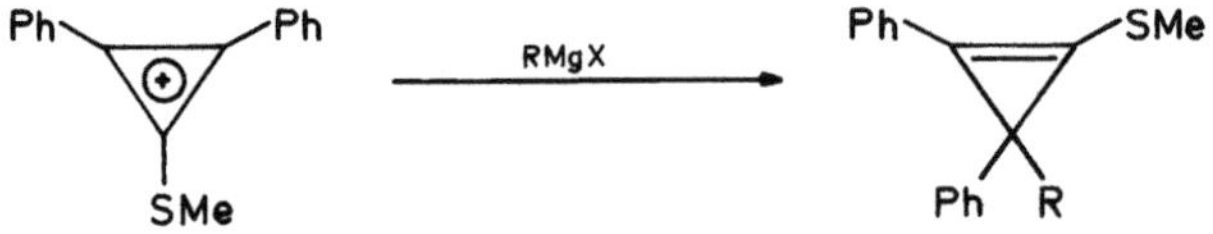

70

In the case of the corresponding thio-derivative, the reaction with Grignard reagents is not completely regioselective, although (70) is the major product [67]. Trapping of cyclopropenium ions with α-lithiodiazoalkenes and with stannylynamines has also been reported [68,69]:

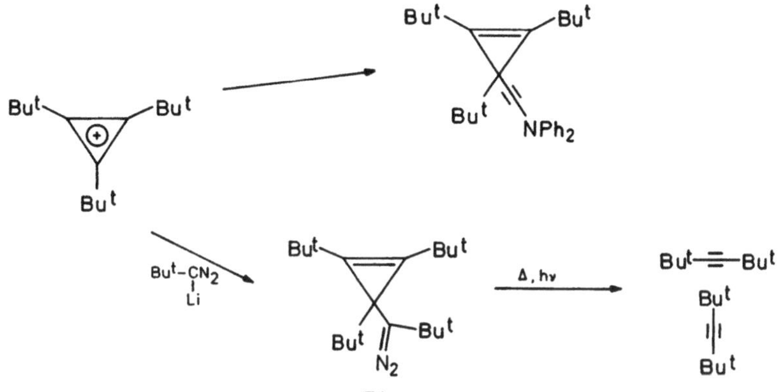

The cyclopropenyl diazocompounds (71) are thermally very stable but decompose under photochemical conditions to produce two alkynes, apparently by fragmentation of a cyclopropenylcarbene [68].

2 Functionalisation of the Cyclopropene Double Bond: Metallation at the 1-Position

Many of the routes to cyclopropenes described in section 1 lead to compounds with hydrogen or halogen substituents on the alkene; these may often be functionalised by metal-halogen or metal-hydrogen exchange followed by trapping with electrophiles. The acidity of the vinylic C—H bonds of cyclopropene is close to those in acetylene — indeed the vinylic hydrogens of methyl cyclopropene-3-carboxylate exchange when the ester is hydrolysed with aqueous sodium hydroxide [70]. The deprotonation can be carried out quite readily using a metal alkyl or dialkylamide, although addition to the cyclopropene is a potential competing process (see addition reactions below). Another process which can compete with metallation at the double bond is removal of an allylic hydrogen from a 1-alkyl substituent on the cyclopropene and rearrangement first to a methylenecyclopropane and then to a vinylcyclopropane. This process occurs readily with bases such as potassium t-butoxide, as seen later in the reactions of some 1-halocyclopropenes; further examples are the conversion of 1,3,3-trimethylcyclopropene to 2,2-dimethylmethylenecyclopropane by reaction with $KOBu^t$ but to 1-lithio-2,3,3-trimethylcyclopropene by treatment with lithium dicyclohexylamide [71], or methyl lithium [72], and the conversion of (72, X = CO_2R, COR, CH_2OH) to (73) by base [73].

The majority of metallations have involved lithium, and the metallated derivative may be trapped by a variety of electrophiles[1] including aldehydes [74], and allylic halides[72,75], although in some cases such as dimethylacetamide, ring opening can result[76].

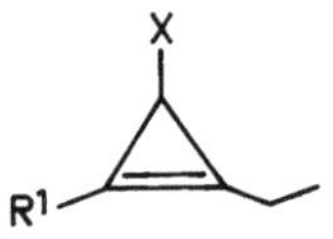

72

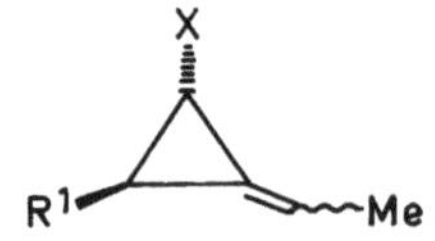

73

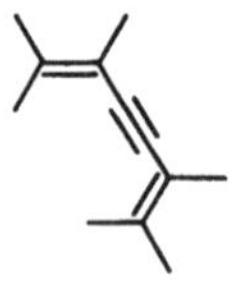

74

Cyclopropen-1-yl sodium derivatives are also readily prepared. Thus reaction of cyclopropene with one equivalent of sodium amide in liquid ammonia leads to 1-sodiocyclopropene which is alkylated by haloalkanes[77,78], reacts with ketones to produce tertiary alcohols and opens epoxides to produce 2-cyclopropenyl-ethanols in moderate to good yields[79]. Moreover, on reaction with two equivalents of base followed by haloalkane, 1,2-dialkylated species are obtained; sequential reactions can also be used to produce unsymmetrically substituted cyclopropenes[78]. Reaction with a deficiency of sodium amide can also cause addition of the cyclopropenyl anion to unreacted cyclopropene, leading to products derived from the 2-cyclopropylcyclopropen-1-yl anion and to 1,2-dicyclopropylcyclopropene[77].

Reaction of a cyclopropenyl lithium with cuprous chloride at —70 °C leads to the corresponding organcuprate, but this is reported to lose copper; thus the 2,3,3-trimethyl-derivative is converted to (74)[80]. The 1-lithiocyclopropenes may however be converted to 1-trialkylsilyl, trialkylstannyl, or trialkylgermyl derivatives, eg.[81,82]:

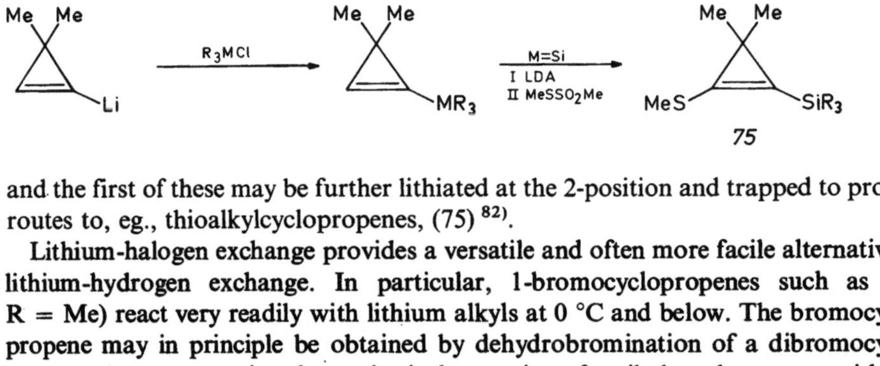

75

and the first of these may be further lithiated at the 2-position and trapped to provide routes to, eg., thioalkylcyclopropenes, (75)[82].

Lithium-halogen exchange provides a versatile and often more facile alternative to lithium-hydrogen exchange. In particular, 1-bromocyclopropenes such as (77, R = Me) react very readily with lithium alkyls at 0 °C and below. The bromocyclopropene may in principle be obtained by dehydrobromination of a dibromocyclopropane, but an attractive alternative is the reaction of a trihalocyclopropane with two equivalents of methyl lithium, followed by trapping with an electrophile[39]:

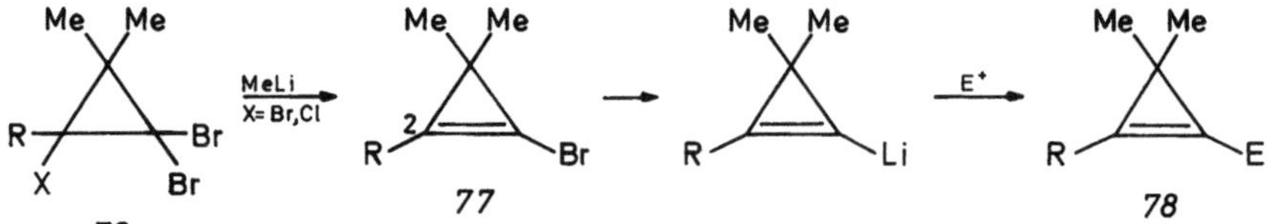

76 77 78

The overall yields of (78) from (76) are generally about 60–80%. The lithiation reaction is successful even with a hydrogen at C-2, as in the formation of (78, R = H,

E = CO_2H). The lithiation of 1-bromo-2-trimethylsilylcyclopropene has also been reported, providing a simple route to 1,2-bis-trimethylsilylcyclopropene after quenching with chlorotrimethylsilane [44]. The chlorocyclopropene (79, R = Me), which may be obtained from the corresponding 1,1-dichloro-2-bromo- or 1,1,1-trichlorocyclopropane and methyl lithium, also undergoes lithium-halogen exchange on further treatment with methyl lithium, but the reaction is slower, requiring 40 min. at 20 °C [39]; in this case, the monochloride (79, R = H) reacts instead by lithium-hydrogen exchange, and the ring opens with loss of lithium chloride to produce carbene (80) [84]. Although the related 1-bromo-2-chloro-3,3-dimethylcyclopropene (79, R = Br) is unstable at ambient temperature, ring opening to two isomeric vinyl carbenes (see later), it reacts with methyl lithium at or below ambient temperature in the presence of an alkene to produce a cyclopropane apparently derived by the addition of the same carbene (80) to the alkene. If it is assumed that this reaction is initiated by lithium-bromine exchange, there are a number of possible ways by which (79, R = Li) can rearrange to the carbene, eg. elimination to produce a formal cyclopropyne, or cleavage of either 1,3- or 2,3-bonds with elimination of lithium chloride. A ^{12}C labelling study indicates that it is the 1,3-bond which breaks [83,84]. A similar study of the generation of the carbene from the reaction of (79, R = H) with methyl lithium also shows cleavage of the 2,3-bond, the label appearing at the central allenic carbon on trapping [39].

The conversion of 2,3-disubstituted 1,1-difluorocyclopropanes to 1-alkyl-2,3-disubstituted cyclopropenes may be achieved by treatment with two equivalents of alkyl lithium at —70 °C. The reaction apparently proceeds by initial lithium-hydrogen exchange and loss of lithium fluoride to generate a 1-fluorocyclopropene. It is suggested that this undergoes rapid reaction with the lithium alkyl at the carbon bearing fluorine; an alternative process which may occur, particularly when the 2-substituent is phenyl, would be an addition to the double bond followed by elimination of fluoride ion. [85] In the case of the difluoride (81, R = F), an alternative reaction occurs with butyllithium and (82) is isolated in moderate yield. It seems likely that an intermediate 1-fluorocyclopropene reacts by lithium-hydrogen exchange at C^2 to produce (83). Labelling studies are consistent with the ring opening of this species by cleavage of the 2,3-bond to produce (84); it is proposed that the ring opening is brought about by attack of the alkyl lithium at C^3 followed by, or concerted with, loss of fluoride ion, but analogy with the results described above for (79, R = Br, H) suggests that the lithio-species may in fact lose lithium fluoride directly to produce the carbene (85), which could be trapped by the alkyl lithium to produce the lithium salt of (82) [85]. The related monofluoride (81, R^1 = Ph, R = Me) is converted to 1-lithio-2-methyl-3,3-diphenylcyclopropene by reaction with butyl lithium, but trapping by carbon dioxide is not very efficient [86].

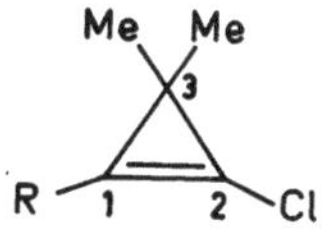

79

80

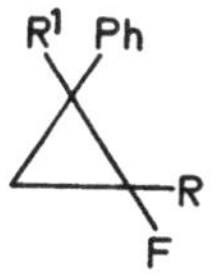

81

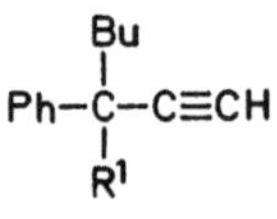

82

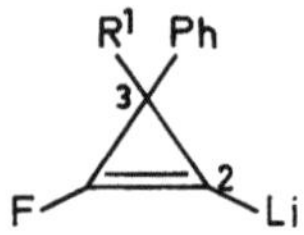

83

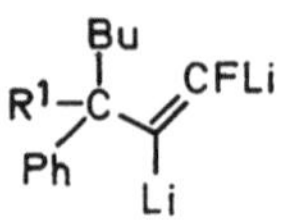

84

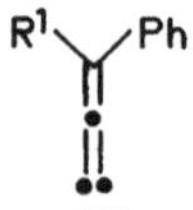

85

An alternative method of vinylic substitution involves the reaction of 1-benzenesulphonylcyclopropenes with an alkyl lithium:

The reaction generally proceeds in good yield and could involve an addition-elimination, though the regiochemistry appears to be incorrect for this. An alternative would be an $S_{RN}1$ process involving (86). The reaction has been applied to the preparation of (87) from the silacyclopropene; photolysis then leads to the novel tetracycle (88) through an intramolecular [2+ 2]-cyclo-addition [87].

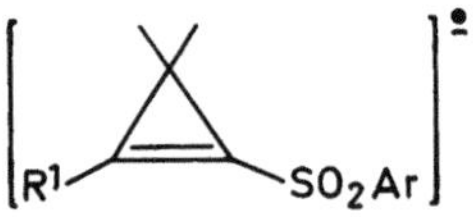

86

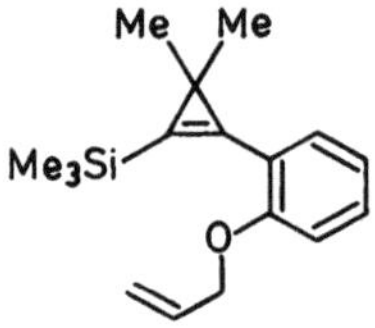

87

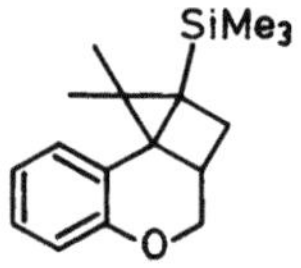

88

3 Thermal Reactions of Cyclopropenes

As might be expected from their inherent strain, many cyclopropenes undergo rearrangement, dimerisation or even polymerisation under relatively mild conditions. The conditions required for reaction are, however, very variable and some cyclopropenes, such as 3,3-dimethylcyclopropene, are stable at relatively high temperature (150 °C in this case). Three main reactions are described below — the ene-reaction, [2+ 2]-dimerisation, and rearrangement to vinylcarbenes.

3.1 The Ene-Reaction

Cyclopropenes having a hydrogen at C^3 often undergo a particularly facile dimerisation by an ene-type reaction. Thus cyclopropene itself has long been known to undergo dimerisation to cyclopropenyl-cyclopropane on standing at —25 °C; the dimer is converted to oligomers at longer reaction times [88]. The cyclopropene fulfils the roles of both ene- and enophile:

The presence of a methyl-group at C-1 confers little selectivity, a mixture of three dimers being obtained in low yield (18 %). The cyclopropene does, however, act as the ene-component in a facile reaction with perfluorobutyne or dimethyl acetylenedicarboxylate at —30 °C, which can proceed with explosive violence at higher temperatures. The initial products are the vinylcyclopropenes (91, X = CF_3, CO_2Me), but these are reported to be unstable at 25 °C [89]. Reaction of the cyclopropene with t-butylcyanoketene leads initially to the ene-product (92), which can either react further or be trapped as an ester by reaction with additional ketene. A similar reaction occurs with perfluorodimethylketene, although in polar solvents such as acetonitrile an additional product is (93); this apparently arises by a polar addition leading initially to (94) which can rearrange to a cyclopropenone and then lose carbon monoxide [90]. In the case of 3,3-dimethylcyclopropenes, which cannot act as the enophile, reaction with t-butylcyanoketene leads predominantly to (95), though once again a competing dipolar pathway is observed, in this case leading to, eg., (96) [91].

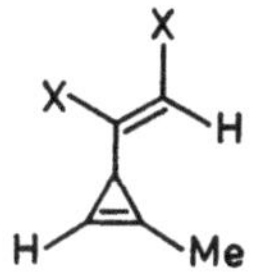

91

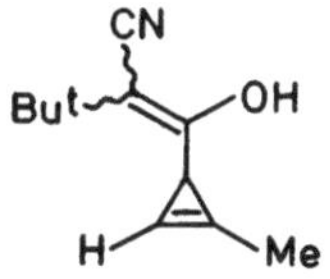

92

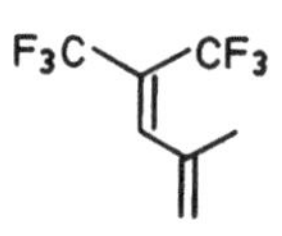

93

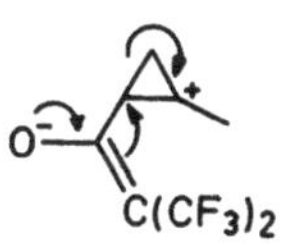

94

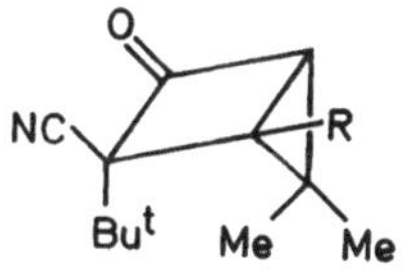

95

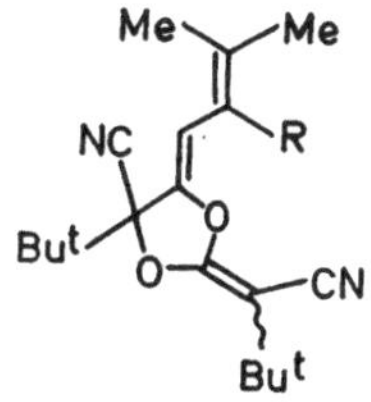

96

1-Methylcyclopropene is also trapped in moderate yield by ene-reaction with bis(trifluoromethyl)thioketene [92]. Tetracyanoethylene can also act as the ene-component in reactions with cyclopropenes. Thus with cyclopropenes bearing an allylic ring hydrogen such as 1-methylcyclopropene the only product is (97), whereas when no such hydrogen is present products apparently derived by trapping of a ring-opened vinylcarbene or diradical, eg., (98) from 1,3,3-trimethylcyclopropene, are isolated [93].

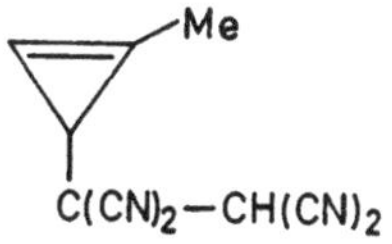

97

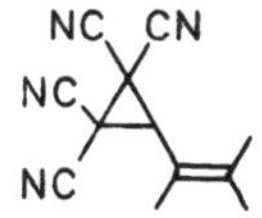

98

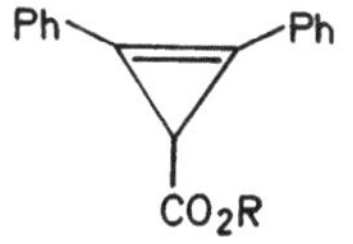

99

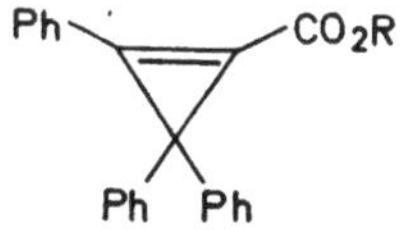

100

Other cyclopropenes have also been shown to act as the enophile or the ene; thus (99) is converted to (100) in reasonable yield by reaction with benzyne [94], and (101) is converted to the diastereoisomers of (102) on heating with trans-2-butene for 11 h at 80 °C. The major product is believed to be (a), resulting from pseudo-exo-approach as in (103) [95].

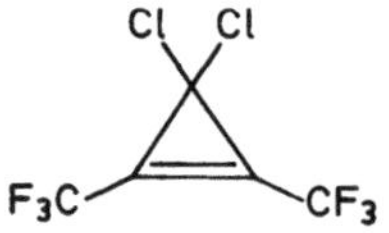

101

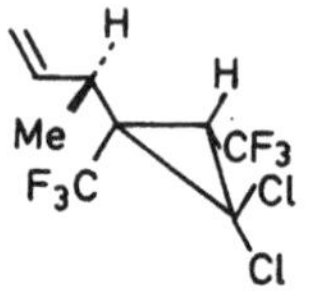

102a

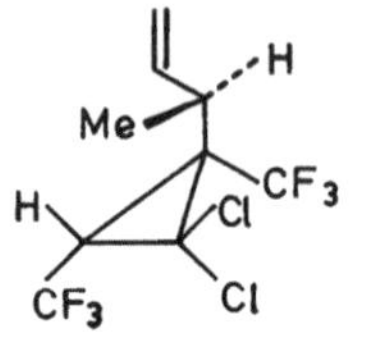

102b

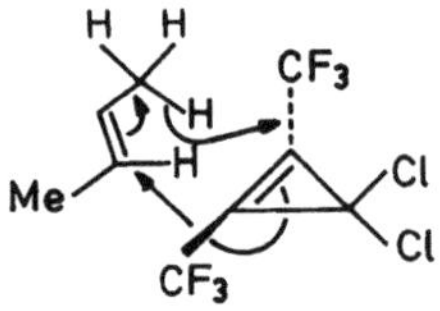

103

In other cases the rate of ene-dimerisation is so high that trapping by external enes or enophiles is inefficient. Thus the diarylcyclopropene (104, R = H) is not trapped by 3-phenylpropene, and leads to only low yields of ene-products analogous to (97) with tetracyanoethene, dibenzoylacetylene or dimethyl acetylenedicarboxylate. The major product in each case is an ene-dimer; this was possibly originally identified as (106) [96], but is now characterised as (105, R = H), and is obtained essentially quantitatively when the cyclopropene itself is warmed above —78 °C. Rate studies reveal a low E_a for the process, but a large negative ΔS^{+}, in agreement with a concerted ene-process. Comparison with the 3-^{2}H-cyclopropene reveals a large isotope effect (3.1 at —30 °C) and the formation of a product with the two H-substituents cis, that is (105, R = ^{2}H) [97]. The ene-reaction of such 1,3-disubstituted cyclopropenes represents a rather novel process in that, if an *exo*-transition state is involved as in (104) → (105), the process can only occur if the two different components are enantiomers. In the case of the cyclopropenes above, with two identical aryl-groups, it is not possible to confirm this analysis. However, the ester (107) and acid (108) form a single dimer on standing at 20 °C. This has been shown by crystal structure analysis to be (109), confirming the exo-nature of the transition state, at least when the ene component has geminal-substituents at C^3 [98].

The acid (108) itself undergoes dimerisation at 0–20 °C, leading to two ene-dimers: the major dimer was shown by crystallography to be (110), consistent with an *endo*-transition state in an ene-reaction. The minor dimer is characterised as (111) on the basis of spectroscopic evidence, the regiochemistry at the cyclopropane in this case indicating an *exo*-transition state geometry [98].

There are also examples of the intramolecular ene-reaction involving cyclopropenes. Thus, while (112, R^1 = Me, R^2 = H) undergoes an intramolecular [2+ 2]-cyclo-

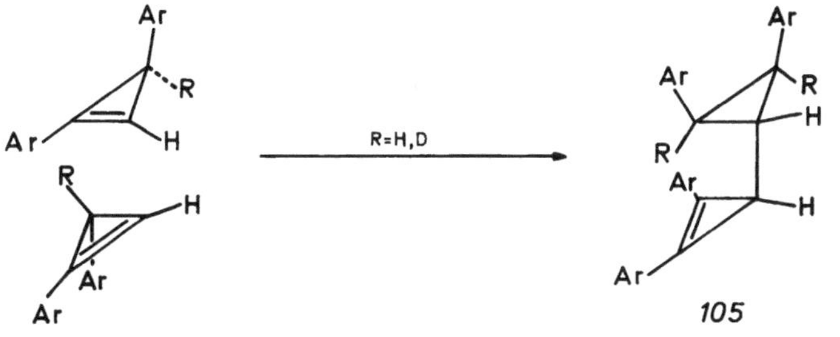

104 105

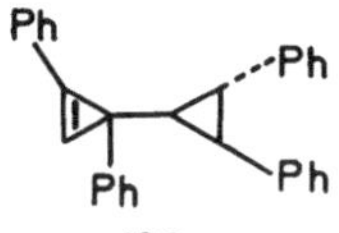

106

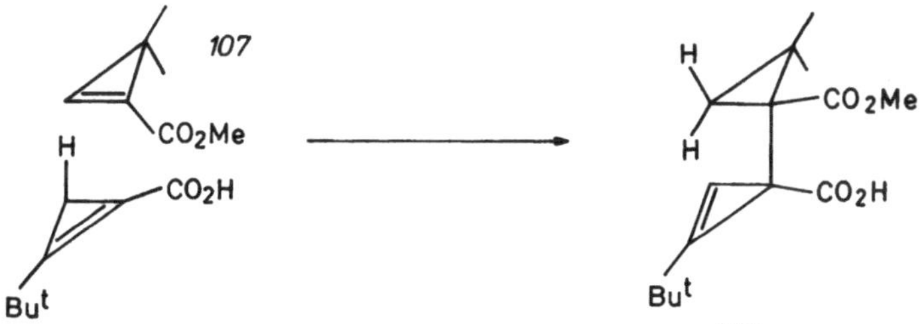

108 109

addition, the isomer (112, R^1 = H, R^2 = Me) is converted to (113), and the homologue (114) leads to (115)[99,100].

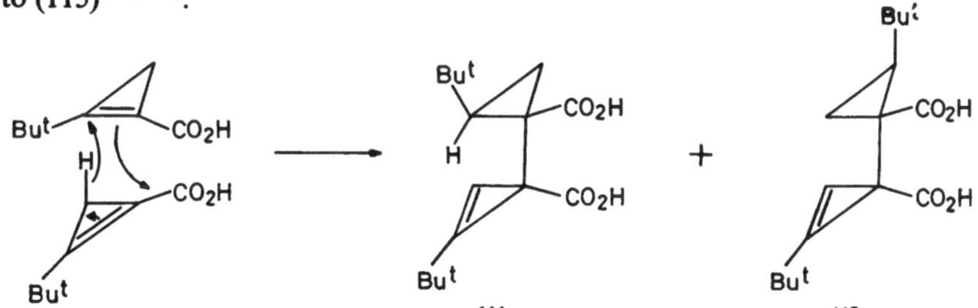

111 110

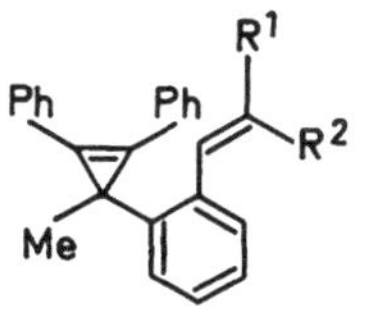

112

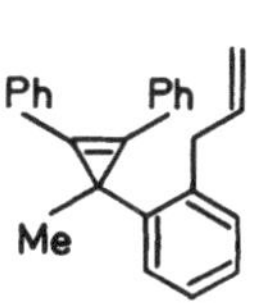

114

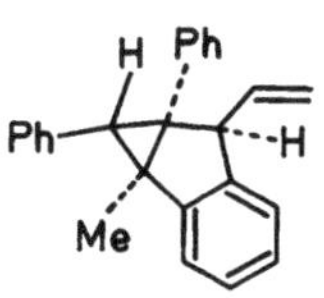

113

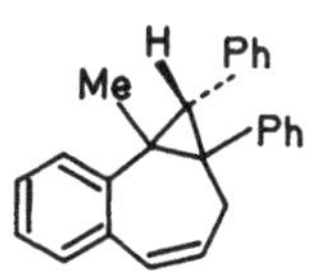

115

3.2 [2+ 2]-Cycloaddition

Although many cyclopropenes dimerise very readily by the ene-process described above, an alternative mode of dimer formation, a [2+ 2]-cycloaddition, may occur under thermal, metal catalysed or photochemical conditions; this is particularly common when the ene-reaction is slow or when it is blocked by 3,3-disubstitution. Thus 3,3-dicyclopropylcyclopropene is converted to (116, R = R^1 = cyclopropyl) on heating to 70–100 °C or in the presence of boron trifluoride [101 a)], while the acetal (116, R, R^1 = $O(CH_2)_2O$) is obtained quantitatively by [2+ 2]-dimerisation of the corresponding cyclopropenone acetal at 0 °C in methanol — although at higher temperatures or with related acetals ring opened products are isolated [101 b)]. Moreover, 3-methylcyclopropene is converted to (116, R = Me, R^1 = H) in near quantitative yield on brief contact with a zeolite at —30 °C [102)], and vinylcyclopropene undergoes the [2+ 2]-dimerization to give (117) even at —60 °C; the facility of the last reaction is explained in terms of relief of ring strain, while the regioselectivity is as expected for a diradical intermediate [26)].

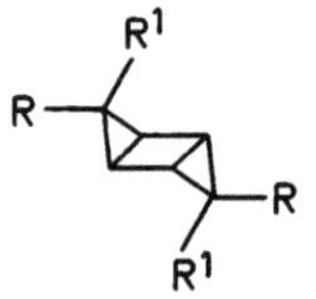

116

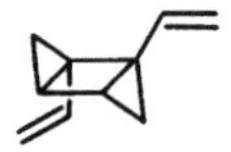

117

In other cases the thermal reaction of the cyclopropene leads to ring opening, but the [2+ 2]-cycloaddition can be brought about by photolysis [103)], or by the presence of a metal salt, eg. [4,93,104,105)]:

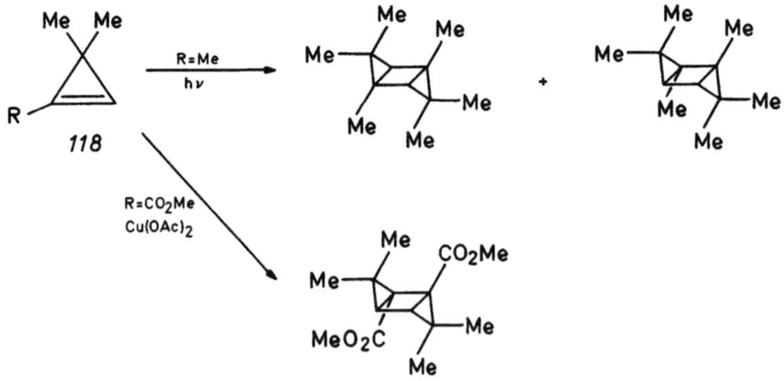

118

The outcome of these reactions seems to be somewhat difficult to predict, as the presence of a copper salt during the reaction of 3,3-dialkylcyclopropenes leads to trienes (see below) [103)], which are the products when (118, R = CO_2Me) is allowed to react in the absence of a catalyst [105)]!

Cyclopropenes may also be made to undergo [2+ 2]-addition to other alkenes; thus 3,3-dialkylcyclopropenes add to norbornadiene in the presence of $Ph_3P \cdot CuCl$ at low temperature to produce bicyclo[2.1.0]pentanes; at higher temperature the product is apparently derived by addition of a vinylcarbene $R'_2C = CH—CH$: to the diene [106]. Photochemically induced intramolecular [2+ 2]-addition is also successful [107 a, c].

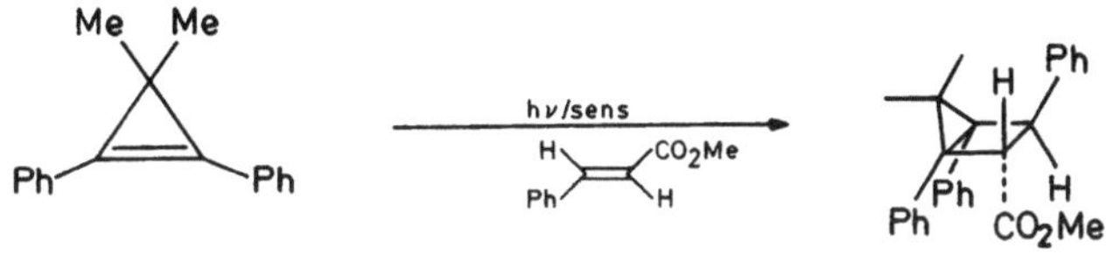

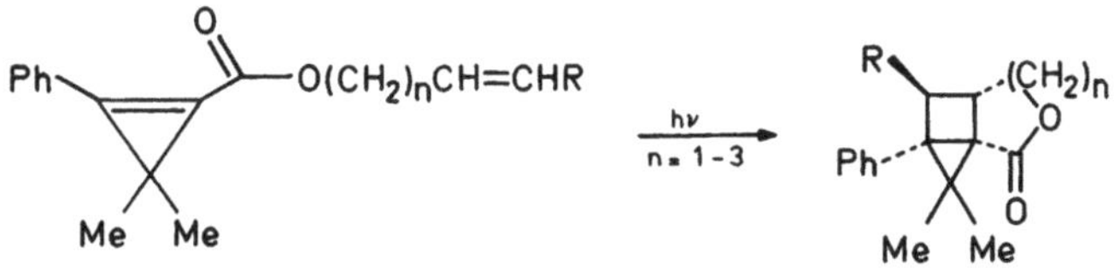

In general, the addition is sensitive to steric factors and the approach geometry is such that interactions are minimised, while the regiochemistry may again be explained in terms of the formation of the more stable biradical intermediate. In other cases a thermal intramolecular addition occurs, eg. [107 b]:

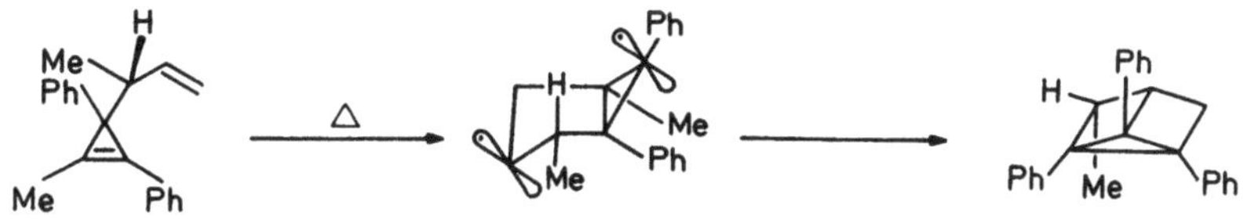

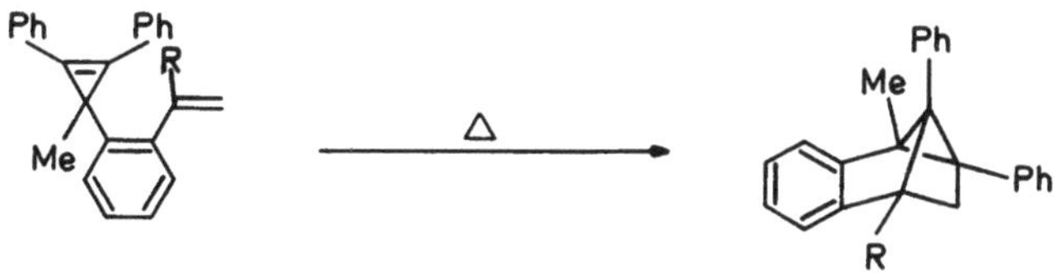

3.3 The Cyclopropene — Vinylcarbene Interface

The ring opening of cyclopropenes to species which behave like vinylcarbenes has long been established in a range of systems; the reverse reaction of ring closure of vinylcarbenes provides one of the basic routes to cyclopropenes (see above). However, it is difficult to be certain that the intermediates in these reactions are always vinylcarbenes, and if so that the exact nature of these species is always the same. Thus vinylcarbenes have a number of possible electronic states such as non-planar singlets and triplets (122), and planar singlets and triplets, eg. (123). Although calculations show the planar triplet is the most stable state it is unlikely to be the first formed intermediate in many reactions which formally lead to vinylcarbenes. In addition, the use of product ratios in analysing details of these reactions is complicated by the reversibility of the cyclopropene-carbene process [108]. The emphasis in this review will be the outcome of the reactions — in particular those which may have synthetic applications — and not the detailed nature of the intermediates, which merits a review in itself!

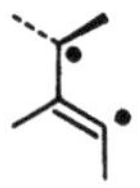

122

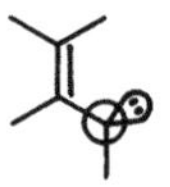

123

The thermal rearrangements of a variety of alkylcyclopropenes (124) lead to ring opening and the formation of alkynes, dienes and in some cases cyclopropanes, which may be explained in terms of known carbene reactions resulting from (125) or (126) [109]. Thus the products of thermal rearrangement of (127) in an acid-free system are consistent with opening to both vinylcarbenes (128) and (129). However the large negative entropy of activation is inconsistent with a simple ring opening and may suggest either bond cleavage concerted with rearrangement, or the formation of a more structured intermediate which rearranges to the isomeric carbenes; formally the major products apparently arise from the more stable carbene (128) [110]. Gas phase kinetic analysis of the thermal rearrangement of 1-methylcyclopropene, however, indicates that all products arise via 1,2-shifts in a diradical-like intermediate, and that the methyl-group deactivates the ring to rearrangement [111].

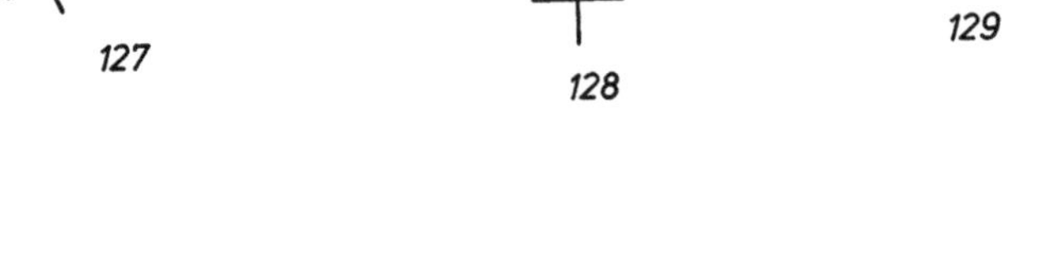

Clearly the use of these reactions in synthesis requires the formation of a single carbene, (125) or (126) from the cyclopropene, and also demands that other thermal reactions such as the ene-reaction or [2+ 2]-cycloaddition do not compete effectively. Although the above alkylcyclopropenes lead to mixtures of products, this is not the case with many other substituents. The regiochemistry of ring opening of aryl-substituted cyclopropenes has been particularly extensively examined both under thermal and photochemical conditions. The details of the photochemical reactions will not be examined here, but some cases which bear on the thermal processes will be discussed. The major product of photolysis of (130) is (131), derived by an overall insertion of the carbene (133) into an adjacent C—H bond of the aromatic ring [112-4]; in contrast, pyrolysis of (130) leads exclusively to (132), apparently derived from (134). Kinetic and product analysis of the thermal decomposition of vinyldiazomethanes (135), leading to 3H-pyrazoles and cyclopropenes, has been used to show that carbene (134) is more stable than (133) [108]. However, product ratios are complicated by return to the cyclopropene. Thus the optically active cyclopropene (136) racemises 2.5 times as fast as it is converted to product, the furan (137); the latter is apparently derived entirely from the singlet vinylcarbene (138), formed by cleavage of bond b, rather than its regioisomer [115].

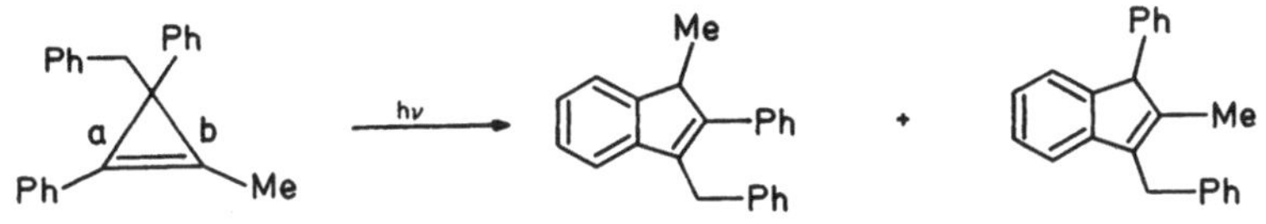

130 131 132

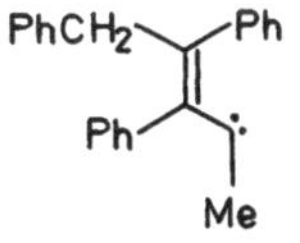

133

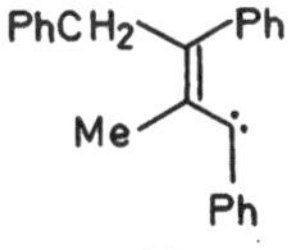

134

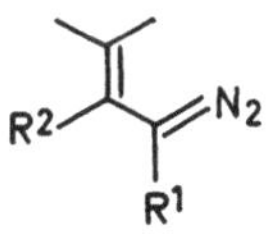

135

136 137

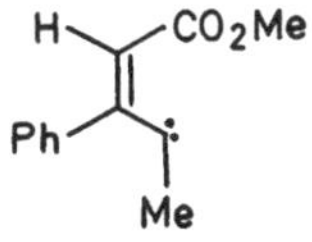

138

The diaryl substituted systems (139) have been shown to undergo regioselective cleavage of the cyclopropene σ-bond bearing the more electron donating substituent under direct photolysis [116].

Aryl-substituted cyclopropene esters and aldehydes have also been examined in detail. Photolysis of (140) in methanol leads to cleavage of the $C^1—C^3$ bond to produce (142); this undergoes a 1,2-hydrogen shift to produce the corresponding ketene which is in turn trapped by methanol to produce (143) [117]. The ester (141) is thermolysed to produce indene (144), kinetic analysis indicating a non-polar transition state and a vinylcarbene intermediate [118]; once again, photolysis of (146) leads instead to the cyclopropene (141) together with the isomeric indene (145) [118,119].

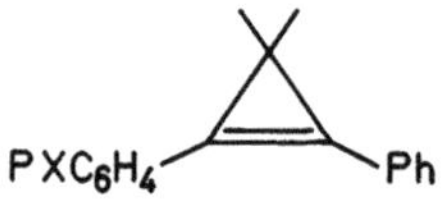

139

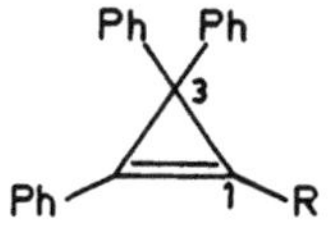

140: R=CHO
141: R=CO2Me

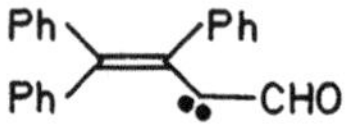

142

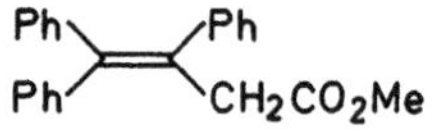

143

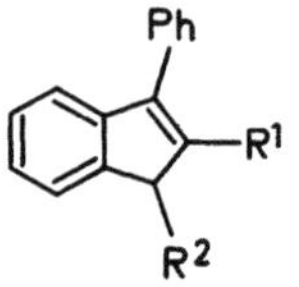

144: R^1=Ph, R^2=CO_2Me
145: R^1=CO_2Me, R^2=Ph

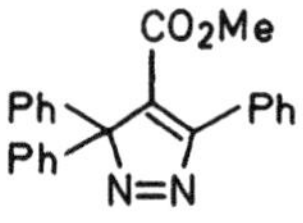

146

The cyclopropene — vinylcarbene rearrangement seems to occur particularly readily when there are geminal substituents at the 3-position. 3,3-Dimethylcyclopropene is reported to undergo thermal reaction in the presence of alkenes to produce adducts of (147, R = Me) [120], while the 3-methyl-3-phenyl compound ring opens at 180 °C to the vinylcarbene (147, R = Ph) which is trapped in low yield by 2,3-dimethylbut-2-ene. An additional product in the latter reaction is the indene (148), derived by a formal insertion of the corresponding (Z)-carbene into a C—H bond of the benzene ring [121]. Ring opening of tetrachlorocyclopropene to (149) occurs on heating to 180 °C, and the carbene is readily trapped by alkenes [122]; this reaction is described in more detail elsewhere in this series [123]. In other cases the ring opening occurs at much lower temperature. Photolysis of the pyrazole (150, R = Et) at —20 °C leads to an unstable intermediate cyclopropene (151) which can be characterised by formation of diasteroisomeric adducts (152) with diazopropane. However, if the cyclopropene is allowed to stand at 5 °C in the presence of a diene such as furan, the product is not a Diels-Alder adduct but instead is a cyclopropane derived by addition of (153, R = Et) to one double bond [124].

On standing for 18 h at 20 °C in the absence of an alkene, cyclopropene (151, R = Et) rearranges to the vinylsulphine (154), which can be trapped by cycloaddition to diazopropane; photolysis of (150, R = 4—MeC_6H_4) in a similar way leads to a high yield of (155), and an intermediate vinylcarbene (153, R = 4—MeC_6H_4) may be trapped by ethyl vinyl ether. In each case the intermediate vinyl carbene apparently

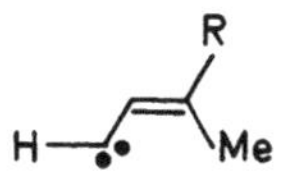

147

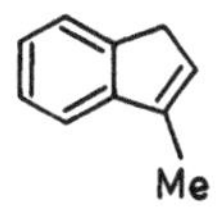

148

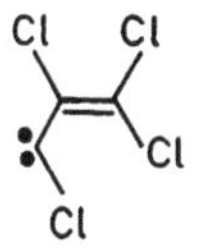

149

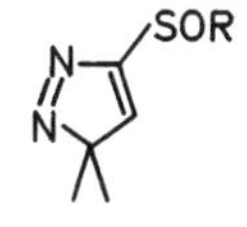

150

151

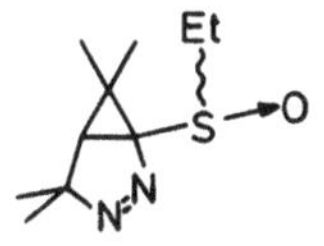

152

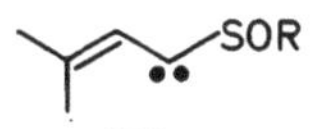

153

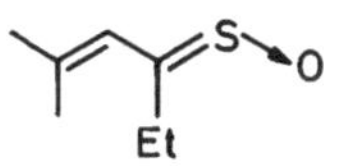

154

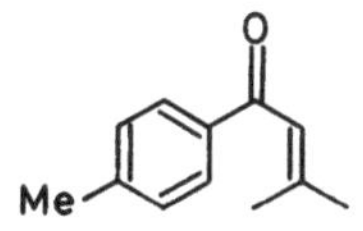

155

rearranges to the sulphine, which in the latter case loses sulphur to give the isolated product [125]. Photolysis of (156) does not lead to an isolable cyclopropene; the two major products, (157) and (158), appear to be derived instead from the vinylcarbene (159); indeed this can be intercepted surprisingly efficiently by addition to electron poor alkenes [126].

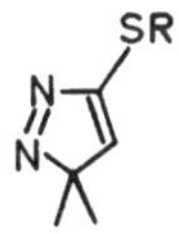

156

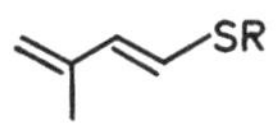

157

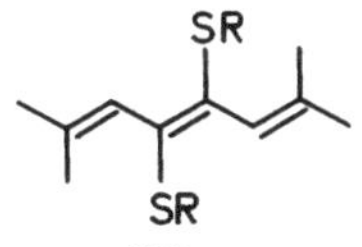

158

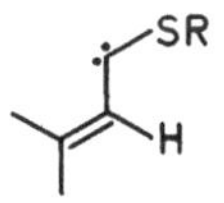

159

Moreover, trapping of the carbene (159, R = pTol) by an allyl thioether leads largely to (160); removal of the protecting group provides a simple route to artemesia ketone (161) in 65% yield from (156, R = pTol) [126]:

156
R=pTol

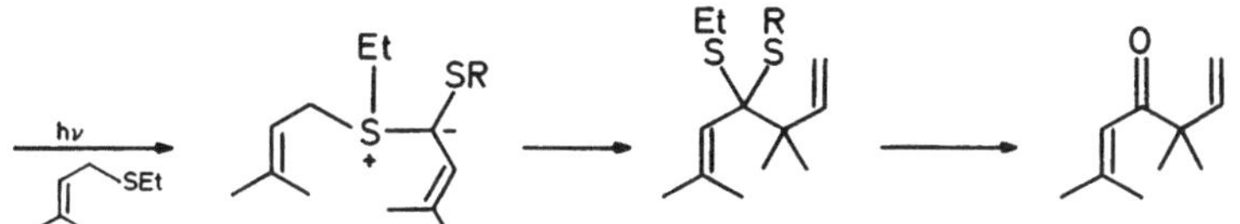

160 161

Addition of (159, R = Et) to methyl β,β-dimethylacrylate followed by Raney nickel desulphurisation also provides a convenient route to *cis*-chrysanthemic acids [127].

Although a 1-(alkylthio)cyclopropene could not be isolated in the above reactions, the corresponding silylated derivative (162) does ring open either on heating or on photolysis, leading to an allene; the reaction may involve a 1,2-silyl-shift in an intermediate carbene (163), though in this case the latter could not be trapped by added alkene [82].

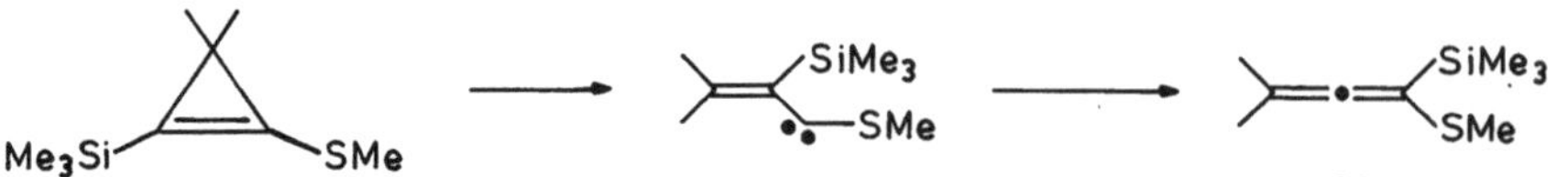

162 163

While the photolysis of (164) can lead to reasonable yields of acylcyclopropenes, no cyclopropenes are detected from the corresponding reactions of (165); however, if the photolysis of (165, R = Me, R′ = H) is carried out in the presence of furan, a 2-oxabicyclohexane is isolated. This is apparently derived by addition of the carbene (166) to the 2,3-bond. The oxabicycle in turn rearranges to a single triene (167) [128].

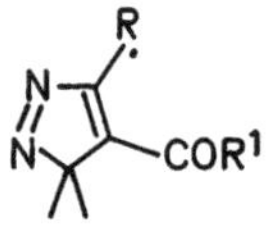

164

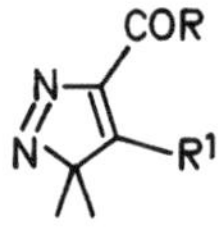

165

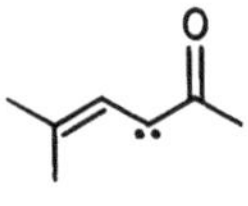

166

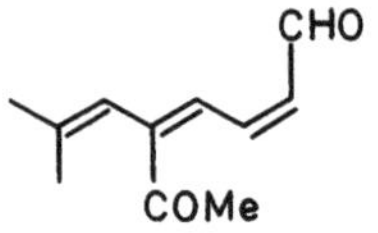

167

Photolysis of (168) in the presence of furan leads to a mixture of products, the major one being the Diels-Alder adduct of the latter with 1-nitro-3,3-dimethylcyclopropene. Two other products, (169) and (170) are apparently derived by trapping of the carbene (171) by furan, the former apparently by a 1,3-dipolar addition. The pyrazole (172) is converted to (173) on photolysis, the intermediate carbene again apparently preferring to rearrange rather than cyclise to a cyclopropene [129].

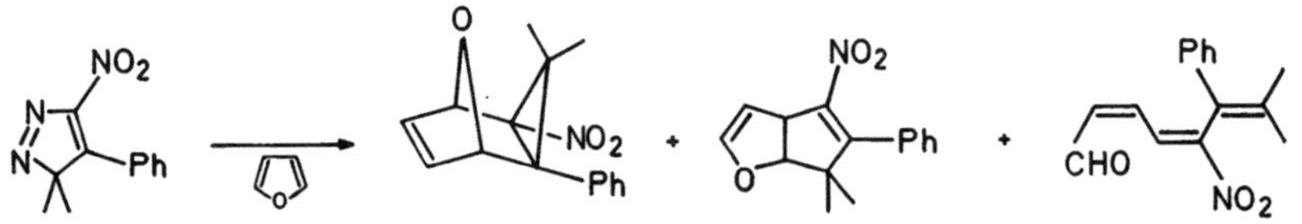

168 169 170

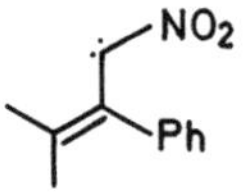

171

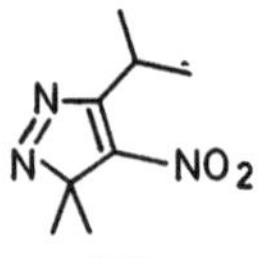

172

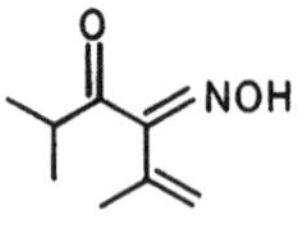

173

Reaction of the tetrachloride (174, X = Cl) with methyl lithium at 0–20 °C in the presence of alkenes leads to the adducts (176) derived by addition of the dichloro-vinylcarbene (177, X = Y = Cl); this is derived by rearrangement of the dichlorocyclopropene (175, X = Cl), which can be trapped by addition of bromine at lower temperature. The same products are obtained when (174, X = H) is treated with methyl lithium. Presumably a lithium-hydrogen exchange occurs more rapidly than lithium halogen exchange; loss of lithium chloride then leads to the same cyclopropene, (175, X = Cl). The bromochloride (175, X = Br), obtained in a similar manner from the 1,1-dibromo-2,2-dichlorocyclopropane, also ring opens, but little regioselectivity is observed in the formation of (177, X = Cl, Y = Br) and (177, X = Br, Y = Cl) [83,84]. The corresponding monohalocyclopropenes, eg. (175, X = Cl, Y = H) also rearrange at ambient temperature, in this case to produce haloalkynes [38]:

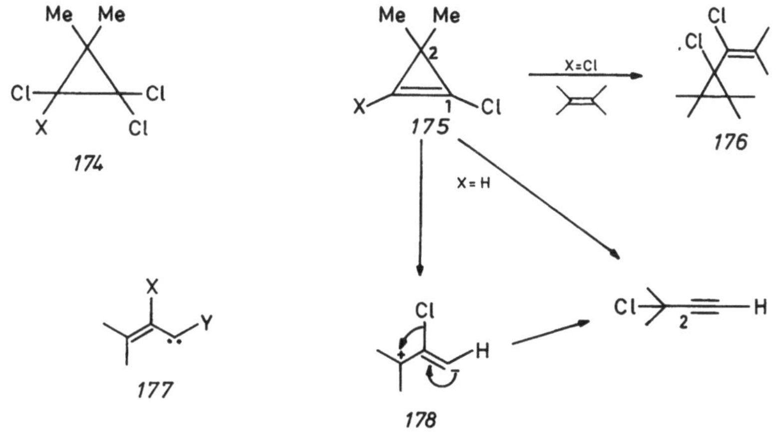

The reaction could be explained in terms of ring opening to a carbene (177, X = Cl, Y = H) followed by a 1,2-chlorine shift as in (178); although the carbene could not be trapped by added alkenes, a labelling study indicated that C^1 became C^2 of the alkyne, confirming that the C^2—C^3 bond of (175, X = H) is broken [39].

Reaction of (180, X = Cl) with methyl lithium in the presence of alkenes at ambient temperature leads to apparent carbene adducts (183), in this case derived from ring opening of (181) to the highly functionalised isoprenoid carbene (182). Surprisingly, the bromide (180, X = Br) reacts by a different course, leading to (184), apparently through initial lithium-bromine exchange followed by 1,3- rather than 1,2-elimination of LiCl [130].

The 3,3-disubstituted ester (185, R = H) also rearranges at or below ambient temperature. The major product is the triene (186), which at first sight appears to be a dimer of the carbene (187) derived by cleavage of the 2,3-bond. However, examination of the mother liquor reveals a second dimer, (188), which could be obtained by addition of (187) to the cyclopropene [105]; a similar dimer, and indeed related trienes, have been isolated in the photochemical reactions of (185, R = Ph), although in this case the C^1—C^3 bond is broken [131]. On standing at 0–20 °C, (188) is converted to (186). It is not certain that all (186) is derived in this way, or whether the *endo*-carbene adduct isomeric with (188) is produced and rearranges rapidly; either intermediate would avoid the need for the unlikely carbene-carbene dimerisation.

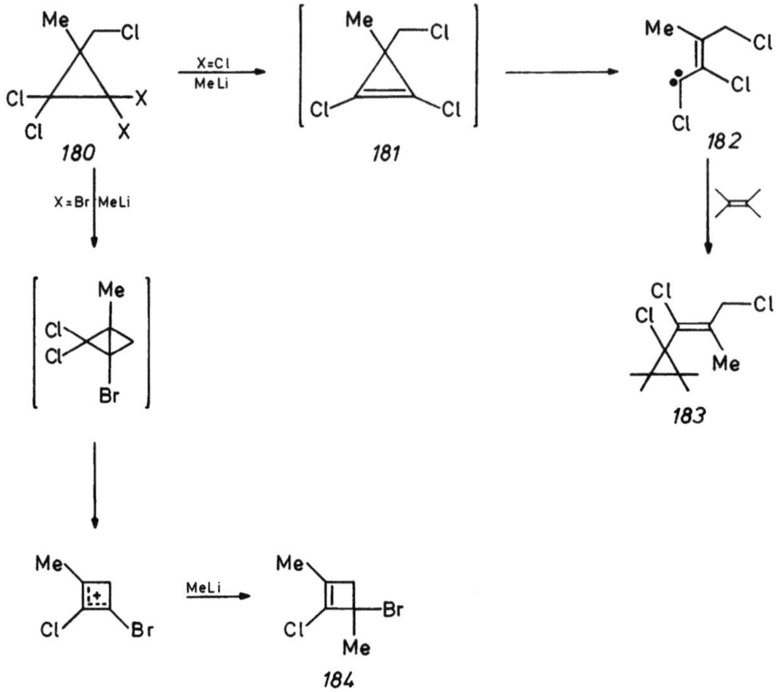

180 181 182 183 184

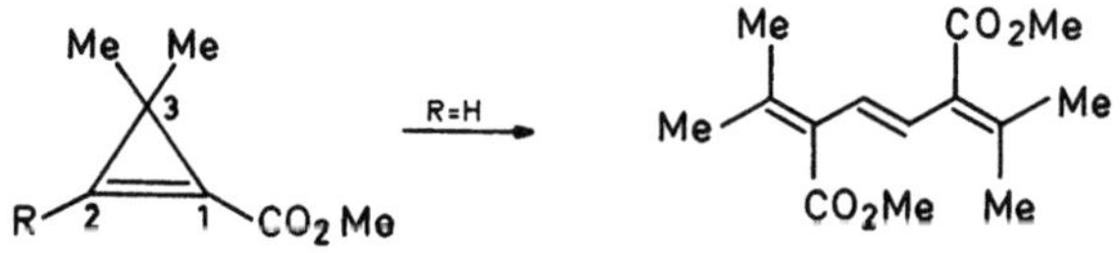

185 186

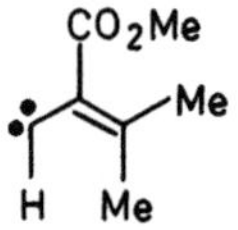

187

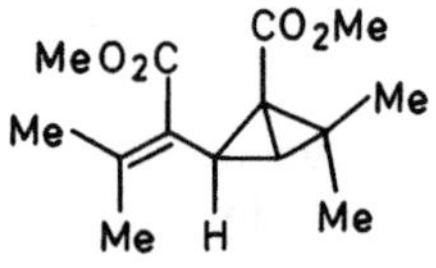

188

The thermolysis of cyclopropenone acetals at 70–80 °C, generally in benzene solution leads to a ring opening which is formally described by the scheme below:

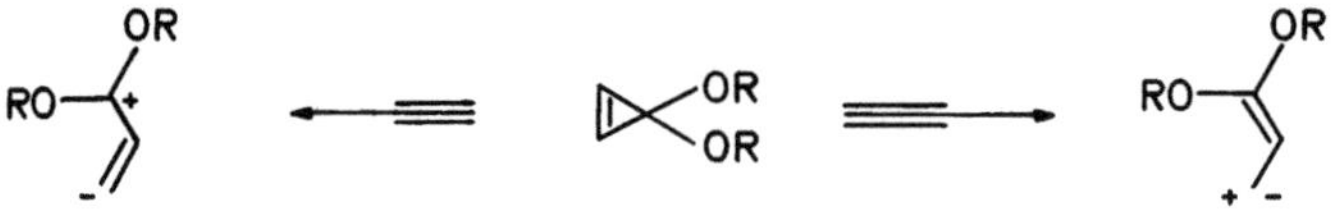

189

In the presence of an alkene having two electron withdrawing substituents at the 1-position, a [3+ 2]-cycloaddition is observed leading to cyclopentenone acetals (190)[132]:

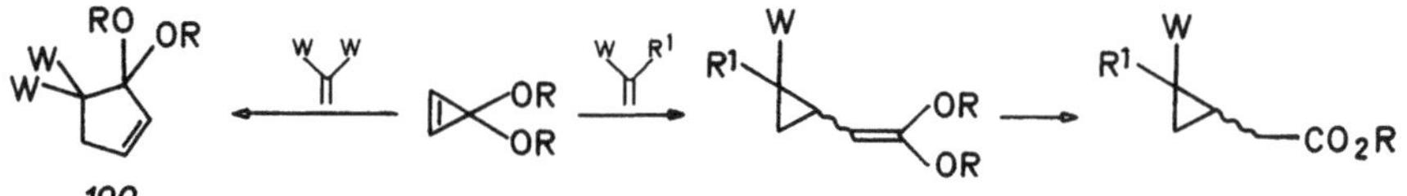

190 191

However, when the alkene has only one electron withdrawing substituent, a complete change in reactivity is observed and the cyclopropyl ketene acetal derivatives (191) are produced. These are converted directly to esters either by chromatography over silica gel or by treatment with acid, and in each case the predominant isomer of the ester has the *cis*-stereochemistry of ester and electron-withdrawing groups. It is suggested that a cyclopropane may also be the primary product in the formation of (190), but that the ring closure is reversible under the reaction conditions when two electron withdrawing groups are present, and the cyclopentene is formed under thermodynamic control[133]. The dipolar form (189) may also be trapped by reaction with aldehydes or ketones to produce butenolide orthoesters, which may further be transformed to butenolides or furans with acetic acid or to γ-ketoesters (R = H) with hydrochloric acid[134]:

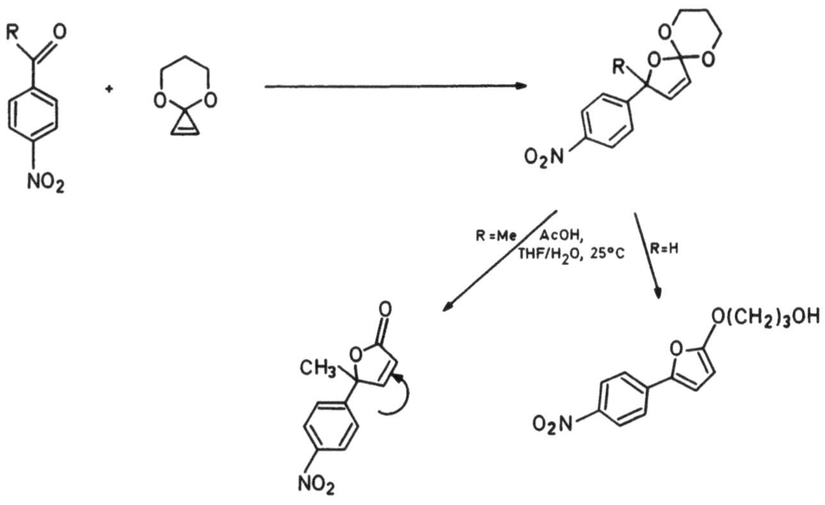

The ring opening of cyclopropenes may also be induced by metal salts or complxes. 3,3-Dimethylcyclopropene is converted to adducts (192) by treatment with $Ni(COD)_2$ in the presence of electron poor alkenes; with diethylmaleate the reaction proceeds with predominant retention of stereochemistry[4,135]. Other 3,3-disubstituted compounds are converted to adducts in good yield by reaction with $(EtO)_3P \cdot CuCl$ in the presence of alkenes at —40 to 20 °C. In the absence of a trap, a triene (193) is isolated[103]:

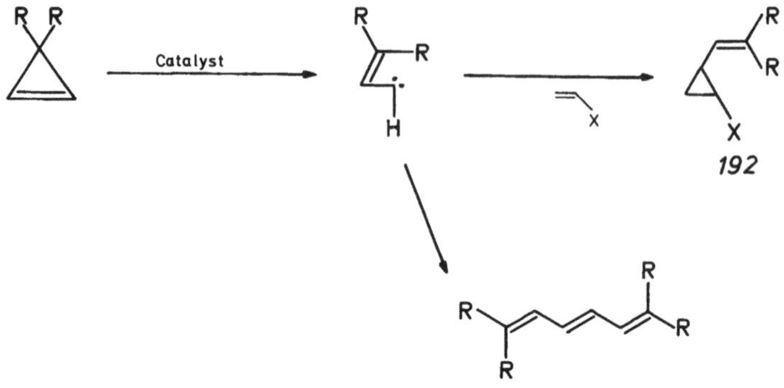

192

193

This is explained in terms of dimerisation of the carbene, or a related carbenoid. However, formation and rearrangement of a bicyclo(1.1.0)butane related to (188) must also be considered.

Ring opening of 1,3- and 1,2-disubstituted cyclopropenes has also been examined. The ester (194) rearranges on heating to 98 °C in the presence of copper to give furan (195); the less substituted cyclopropene single bond appears to be cleaved to produce a carbene-metal derivative, which cyclises to the ester group [136]. A similar photochemical transformation of a cyclopropene-3-ester to a furan has already been described [115].

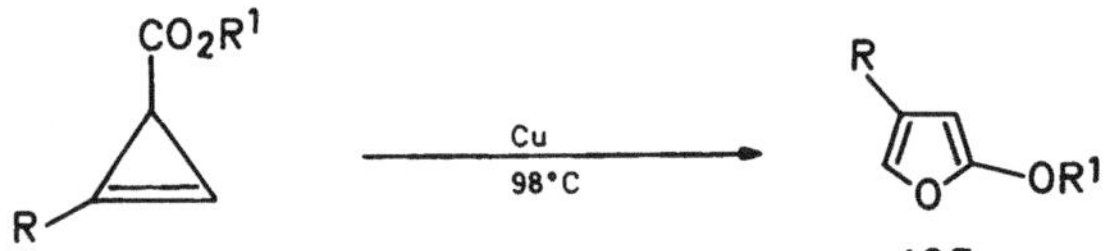

194 195

$XCH{=}CHRCH_2CO_2R^1$

196

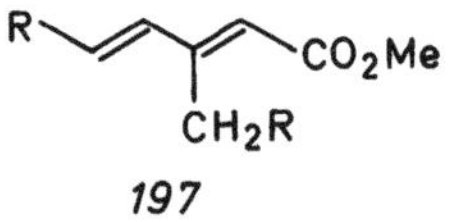

197

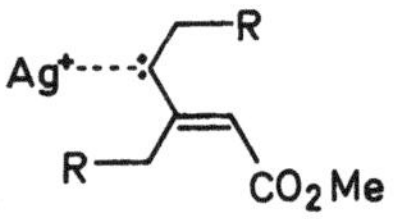

198

Reaction of (194, R = Pr) with cuprous chloride induces a similar ring opening and the carbenoid can be trapped by alkenes [137]. If the reaction is carried out with an organic or inorganic acid HX present in place of the alkene, reasonably high yields of the (E)- and (Z)-alkenes (196) are isolated; no rearrangement occurs in the absence of the copper chloride [138]. Silver ion induced ring opening can also occur, 1,2-dialkyl-3-carbomethoxycyclopropenes being converted to dienes (197) in reasonable yield, apparently through the silver-carbenoid (198) [139a]. In other cases intramolecular trapping may occur [139b]:

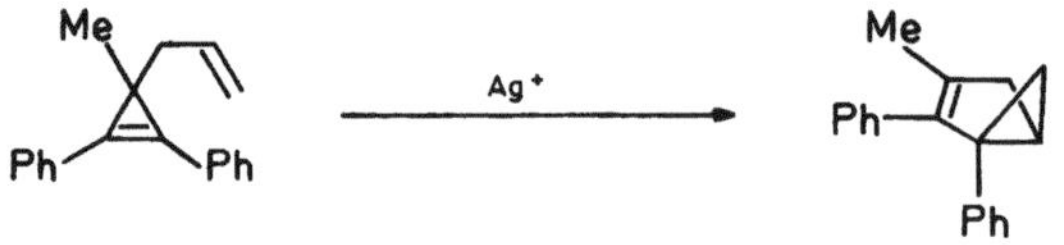

These reactions are discussed again in Section 5i.

Rearrangement of cyclopropenes to vinylcarbenes can also occur at low temperature when they are fused to small or medium rings. Treatment of (199, X = Z = H, Y = Cl) with base in THF leads to the ether (200) which can be explained in terms of trapping of an intermediate carbene, (201), derived either by rearrangement of a cyclopropene or by a direct fragmentation. Evidence for the former route was obtained by addition of methane thiol to the reaction, when (199, X = Z = SMe, Y = H) was isolated; the regiochemistry of the addition of the second thiolate to the (presumed) intermediate cyclopropene (202) presumably occurs because the alternative benzylic carbanion cannot easily become planar [140, 141]. Evidence for the initial cyclopropene formation is obtained from the reaction of (199, X = H, Z = Me, Y = Cl) with base in the presence of thiolate anion, which leads to a single diastereoisomer of (203) in high yield. A most interesting observation was that the monochloride (204) failed to react with potassium t-butoxide even after extended times, presumably because the elimination in these systems requires a syn-planar arrangement of leaving groups [141]; however, a later report has shown that (204) does react with base in THF-DMSO, although only low yields of product were isolated [142].

These reactions are discussed again in Section 51.

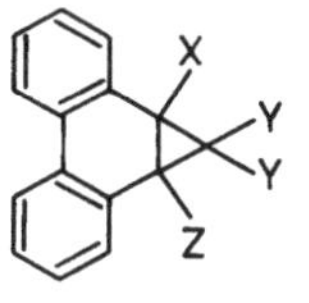

199

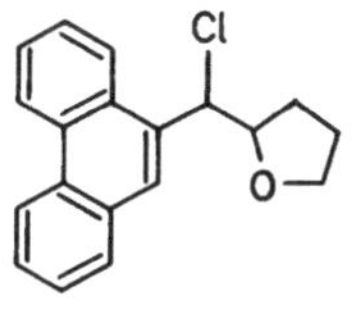

200

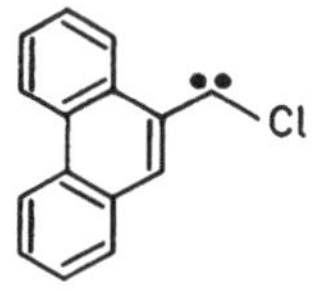

201

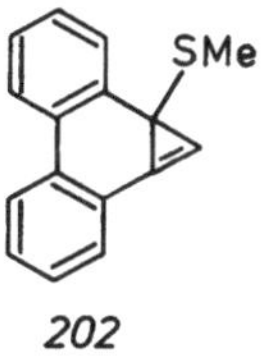

202

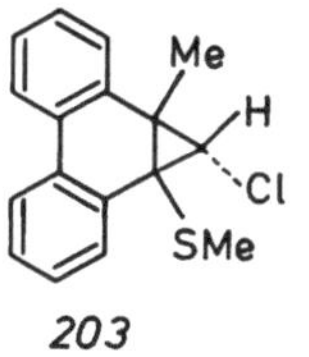

203

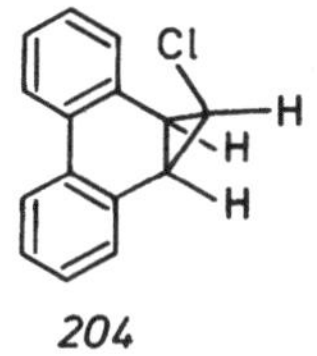

204

Treatment of (205) with a large excess of potassium t-butoxide in THF leads to ethers and related compounds; their formation may be rationalised in terms of the formation of a cyclopropene (206) which can undergo a complex set of rearrangements to naphthylcarbenes which are then trapped by alkoxide ion [141, 143].

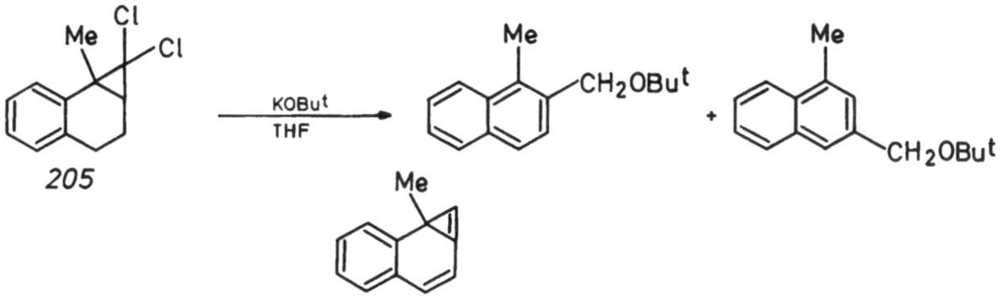

205

206

Although these reactions are of considerable mechanistic interest, the formation of mixtures limits any synthetic application. However the cyclopropene (207) ring opens to (208) which is trapped reasonably efficiently by furan [38]:

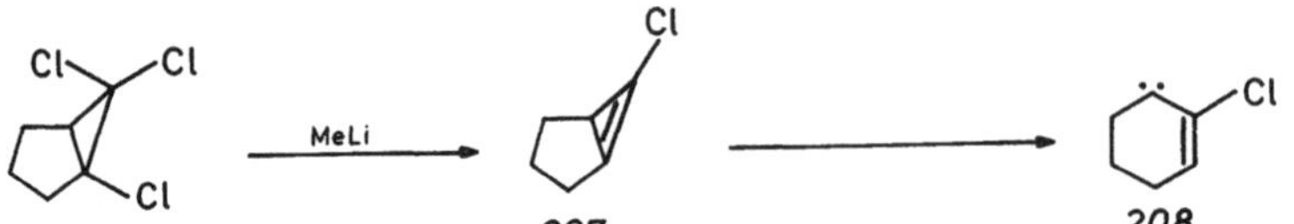

207 208

4 Base Induced Double Bond Migration

Reaction of cyclopropenes with bases such as alkoxide or amide ions often leads to a methylenecyclopropene by removal of an allylic hydrogen and reprotonation [6–9,71] though other reactions such as nucleophilic addition (see Section 5) or metallation at a vinylic position (see Section 2) may compete. Thus the ester (209) is isomerised by KOH to (210), and under more vigorous conditions to (211) [144]:

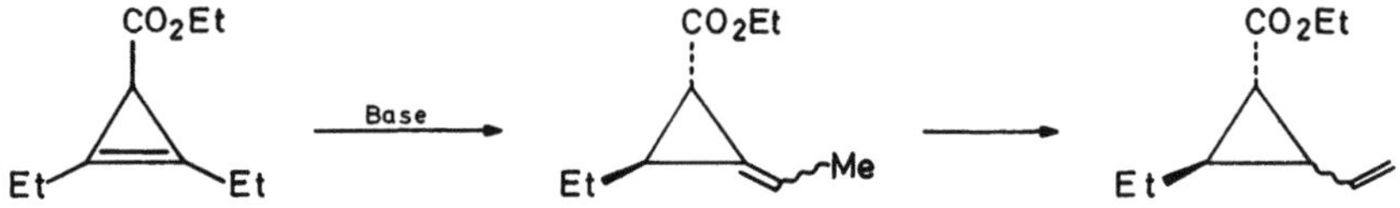

209 210 211

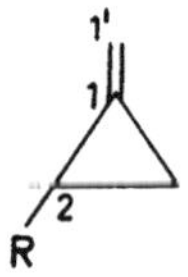

212

The presumed intermediate allylic ions of type (212, R = H) may also be generated from methylenecyclopropane by reaction with eg., butyl lithium; trapping by carbonyl compounds occurs by bond formation from C^2 [145], although when R = $SiMe_3$ trapping by benzaldehyde occurs only at $C^{1'}$, and probably involves an electron transfer process [146].

Many examples of double bond migration appear in the reactions of dihalocyclopropanes with alkoxide ion, in which the halocyclopropene is a presumed intermediate. Thus dehydrohalogenation of 2,3-dialkyl-1,1-dichlorocyclopropanes provides a very simple route to methylenecyclopropanes, which often rearrange on heating, eg. [147]:

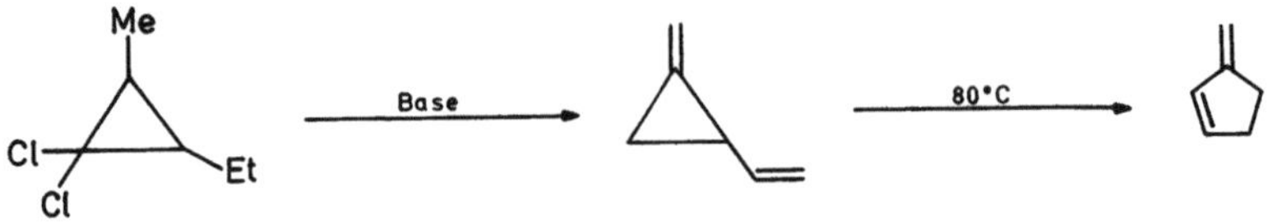

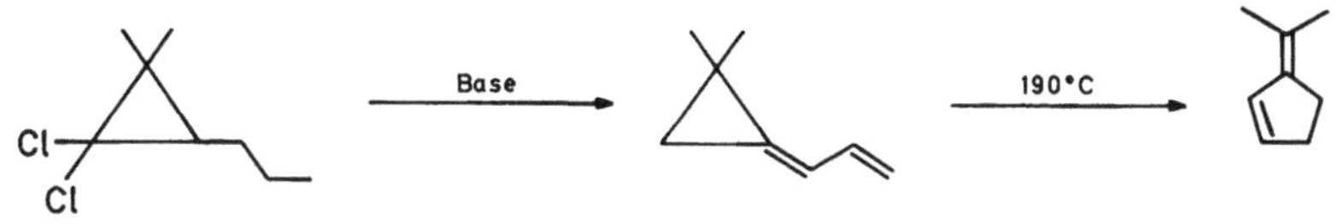

The reaction of the dichlorides (213, n = 3–7) with potassium t-butoxide in DMSO leads to the elimination of two molecules of HCl and the formation of (214, n = 3–7) respectively. In each case the reaction can be explained by a sequence of elimination to a chlorocyclopropene, prototropic shifts, then a second elimination to a cyclopropene followed by prototropic shifts. Thermolysis of (214, n = 3 or 4) leads to (215, n = 3 or 4), although the mechanism of this reaction is rather uncertain; in contrast the larger ring species (214, n = 6 or 7) rearrange to a methylenecyclopentene, (216) [148,149]. In the case of (213, n = 2), the analogous elimination product (214, n = 2) is not observed, presumably rearranging under the reaction conditions to the observed product (217) [149].

213 214

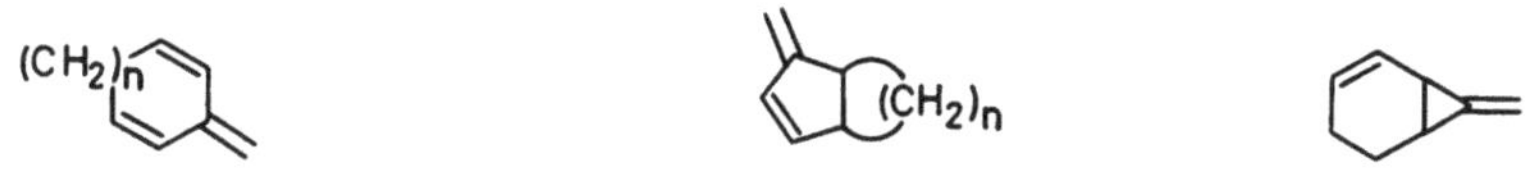

215 216 217

Moreover, when the ring size is reduced further, very complex product mixtures are obtained. Thus (213, n = 1) reacts with potassium t-butoxide in DMSO to produce toluene, cycloheptatriene, ethylbenzene, 2-ethyltoluene, and isomeric 3-ethylidenecyclohexenes [150]. In the same way 7,7-dibromobicyclo[4.1.0]heptane is converted largely to ethylbenzene and 2-ethyltoluene; when deuterated DMSO is used, the methyl-group of the ethylbenzene is almost completely deuterated, and methylene and aromatic positions are also heavily labelled. It is thought that the highly strained intermediate (218) reacts with the solvent as below, further reactions then leading to the eventual products:

218

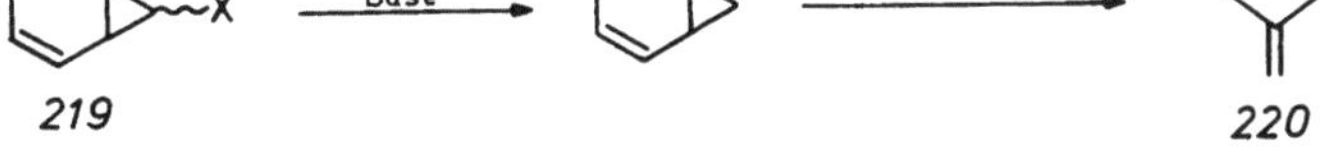

219 220

Although no products analogous to (214) were isolated from these reactions, treatment of (219) with KOBu^t-DMSO did lead to (220), albeit in very low yield [150]. It is important to note, however, that the reactions of chlorobicyclo[4.1.0]heptanes with base are highly dependent on both the nature of the base and the solvent [151]. Thus (213, n = 1) reacts with KOBu^t in benzene in the presence of a crown ether to produce (221) in moderate yield, while the addition of a small amount of DMSO leads to the products described above. Moreover, reaction with Pr^{i}OK leads to products such as (222), apparently derived by a reduction at some stage [151].

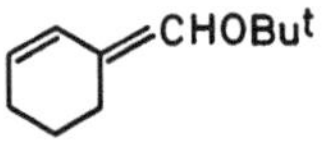

221

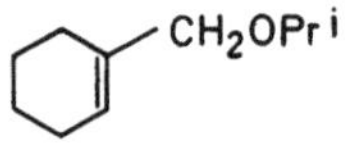

222

The use of these reactions in the preparation of highly strained molecules is illustrated by the reaction of (223) with potassium t-butoxide at low temperature:

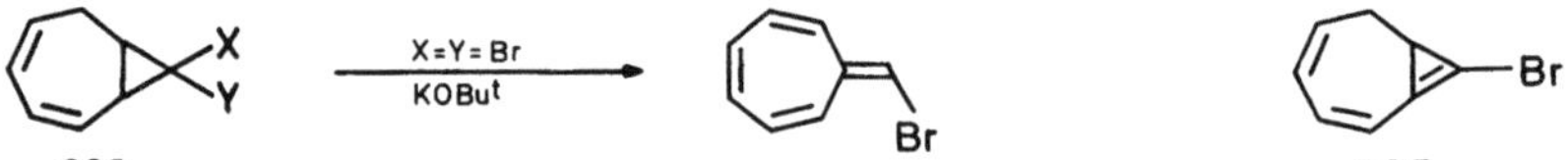

223 224 225

The intermediate bicyclo[5.1.0]octatriene (225) may be trapped as a Diels-Alder adduct, while at elevated reaction temperature, the heptafulvene (224) may likewise be trapped. The chloride (223, X = Cl, Y = H) may similarly be converted to the parent heptafulvene by reaction with the same base in tetraglyme at 90 °C and low pressure [152].

Double dehydrohalogenation of 7,7-dichlorobicyclo[4.1.0]heptenes represented one of the first routes to benzocyclopropenes [2,153]. In a classical experiment, the labelled species (226) was shown to lead to (227), in agreement with the occurrence of a sequence involving 1,2-elimination, followed by prototropic shifts [154]:

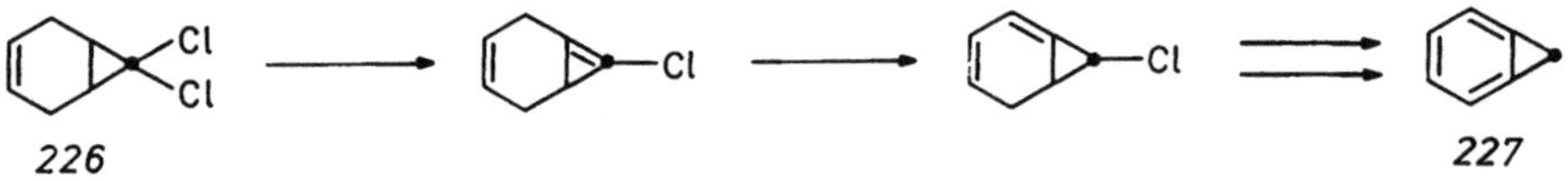

226 227

A number of competing pathways have also been identified [155,156]. Indeed the dehydrohalogenation of (228, X = Br, Cl) is reported to lead to (229). A mechanism is proposed which does not in this case involve a cyclopropene, but instead is initiated by an elimination with rearrangement [157]:

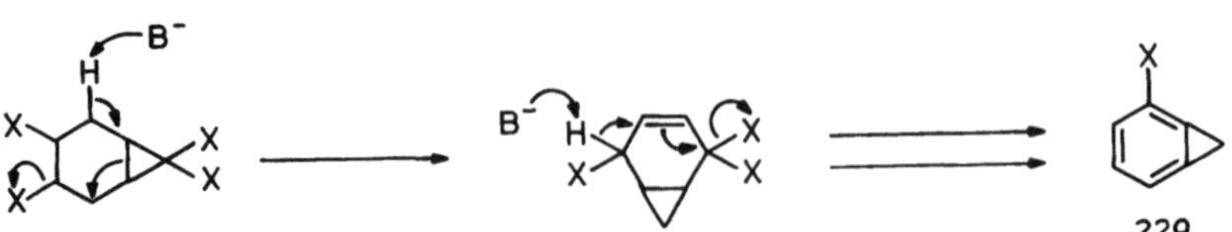

228 229

Similar routes have been described to naphtho(b)cyclopropene [158], and to annelated derivatives [159].

5 Addition

The double bond of cyclopropenes is sufficiently reactive that in many cases attack of either electrophiles or nucleophiles can occur. 1,2-Addition to cyclopropenes can create up to three new chiral centres:

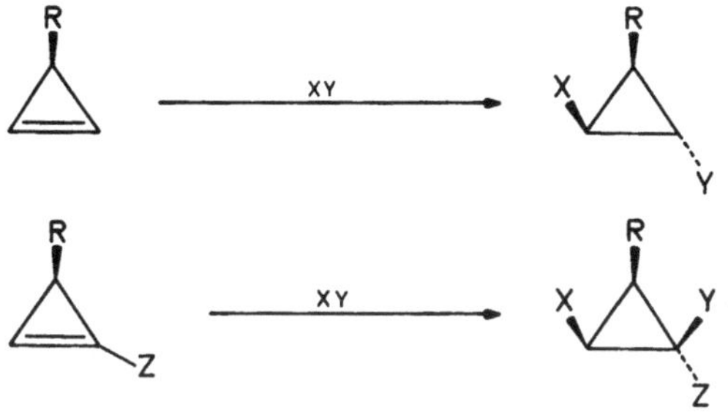

In principle the addition could be controlled by a chiral centre at C^3 or by chiral auxiliaries in the C^1 or C^2 substituents or the addend. This would provide a versatile route to optically active cyclopropanes, but to date no emphasis has been laid on this possibility. Instead, a wide range of additions leading to racemic cyclopropanes has been examined.

5.1 Electrophilic Addition

The addition of an electrophile to the cyclopropene double bond formally leads to a cyclopropyl cation; this may be expected to undergo ring opening to an allyl ion unless it is rapidly trapped by a nucleophile. In some cases, however, electrophilic attack may occur at one of the σ-bonds, leading directly to an allylic cation.

Addition of halogens often occurs without ring opening. Early reports described a *cis*-addition [1]; though the stereochemistry of addition of chlorine to (229) is *cis*- [160], addition of bromine is *cis*- in non-polar solvents in the presence of sunlight but *trans*-in relatively polar ones [161], and addition of bromine to 1,2-dichloro-3,3-dimethylcyclopropene leads to a mixture of (E)- and (Z)-dibromodichlorides [83]. 1,3,3-Trimethylcyclopropene is, however, reported to ring open to 1,3-dibromo-2,3-dimethylbutene on bromination, though 1,2-addition of chlorine occurs with $PhICl_2$ [162]. In general little use has been made of these reactions, although the ring-opening of alcohol (230) on treatment with bromine and intramolecular trapping provides a route to optically active dihydropyrans [130]:

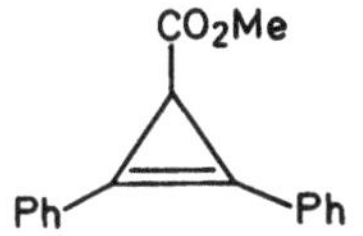

229

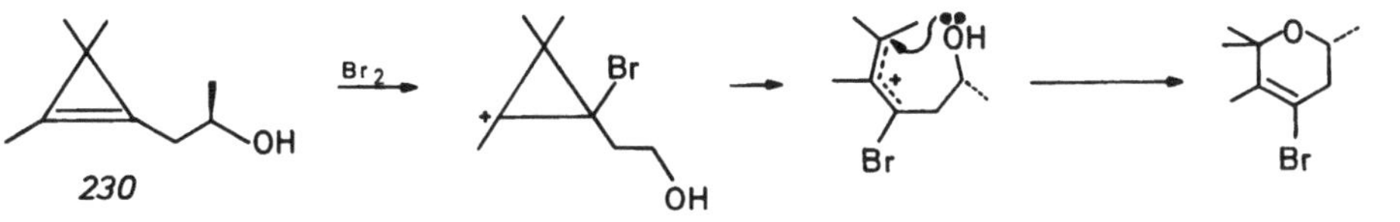

230

1-Methylcyclopropene undergoes ring opening to methallyl chloride and methallyl acetate on reaction with HCl in acetic acid; with phenylsulphenylchloride addition occurs without ring opening, though with low selectivity, producing (231) and its regioisomer [163]. The *trans*-addition of arylsulphenylchlorides to 3,3-dimethylcyclopropene shows a large negative entropy of activation, and is believed to involve a tight ion pair, (232) [164]. In the case of the 1-chloro-2,3,3-trimethyl analogue the product is (233), apparently derived by deprotonation of the intermediate episulphonium ion; the cyclopropane rearranges on standing over neutral alumina, leading to diene (234) [165].

The addition of thiocyanogen to 1-methylcyclopropene is complicated, though the *cis*-adducts (235, X = SCN) and (235, X = NCS) can be isolated [163]. Addition of nitrosyl chloride is also reported, leading to (236) from (229) [166].

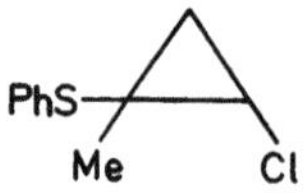

231

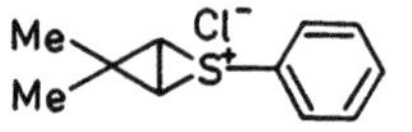

232

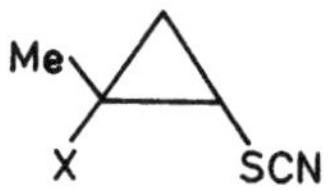

235

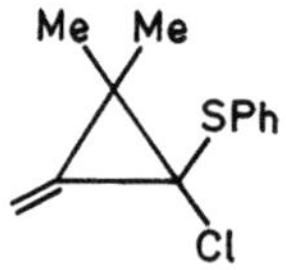

233

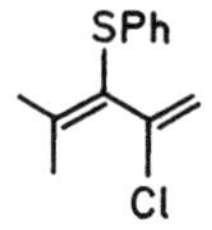

234

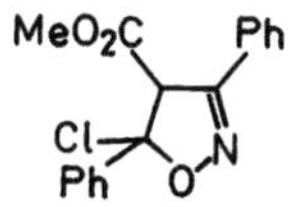

236

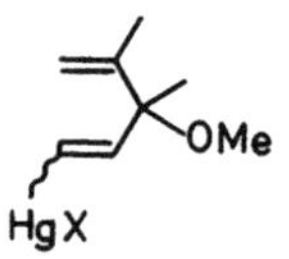

237

Oxymercuration of cyclopropenes can also occur without ring opening, the major product from (229) having (E)-stereochemistry [167]. However, in other cases ring-opening does occur, eg., 3-methyl-3-isopropenylcyclopropene is converted to (237) with mercuric acetate in methanol, presumably by solvolysis of an intermediate ring opened allyl cation [168]. One again, intramolecular trapping of intermediate allyl cations can lead to cyclisation, eg., to furans [130]:

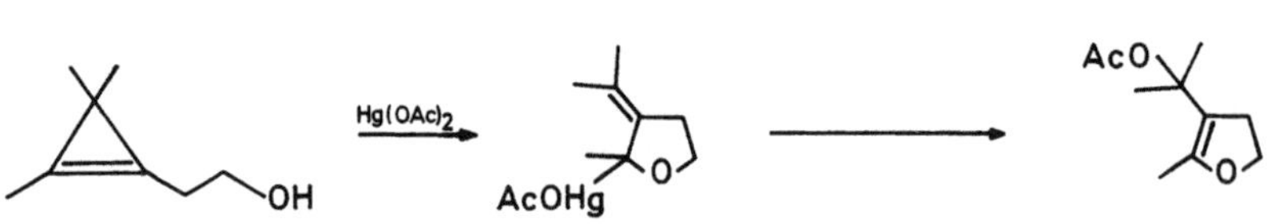

the intermediate organomercury compound apparently undergoing solvolysis. Reaction of 1-methylcyclopropene with diborane in pentane leads predominantly to the 2-methyl-substituted borane, although a small amount of the 1-methyl system is also formed [169]. Reaction of either 1- or 3-methylcyclopropene with tetraethyldiborane, however, proceeds cleanly to (238), which can be converted to essentialy pure *trans*-2-methylcyclopropanol by oxidation with trimethylamine-N-oxide [9,169]. Triallylboranes such as (239) also lead to *cis*-addition with allylic rearrangement, in good yield [170].

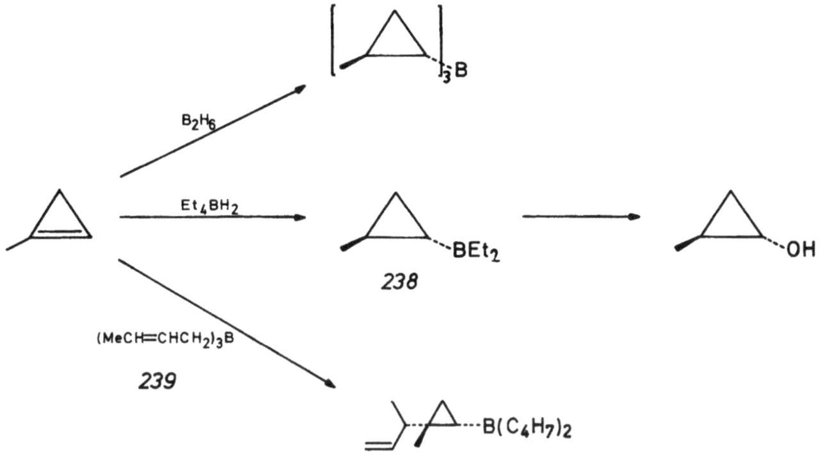

238

239

3,3-Dimethylcyclopropene undergoes [2+ 2]-cyclodimerisation catalysed by reagents such as boron trifluoride or triethylaluminium etherates, but treatment with triethylaluminium in pentane leads to (241), presumably by formation of an intermediate such as (240) followed by trapping with a second equivalent of cyclopropene [171]. Reaction of 1,3,3-trimethylclopropene with tri-isobutylaluminium in hexane also leads to ring opening:

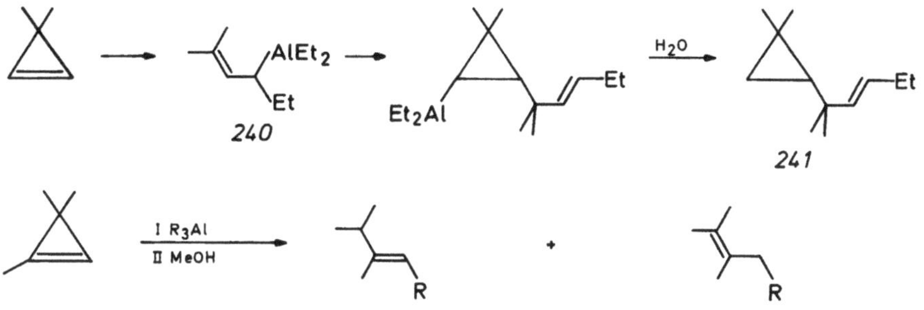

240

241

242

243

The process is believed to occur by addition to form the more stable cation, (242), which undergoes a cyclopropyl-allyl rearrangement followed by a 1,2-shift of an R-group to produce an allyl-substituted dialkyl aluminium, (243), as in the above case. This is trapped by added water, or undergoes an allylic rearrangement before being trapped [172].

Silver-ion induced isomerisation of cyclopropenes gives products which contrast sharply with those derived by photolysis. Thus reaction of (244) with catalytic quantities of silver perchlorate in benzene leads to a quantitative yield of (245), whereas the photochemical process leads to the 1-methyl-2-phenyl regioisomer. The alcohol (246) is converted to (247) in good yield by reaction with Ag^+ while (248) leads to the bicyclo[3.1.0]hexene — although other 3-allylcyclopropenes give more complex product mixtures including bicyclohexenes.

The final bicyclic product is obtained with complete retention of the stereochemistry about the double bond of the allyl group, and a mechanism is proposed in which the silver ion attacks one of the ring σ-bonds to produce the more stable cation [173,174]:

5.2 Hydrogenation

The hydrogenation of cyclopropenecarboxylic acids leads to good yields of the saturated acid. Two elegant routes to *cis*-chrysanthemic acids have been reported based on this process. In the first, the ester (249) is reduced by either diimide or nickel boride and hydrogen [49]:

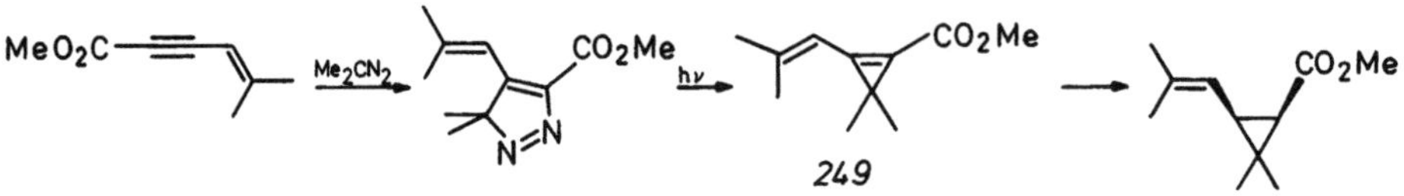

In the second, the use of diazopropane is avoided; the alcohol (250) is converted to the pyrazole (251) by a two step sequence of reaction with hydrazine in acetic acid and then oxidation with manganese dioxide. Photolysis then leads cleanly to the cyclopropene, without any interference by cyclisation of an intermediate carbene to the alcohol group; the product is hydrogenated directly, the conversion of (251) to (252) occurring in 94% yield. Elimination of the elements of water leads to *cis*-chrysanthemic acid in good overall yield [175]:

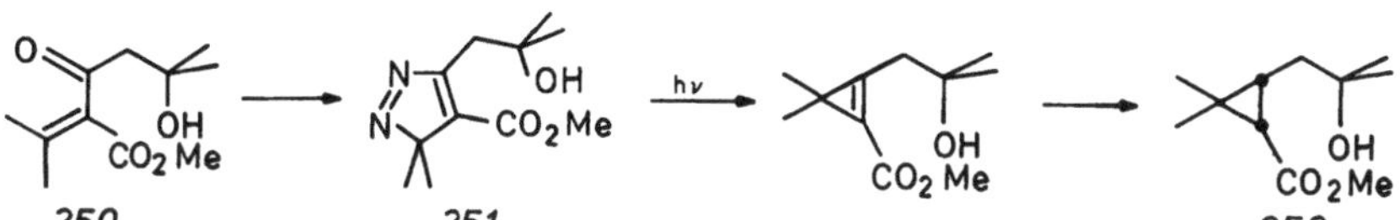

250 251 252

The reaction can also be used to produce *cis*-2-substituted halocyclopropanes, 1-chloro-2-phenylcyclopropenes undergoing efficient hydrogenation using palladium on calcium carbonate by *cis*-addition of hydrogen with no evidence for competing hydrogenolysis of the C—Cl bond [27].

5.3 Hydride Reducing Agents

Lithium aluminium hydride reduction of 1,2-disubstituted cyclopropene-3-carboxylates occurs initially at the ester group, but with additional reagent good yields of *cis*-1,2-disubstituted-*trans*-3-methanols are obtained [176]. The reduction of the double bond is regioselective, leading to the more stable carbanion, and the attack of the hydride ion exclusively *cis* to the 3-substituent may be explained in terms of initial formation of an alkoxyaluminium complex followed by intramolecular hydride transfer [176]. In the case of cyclopropene-1-carboxylates, direct reduction to the saturated alcohol occurs; thus (253) is converted to the *trans*-alcohol [177]:

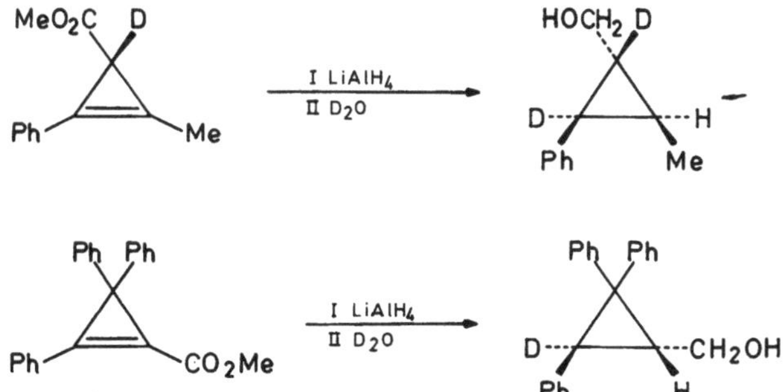

253

5.4 Nucleophilic Attack

Alkoxides and thiolates

Although the addition of alkoxides and thiolates and indeed of methanol itself to simple cyclopropenes has been reported, eg. [178]:

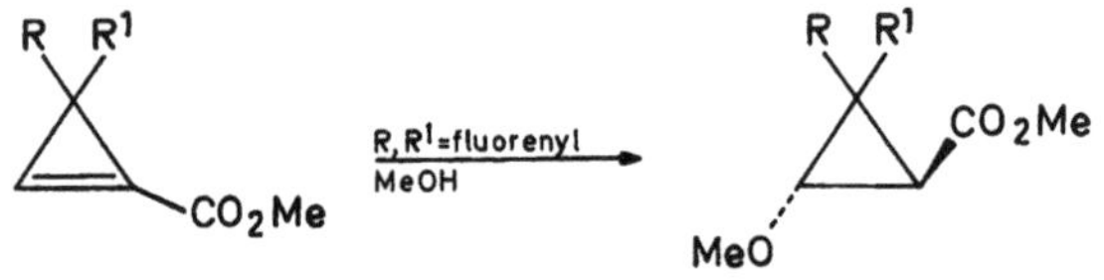

the most common examples of addition to cyclopropenes are reactions in which dihalocyclopropanes are treated with strong base in the presence of a good nucleophile. Thus reaction of (254) with potassium isopropoxide in DMSO leads predominantly to the diether (255). Apparently an initially formed 7-chlorobicyclo[4.1.0]hept-6-ene is trapped by attack of alkoxide at C^6, before a prototropic shift (see Section 4) can occur. The derived anion is protonated and then undergoes a second dehydrohalogenation — addition sequence. The regiochemistry of the second addition is presumably controlled by the stability of the intermediate anion [179]. When the reaction is carried out with potassium isopropoxide in [D_6]-DMSO the product diether is doubly deuterated at C^7 [180].

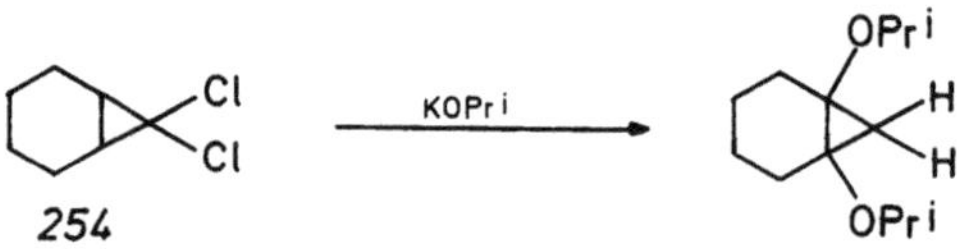

254 255

In other cases the product is an unsaturated ether [179]:

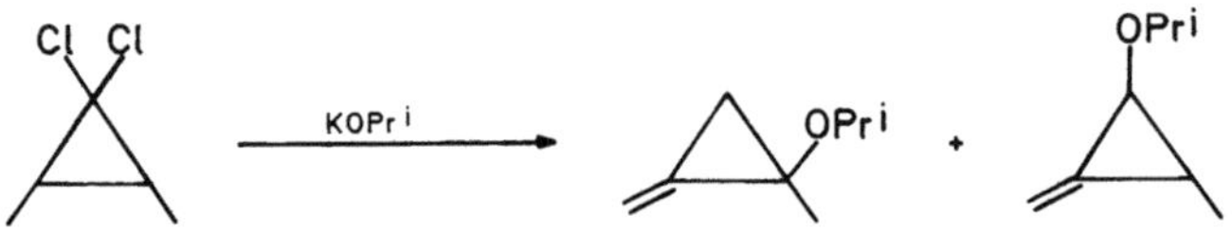

but the exact sequence of steps is not clear. Good evidence for the formation of cyclopropenes in these reactions is seen in the conversion of (256) to (258), the observation of the intermediate (257) by n.m.r. and its trapping with cyclopentadiene at low temperature [181]. The presumed chlorocyclopropene intermediate from the reaction of 1,1-dichloro-2-methylenecyclopropane with potassium t-butoxide in THF (evidence for which is obtained by trapping with thiolate ion) is converted into the enyne (259) in moderate yield. In this case attack of butoxide at C-2 of the 1-chloroalkene followed by loss of chloride ion may produce (260) which is then rearranged to the acetylene [182].

The regiochemistry of addition of nucleophiles to cyclopropenes contained in a bicyclic ester skeleton has been examined in some detail. Generation of (261, R = Me, n = 1) in the presence of t-butoxide leads to (262, R = Me) by attack of the nucleophile at the ring junction, whereas in the cases of (261, R = Me, n = 2 or 3)

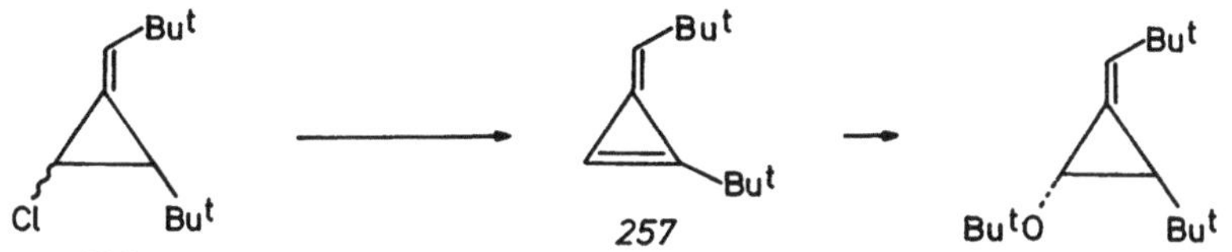
256 257 258

259 260

attack occurs at the benzylic position leading to (263, R = Me). Calculations suggest that the regiochemistry is steric in origin. The larger ring cyclopropene (261, R = Me, n = 3) can in fact isolated [183]. Reaction of (261, R = H, n = 2), generated in the same way, with t-butoxide leads to attack at both ends of the alkene, whereas (261, R = H, n = 3) only produces the 8-ether (263, R = H, n = 3) [184].

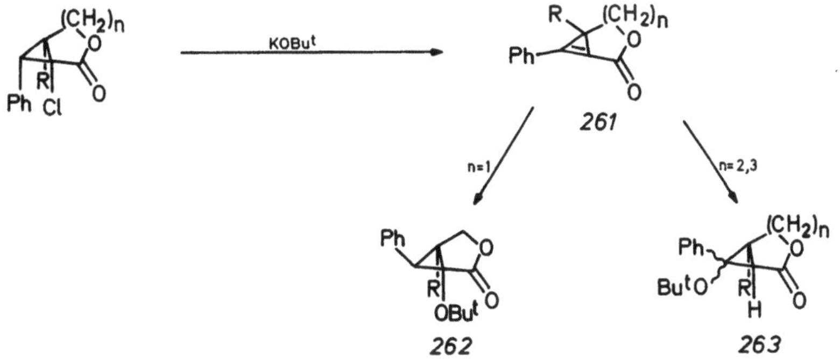
261 262 263

Addition of alkoxides to (264, R = $SiMe_3$) apparently occurs by attack on silicon to produce (264, R = H) followed by addition to the alkene. However, attack by thiolates occurs initially on the double bond to produce (E)- and (Z)-isomers of (265); subsequent attack on silicon leads to anions which invert rapidly owing to the presence of the adjacent sulphone and lead to the same (E)-cyclopropane derivative [185].

264 265

An interesting example of nucleophilic addition occurs in the reaction of (266) with methanol in the presence of triethylamine, which leads to (267) (or possibly the corresponding isomer with the methoxy group *endo*) [186]:

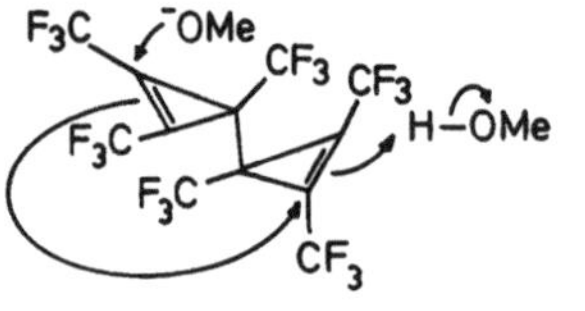

266

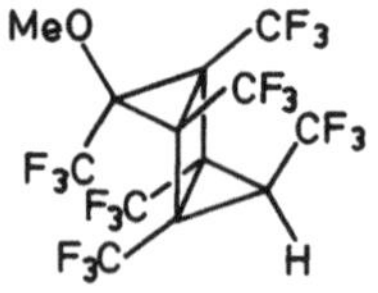

267

Nitrogen and Phosphorus Nucleophiles

Early reports of the reactions of cyclopropenes with amide ions indicated complex products derived by addition of an intermediate 2-aminocyclopropylanion to unreacted cyclopropene [1]. However, amines themselves can add to cyclopropenes:

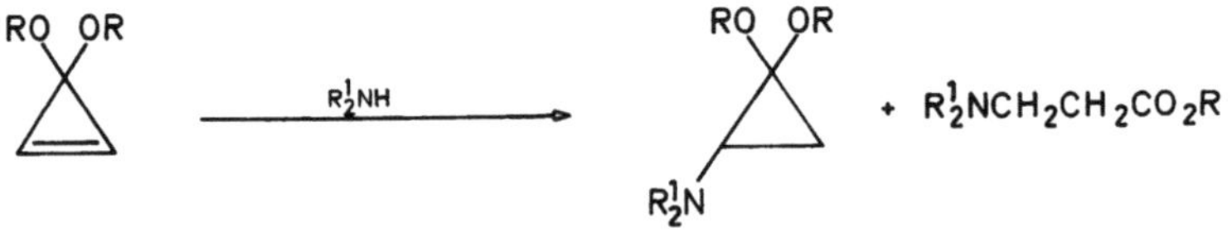

With diethylamine, the cyclopropane is the major product but with dipropyl or diphenylamines increasing amounts of ring-opened product are isolated [187 a)]. Reaction of 3-acetyl-1-methylcyclopropene with butylamine is also reported, leading to a mixture of 2,4- and 2,5-dimethyl-N-butylpyrroles, though the mechanism of this process has not been examined in detail [187 b)].

Treatment of tetrachlorocyclopropene with pyridine leads to the two indolizines (268) and (269) in essentially quantitative overall yield:

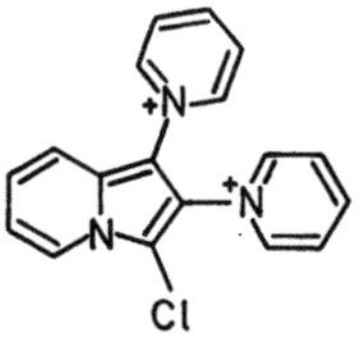

268

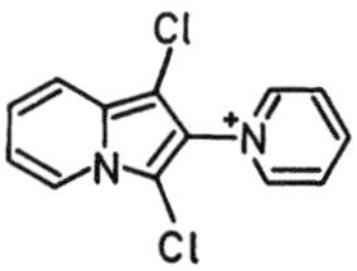

269

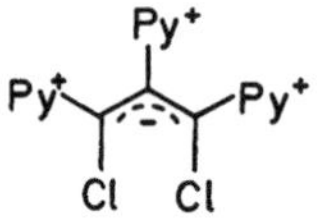

270

The reaction apparently proceeds though displacement of two of the halogens by pyridine, followed by nucleophilic attack at the double bond and opening of the resulting cyclopropylanion to (270); evidence for the intermediacy of such ions was obtained when the 2- and 6-positions of the pyridine were blocked. With pyridines bearing electron withdrawing groups the reaction stops after diplacement of one of the 3-chlorines [188)], while with electron releasing substituents the (tetrapyridinium)-cyclopropene tetra-cation is formed [189)].

Although the reaction of cyclopropenes with phosphines can lead to ring opening [178 b)], treatment of 1,2-dichloro-3,3-difluorocyclopropene with trialkylphosphites leads to moderate yields of the cyclopropanes (271) [190)].

Carbon Nucleophiles

Addition of organolithium reagents to cyclopropenes can occur, though it is much slower than reaction of an organolithium with an acid or ketone group at the 3-position [191], and is generally not observed in those systems where lithium-hydrogen or lithium-halogen exchange at the vinylic positions or removal of an allylic hydrogen can compete; for example, the reaction of (272) with n-butyl lithium leads only to low yields of the corresponding 1-ethylidene-2-ethylcyclopropane [192]. Addition of phenyl lithium to cyclopropene itself does, however, lead to 2-phenylcyclopropyl lithium with over 99% (Z)-stereochemistry [193], while treatment of (273) with methyl lithium in THF at −65 °C leads to the 1-chlorocyclopropene, apparently by addition of methyl lithium to the alkene followed by elimination of lithium chloride [194].

271

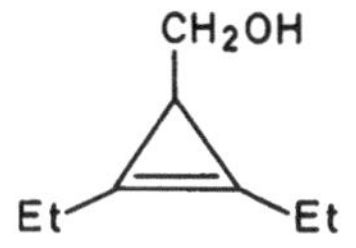

272

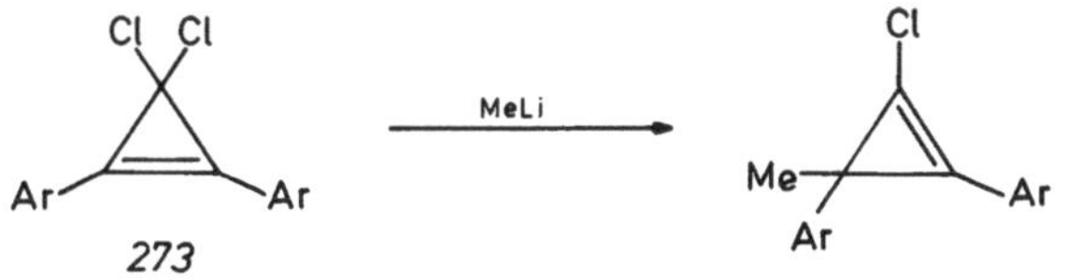

273

Addition of Grignard reagents to cyclopropenes occurs much more readily. Reaction of methyl magnesium iodide with 1-methyl- or 1,3,3-trimethyl-cyclopropenes leads to the Grignard of the more stable anion (274), which may be trapped by a variety of electrophiles in an overall *cis*-addition; in some cases the organomagnesium is trapped by unreacted cyclopropene to produce a dicyclopropyl magnesium halide [195]. Reaction with allylic Grignard reagents leads to a clean *cis*-addition. The reaction with vinylic, γ,δ- or δ,ε-unsaturated Grignards is less satisfactory, and higher molecular weight side products derived by addition of the cyclopropyl Grignard to the cyclopropene can be obtained [196]. Thus 1-isobutenylmagnesium bromide adds to cyclopropenes to produce (275), together with considerable amounts of 1:2 and 1:3 adducts. Quenching of (275) with carbon dioxide or ethyl chloroformate provides an efficient route to *cis*-cyclopropanecarboxylic acids or esters [197]. Addition of allylic Grignards to 1,2-diethyl-3-(hydroxymethyl)cyclopropene also occurs in a *cis*-manner,

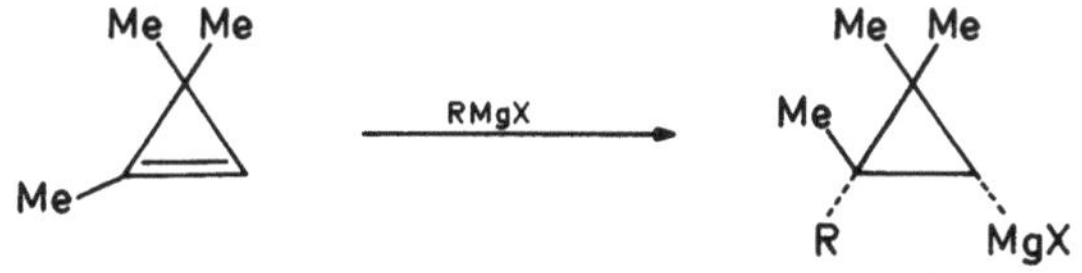

274

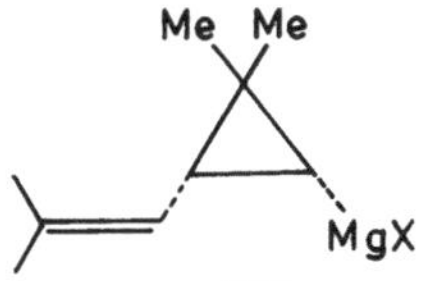

275

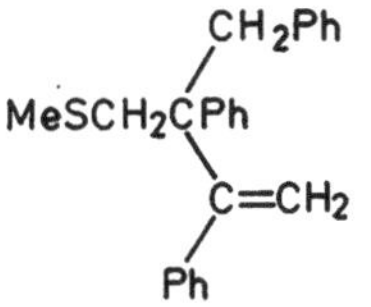

276

from the same face as the 3-substituent and bond formation occurs at the more substituted allylic carbon and at the more substituted cyclopropene carbon [198].

Reaction of dialkylmagnesiums with spiro[2,4]hept-1-ene proceeds with second order kinetics and leads to the *cis*-2-alkyl cyclopropyl magnesium alkyl with α-substituent effects on the rate in the order primary < secondary > tertiary, suggesting a balance between competing electronic and steric effects [199].

In a related reaction, addition of organocuprates, alkyl zinc bromide or di-isobutyl aluminium hydride to 1-trimethylsilylcyclopropenes occurs regio- and stereo-selectively, eg. [12]:

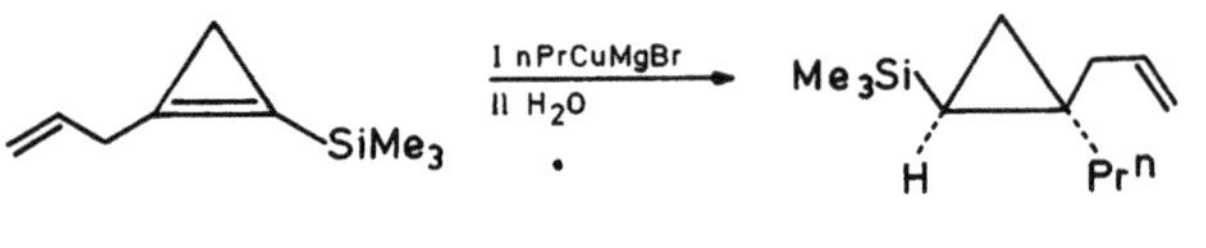

Addition of dimethylsulphonium methylide to triphenylcyclopropene leads to ring opening to (276) [200]. This may be derived by nucleophilic attack followed by ring opening of the triphenylcyclopropylanion to an allyl anion, and intramolecular proton transfer. Addition of t-butylisocyanide to cyclopropenes can also lead to apparent cyclopropyl-allyl anion ring opening, in this case to a vinylketenimine [201].

Reaction of 7,7-dichlorobicycloheptane with base in the presence of malonate, α-cyanoacetate or $^-$CMe(CN)Ph ions leads initially to, eg., (277). Yields are not always high because of further reactions, but in some cases reach over 80% [151, 202].

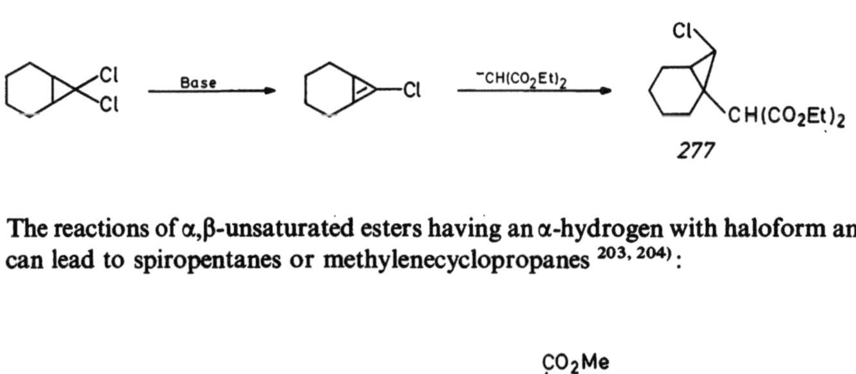

277

The reactions of α,β-unsaturated esters having an α-hydrogen with haloform and base can lead to spiropentanes or methylenecyclopropanes [203, 204]:

and these transformations may also be explained in terms of nucleophiic addition to an intermediate cyclopropene, followed by further additions or, when X = Br, an allylic rearrangement:

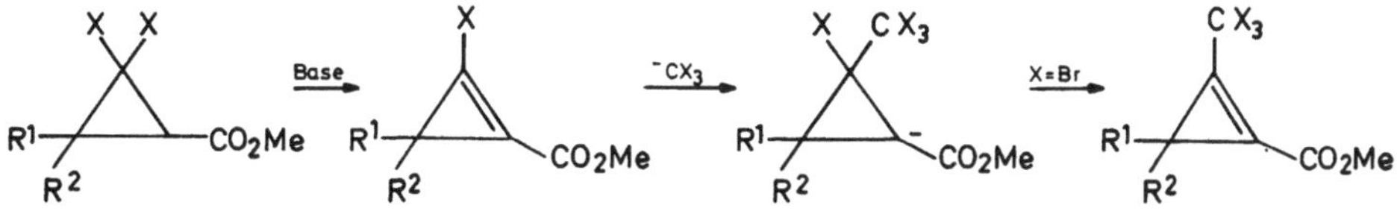

Halide Ions

The reaction of tetrahalocyclopropenes with iodide ion leads to replacement of the vinylic halogens by iodine, presumably by an addition — elimination process [205].

5.5 Photoaddition

The triplet states of cyclopropenes containing a heteroatom at the 2-position of a 3-substituent undergo hydrogen atom abstraction to produce diradicals, which close to produce bicyclo[3.1.0]hexanes, eg. [206]:

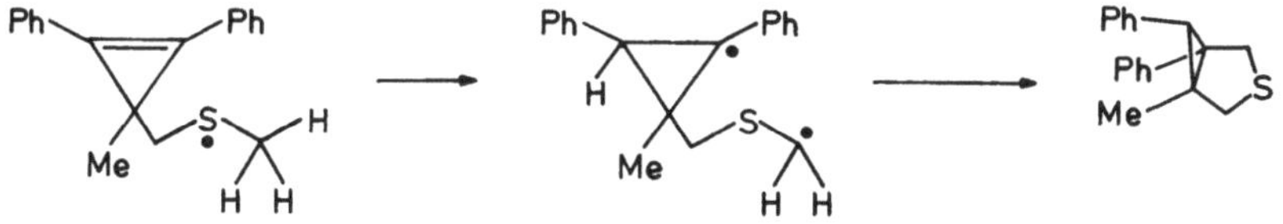

6 Cheletropic Addition

Although some carbenes are reported not to add to cyclopropenes [207], there are several examples of inter- and intra-molecular addition leading initially to the formation of bicyclobutanes. 1,2-Diphenylcyclopropene-3-carboxylates are converted to a mixture of three stereoisomeric bicyclo[1.1.0]butanes by reaction with ethoxycarbonylcarbene generated from the thermolysis of ethyl diazoacetate; an additional product is the diene (278) which is apparently formed by rearrangement of an intermediate zwitter ion [208]. It should be noted, however, that cyclopropenes readily undergo addition to diazo-compounds, and that subsequent transformations may then lead to bicyclobutanes (see Section 8), and that a free carbene may therefore not be involved in the above process.

Reaction of 1,2,3-trimethylcyclopropene with trichloromethyl lithium generated from bromotrichloromethane and methyl lithium at —110 °C produces the cyclobutene (279). This may be explained in terms of intermediate formation of the bicyclobutane (280), followed by cyclopropyl-allyl rearrangement, though the dichlorocyclopropane could not be detected even at —73 °C [209].

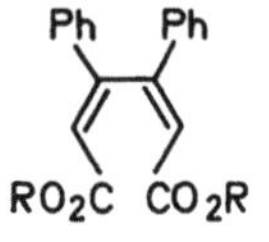

278

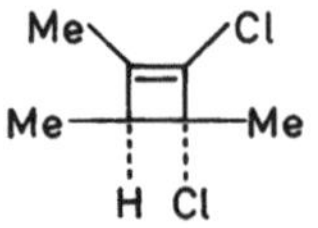

279

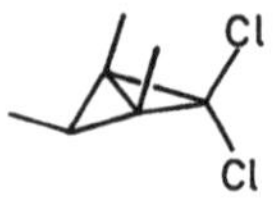

280

It is noteworthy that the reaction of phenylallene with dichlorocarbene under basic conditions has been reported to lead to the pentachloride (281) [210 a)]. However, the structure of the product has now been reassigned as (282), the formation of which appears to involve a similar addition-rearrangement sequence to that described for the formation of (279), the intermediate cyclobutenylcation being trapped by chloride [210 b)].

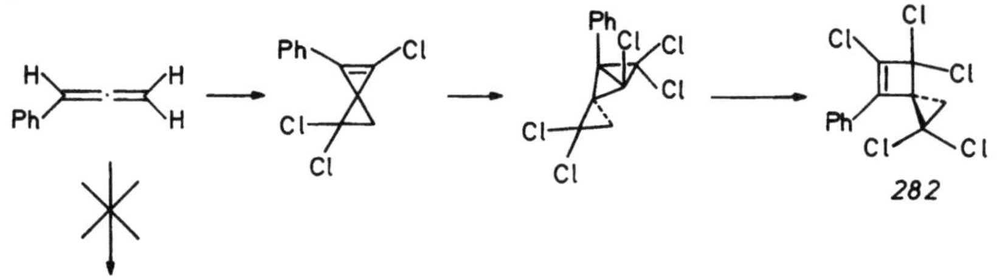

282

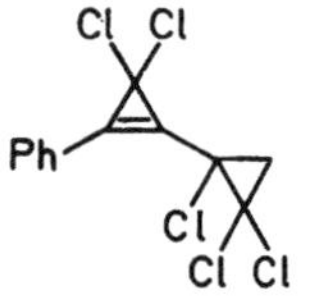

281

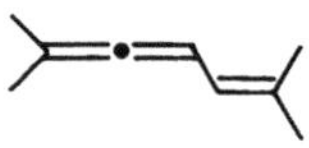

283

284

Addition of dimethylvinylidene or a related carbenoid, generated from the reaction of propanone with a diazophosphonate and base, to alkylcyclopropenes leads largely to trienes, eg. (283) from 3,3-dimethylcyclopropene. Apparently the same products are obtained when the carbenoid is generated from 1,1-dibromo-2-methylpropene and base, but it is not clear whether the reactions involve formation and rearrangement of a bicyclobutane or rather collapse of a polar intermediate such as (284) [211)].

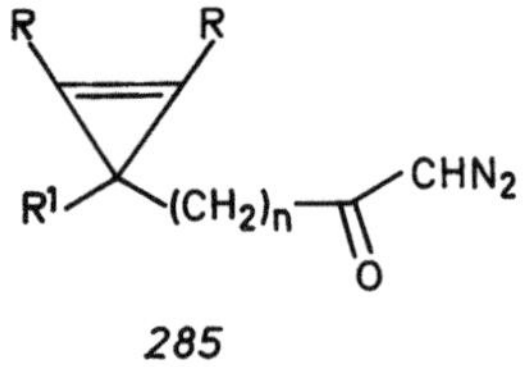

285

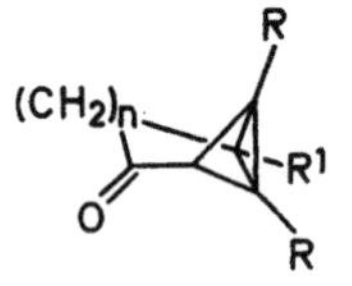

286

Thermolysis of (285, n = 2, R = R^1 = Ph) in the presence of copper powder leads to moderately efficient intramolecular addition of the carbene (carbenoid) to the double bond to produce (286, n = 2) [212)]. Photolytic or catalytic decomposition of the lower homologues (285, n = 0,1, eg., R = Ph, R^1 = H; R = $ClCH_2$, R^1 = H) also leads to intramolecular addition to produce (286, n = 0,1) [213)]. This has been applied in a route to 3,4-fused furans [213 b)]:

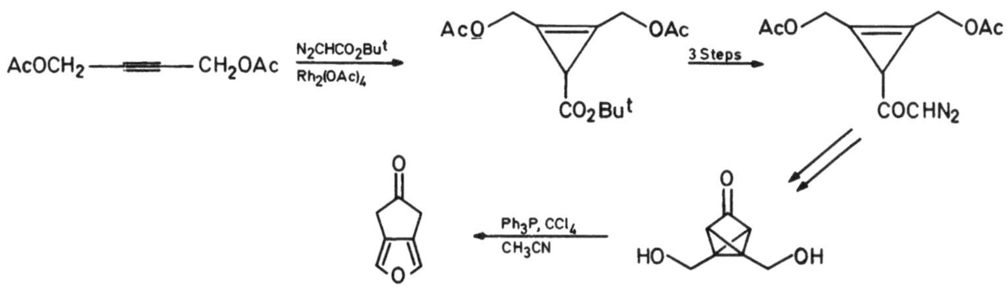

Photolysis of an alkyl azidoacetate in the presence of 1-methyl- or 1,3,3-trimethylcyclopropenes may be explained in terms of an addition of ethoxycarbonylnitrene to the double bond to produce a 2-azabicyclo[1.1.0]butane, but this rearranges under the reaction conditions to the azadiene, although only in low yield [214,215]:

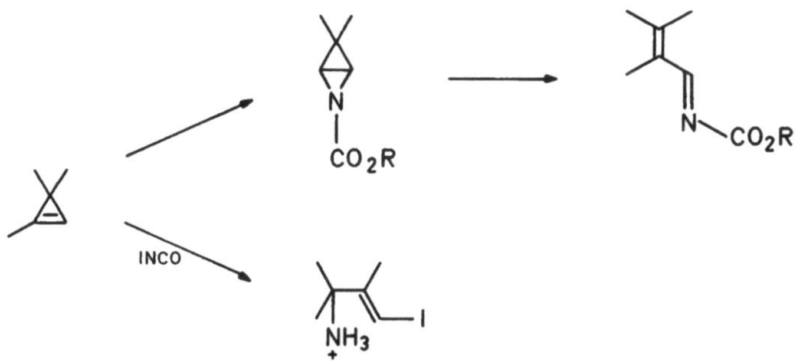

287

No addition occurs without irradiation, and attempts to obtain the azabicycle by addition of iodine isocyanate to the former cyclopropene led only to ring opened product, (287) [215].

7 Diels-Alder Addition

In general, cyclopropanes are good dienophiles, although their inherent thermal instability can lead to side reactions, and the presence of substituents at C-3 causes some steric retardation. Cyclopropene itself cycloadds to a range of dienes including forming an *endo*-adduct with cyclopentadiene; in contrast a 1:1 mixture of *exo*- and *endo*-adducts is formed with furan. The change in stereochemistry is probably explained by the reduction in steric repulsion on replacing the methylene group by an oxygen [216]. Addition to 5-iodocyclopentadiene leads to a rearranged product [217]:

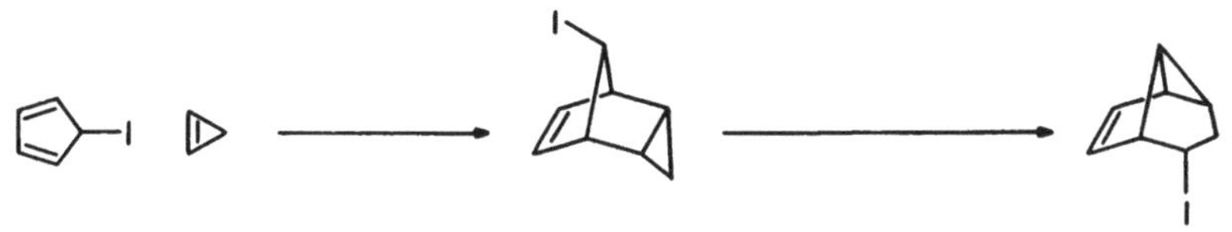

Cyclopropene also adds to less reactive, acyclic, dienes [218], though it is worth noting that the reaction with cyclohexadiene only proceeds in 10% yield [219]. Addition of a range of alkylcyclopropenes to thiophene dioxides leads to cycloheptatrienes, presumably by cheletropic elimination of sulphur dioxide from the intermediate adduct [220]:

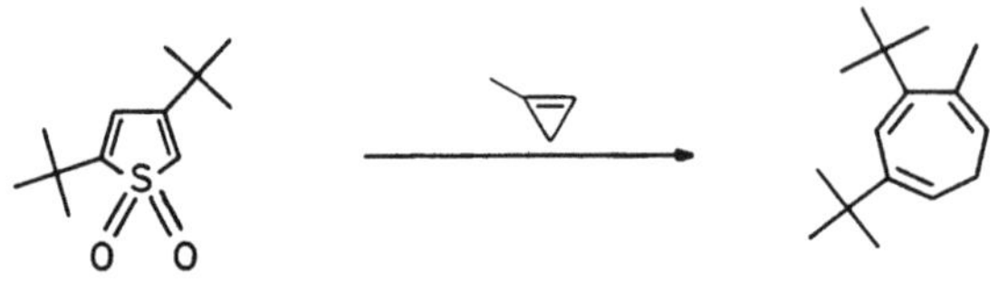

Reaction with cyclopentadienones or their acetals also occurs readily; hydrolysis of the acetals and decarbonylation also leads to cycloheptatrienes, while hydrogenation prior to the elimination produces cyclohepta-1,4-dienes [221, 222]:

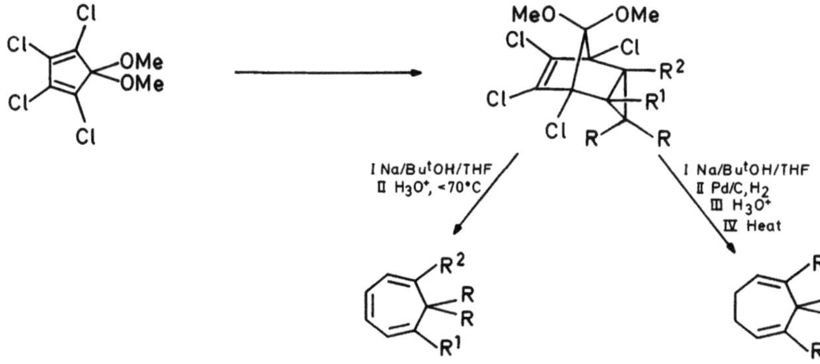

In the same way, addition to 1,2,4-triazines and 1,2,4,5-tetrazenes leads, after loss of nitrogen, to the corresponding aza- or diazacycloheptatrienes, which are in some cases in equilibrium with their bicyclic forms [223]:

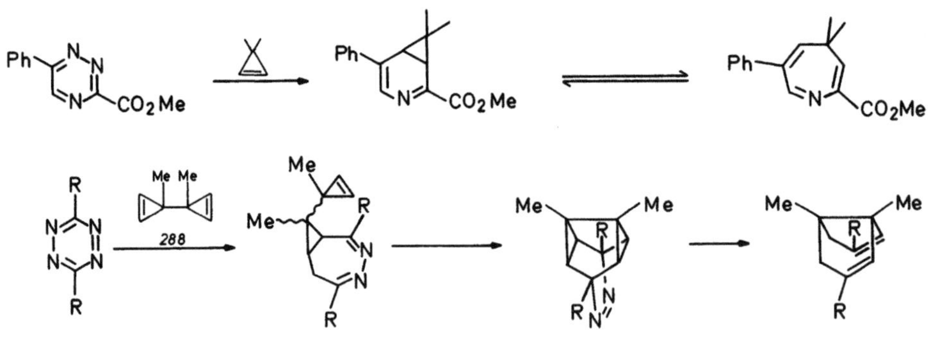

The addition of a second equivalent of cyclopropene can then also occur [224]. In the case of 3,3′-bicyclopropenyls (288), this provides a ready route to semibullvalenes, the epimeric intermediate cyclopropenes equilibrating and allowing intramolecular cycloaddition to occur [225]. Addition also occurs to oxadiazin-6-ones [226], to 5-phenylazotropones (289) [227], and to the tropylium ion in aqueous dioxan, though yields in the latter case are low [228]:

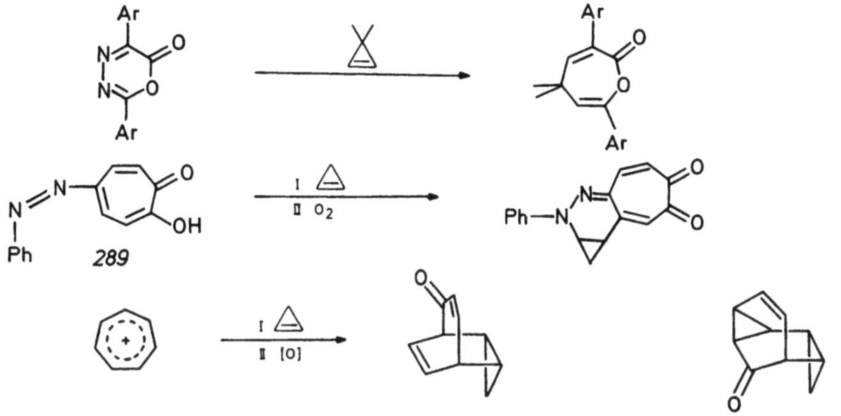

289

Cyclopropenes bearing electron withdrawing groups are, as may be expected, particularly good dienophiles. The cyclopropenes (290, R^1 = H or CO_2Me, R^2 = CO_2Me; R^1 = H, R^2 = $POPh_2$) undergo *endo*-addition to cyclopentadiene and *exo*-addition to furan in the same way as non-electrophilic cyclopropenes [229–231, 53]. Addition of (290, R^1 = H, R^2 = CO_2Me) to isoprene occurs with low regioselectivity, leading to a 2:1 mixture of (291) and its isomer [229, 230].

290

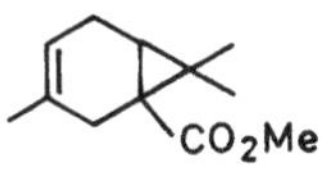

291

Tetrahalocyclopropenes also undergo the cycloaddition [232], and stereochemical studies, the effect of solvent and activation parameters appear to be consistent with a concerted mechanism [233]. In these cases, *endo*-adducts are formed with furan. 3-Chloro-, 1,3-dichloro- and 3,3-difluorocyclopropenes add to cyclopentadiene to give *endo*-adducts [233, 234].

The Diels-Alder reaction has been particularly widely used in the preparation of cycloproparomatics, using either 1,2-dihalo- [41–43], or tetrahalocyclopropenes [235, 236]:

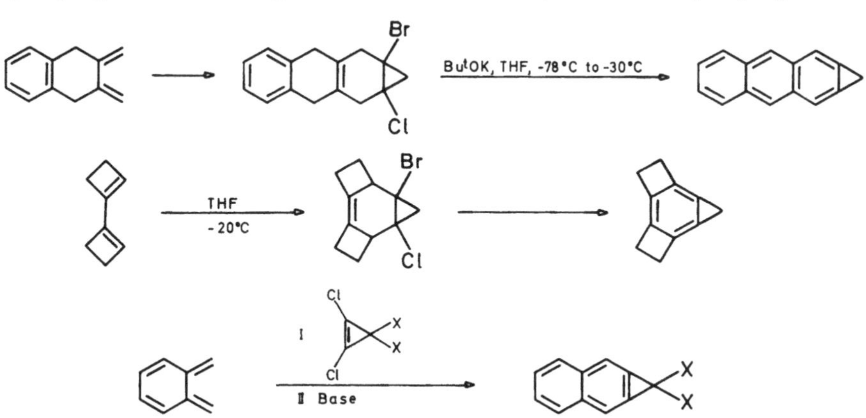

The perfluorinated cyclopropenes (292, X = $C(CF_3)NH$) may be prepared from (293) by 1,3-dipolar addition of diazomethane followed by desulphurisation with triphenylphosphine to produce, (294), and then thermolysis. Addition of (292, X = $C(CF_3)O$) to 2,3-dimethylbutadiene occurs predominantly in an *endo*-manner, the intermediate undergoing an intramolecular ene-reaction to produce (295). In the same way reaction with pyrrole leads to (296), in this case presumably by an intramolecular nucleophilic attack in the initially formed *endo*-adduct [237].

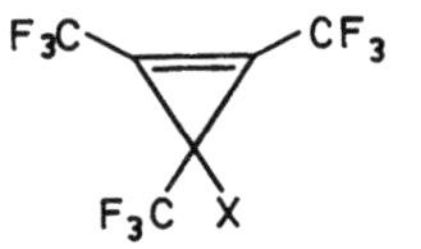
292

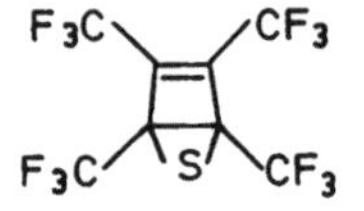
293

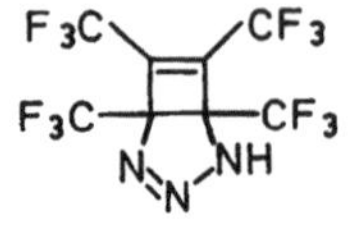
294

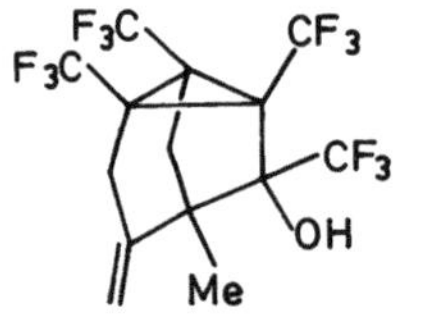
295

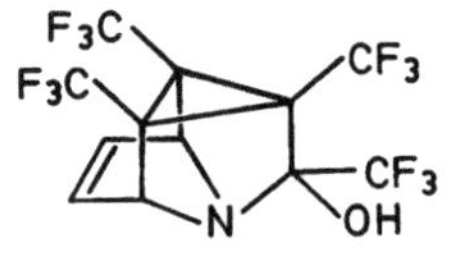
296

8 Dipolar Cycloaddition

8.1 Diazo-Compounds

Diazocompounds readily undergo dipolar cycloaddition to cyclopropenes, leading initially to pyrazolines; however, these are in some cases very sensitive to base and a number of early reports of this reaction indicated that the products were in fact pyridazines, eg. [238]:

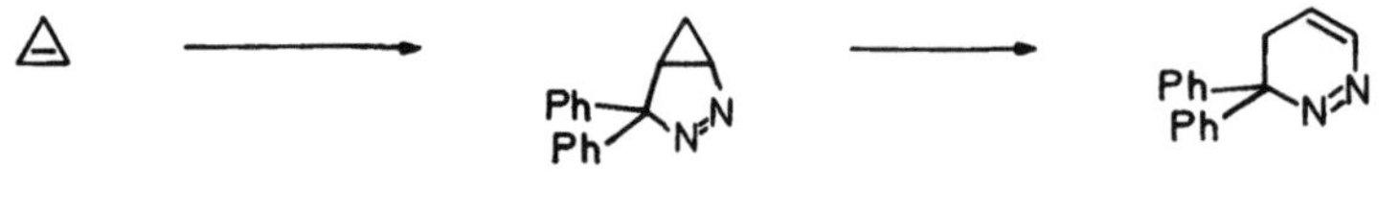

A further complication is the conversion of pyrazolines to diazocompounds, which can be brought about in many cases either by heat or by light, coupled to the secondary decomposition of these diazo-species. Reaction of diazoethane with 3-methylcyclopropene leads to the epimeric diazabicyclo[3.1.0]hexanes (297). Thermolysis of (297a) at 120–190 °C leads to (298a) as the major product, while (297b) leads to (298b), in each case accompanied by isomeric products. The results are consistent with initial rearrangements to the diazo-compounds (299, a and b), followed by loss of nitrogen to produce the corresponding carbenes [239].

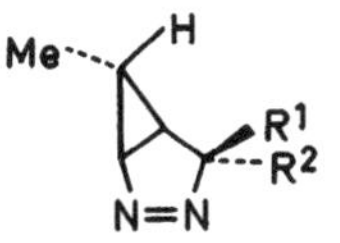

297

(a) R^1=H, R^2=Me
(b) R^1=Me, R^2=H

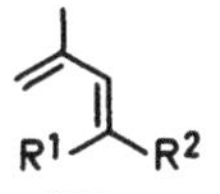

298

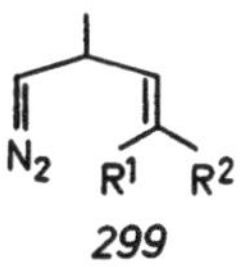

299

Photolysis of various diazabicyclo[3.1.0]hexenes related to (300) leads to mixtures of dienes and bicyclo[1.1.0]butanes, eg.:

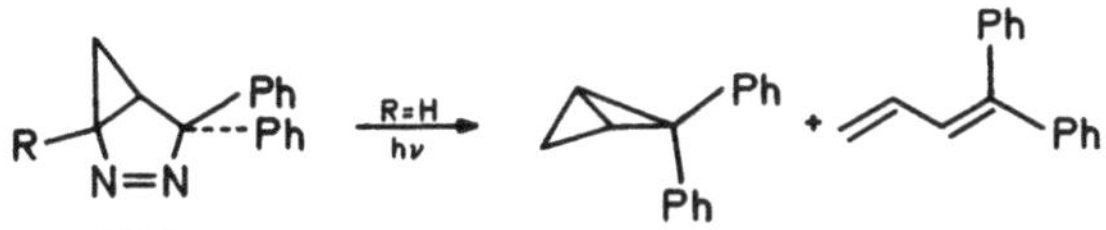

300

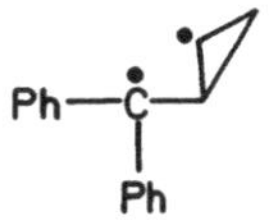

301

The dienes are again explained in terms of a rearrangement to a diazo-compound followed by loss of nitrogen to form a carbene and then 1,2-hydrogen shifts. The bicyclo[1.1.0]butanes are thought to arise through loss of nitrogen to form a diradical (301) followed by cyclisation [39]. It is interesting, however, to note that diazo-compounds related to (299) can be converted to bicyclo[1.1.0]butanes in high yield (see below), and that in other cases thermolysis of pyrazolines has been reported to lead to high yields of bicyclobutanes [241]. Whatever the origin, rearrangement of bicyclobutanes such as (302) using a rhodium complex catalyst provides a ready access to the azulene system [240]:

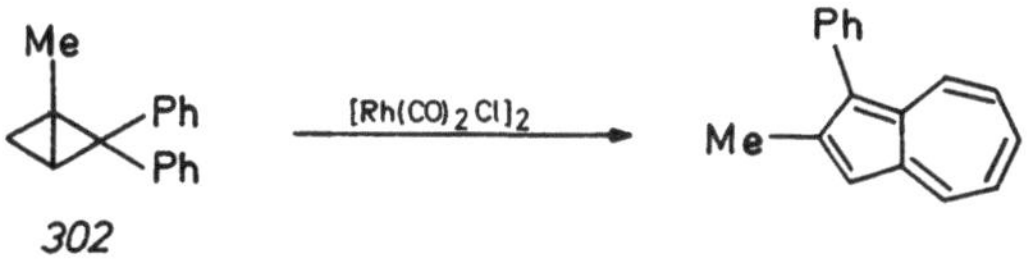

302

Addition of diazomethane to 3,3-dimethylcyclopropene leads to the pyrazolines (303, R = H) but methyl diazoacetate leads to (304) and (305); the former is apparently derived by base induced reaction of the pyrazoline, while the latter may be explained in terms of rearrangement to the diazocompound (306) followed by dipolar addition to the cyclopropene [242].

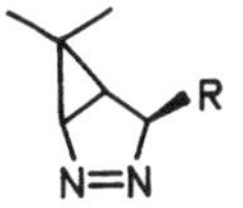

303

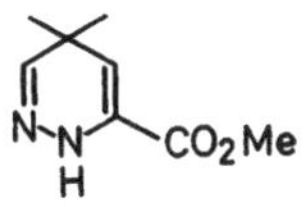

304

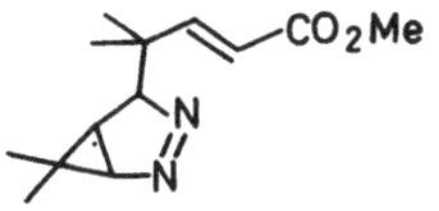

305

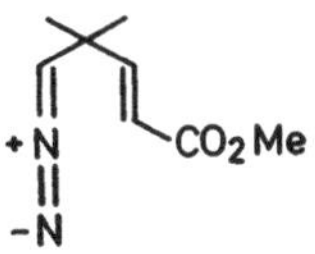

306

In the case of alkyl substituted pyrazolines such as (307), thermal decomposition at 40–70 °C or photolysis leads to complex products which could be explained either in terms of a diradical intermediate or of rearrangement to a diazo-compound followed by loss of nitrogen to produce a carbene. However, on brief heating to 80 °C or on photolysis at —50 °C, the pyrazolines are converted to yellow solutions of diazo-compounds (308) which survive only for a few hours at 20 °C. The fact that the diazocompound can be isolated from the photolysis of (307) at low temperature and leads to the same products on heating is strong support for its intermediacy in the reactions at higher temperature [243].

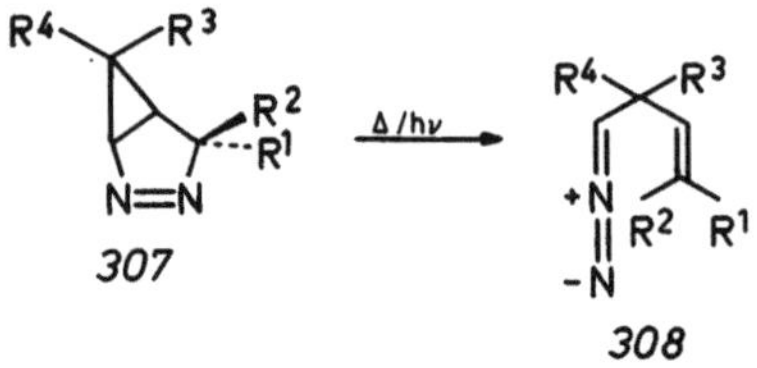

307 308

Diazoalkane addition to cyclopropenes having electron withdrawing substituents at 1- or 2-positions generally leads to the regioselectivity predicted on electronic or frontier orbital grounds:

Ref. 246, 247

310 309 Ref. 244

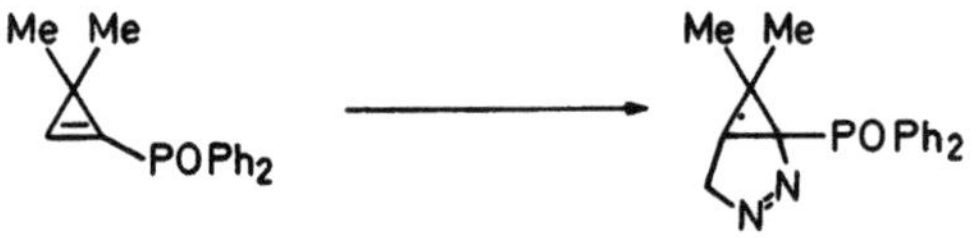

Ref. 245

310 311

although in some cases minor amounts of the regioisomers are isolated. The adducts, and in particular the major isomers, often rearrange to diazocompounds under mild conditions. The phosphonates (309) rearrange in refluxing benzene or toluene, while reaction of (310) with ethyl diazoacetate or $N_2C \cdot PO(OMe_2)$ at ambient temperature leads directly to the corresponding diazo-compounds [244, 245]. All diazomethane adducts in this series rearrange to pyridazines, eg. (311), on reaction with a trace of base [245]. Although the adduct between (312) and diazopropane is reported to be thermally stable[246)], the related ester (313) rearranges cleanly to the diazo-derivative at 70 °C; catalytic decomposition in the presence of rhodium acetate leads to vinyl-migration in an intermediate metal-carbenoid[247)]:

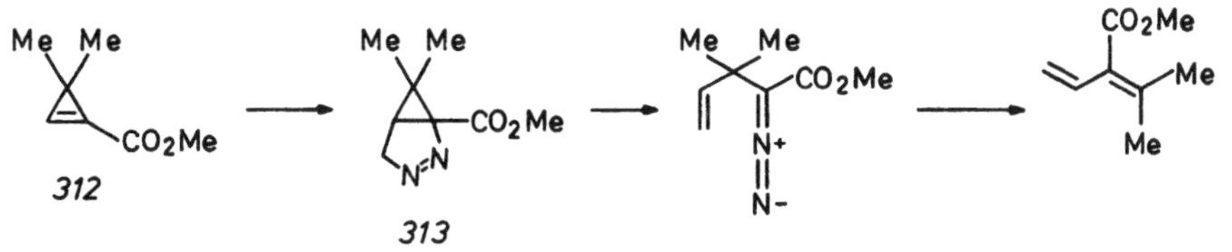

In other cases, however, the diazo-compound is converted cleanly to a bicyclobutane, eg.[247)]:

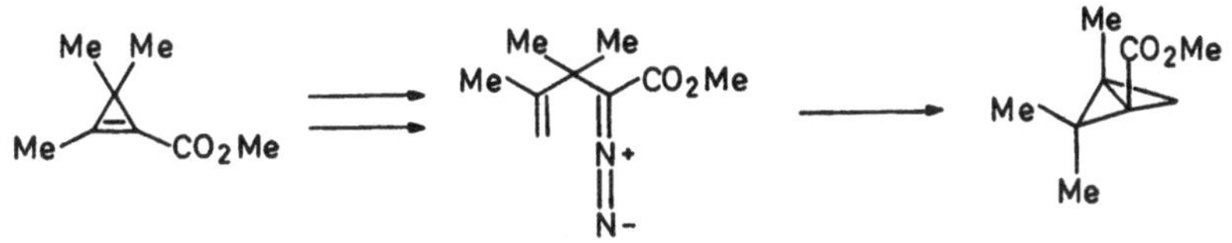

1-Trimethylsilylcyclopropenes also undergo regioselective addition of diazomethane[248, 247)] and rearrangement to the diazo-compound[247)]. Catalytic decomposition of the latter can lead to bicyclobutanes or to isomeric silyldienes, eg.:

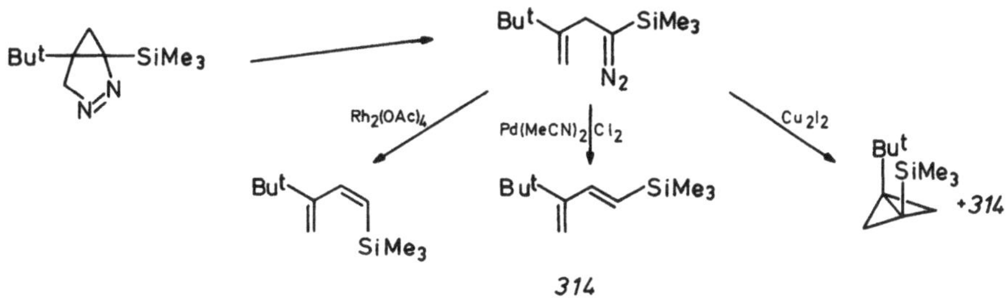

Photolysis of pyrazolines derived from 1,2-bis-trimethylsilyl-cyclopropene-3,3-dicarboxylates has also been shown to provide an efficient route to bicyclobutanes[249)].

In the case of methylenecyclopropenes such as (315), the pyrazoline rearranges to a diazo-compound, which in turn cyclises at the original exocyclic double bond[250)]:

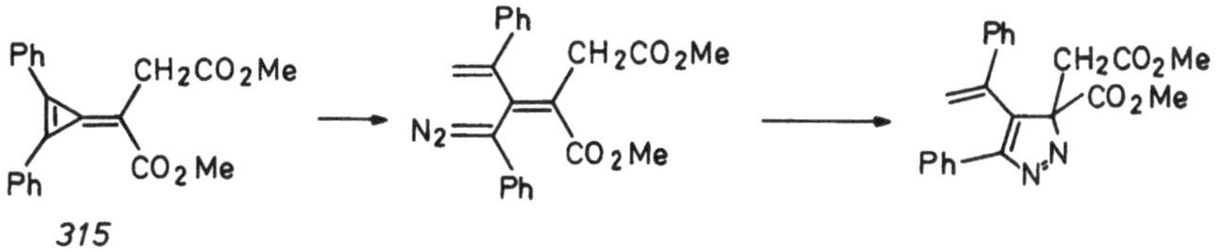

8.2 Azides

The reaction of 3,3-dimethylcyclopropene with phenyl- or p-toluenesulphonylazides leads to 2:1 adducts (316). These are apparently derived by rearrangement of an initial dipolar cycloadduct to give (317), which then reacts with more starting material [242]. Addition of aryl azides to tetrachlorocyclopropene occurs at elevated temperatures and leads to the tetrachlorides (318), apparently derived from initially formed dihydropyridazines by rearrangement with loss of nitrogen [238 b)].

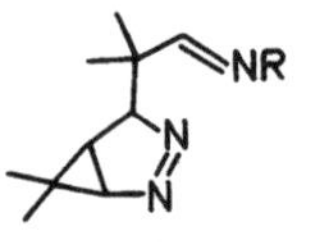

316

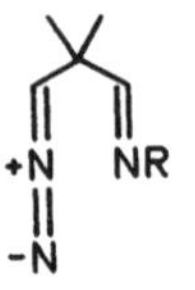

317

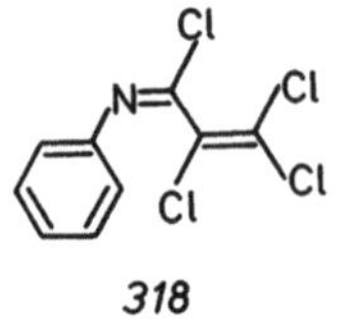

318

8.3 Nitrile Oxides and Nitrile Imines

Cyclopropene itself and the 3,3-dimethyl-derivative add to nitrile oxides or nitrile imines in good yield [251]:

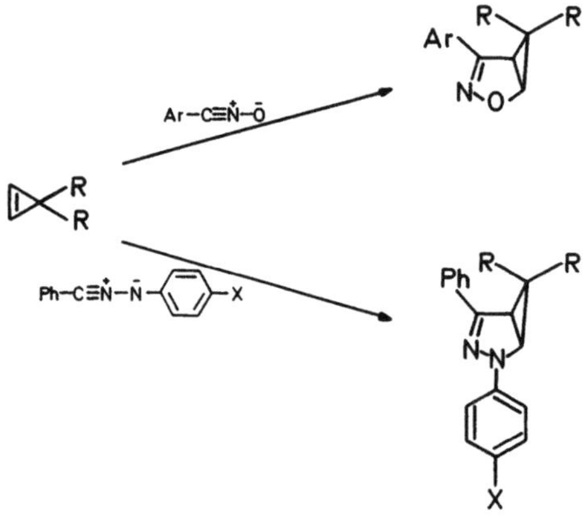

8.4 Immonium Ylides

1,2,3-Triphenylcyclopropene cycloadds to a variety of pyridinium dicyanomethylides (319) to produce indolizines and quinolizines, eg.:

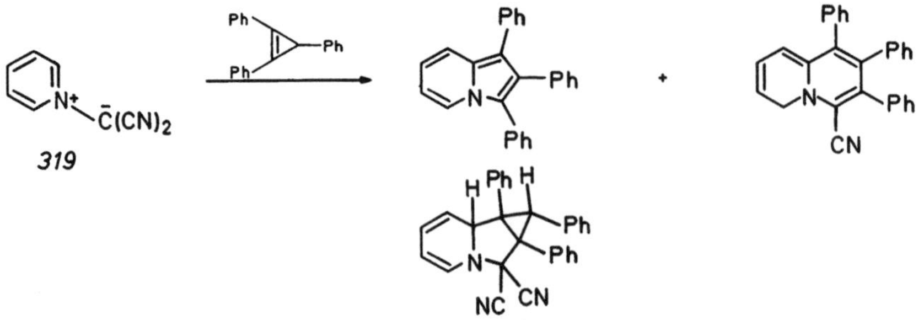

319

320

Both products are thought to be derived from an initial adduct (320) but the detailed mechanisms are very unclear [252]. Similar products are obtained from pyridazinium, phthalazinium and pyrazinium dicyanomethylides [252].

Although some pyridinium ylides and pyridazinium N-oxide do not form adducts with tetrachlorocyclopropene, compounds (321, Z = NCOR, NCO_2R or $C(CN_2)$) do form moderate to low yields of adducts, eg. (322). This compound undergoes an interesting elimination of AcCl on heating, the a–b bond being cleaved to lead, after proton transfer, to (323) [253].

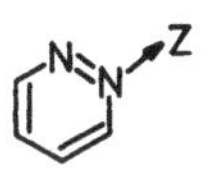

321

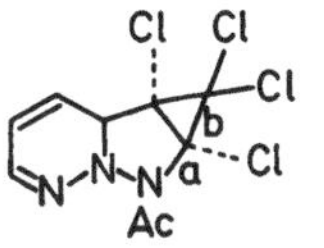

322

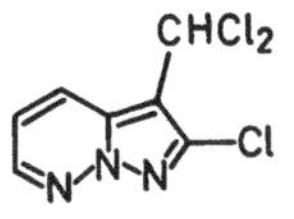

323

Other examples of heterocyclic ring synthesis using cyclopropenes are shown below:

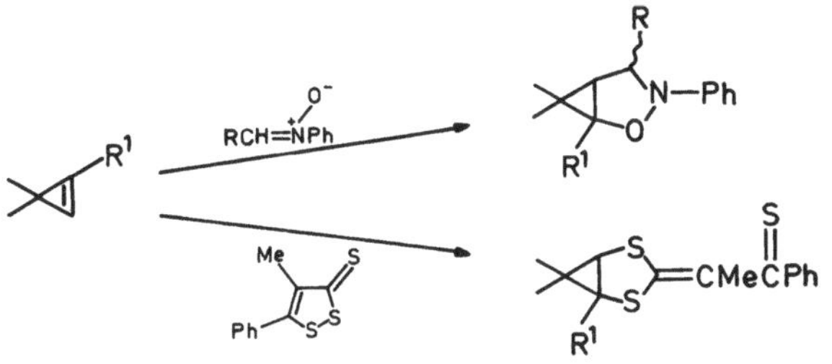

Ref. 254a

Ref. 254b

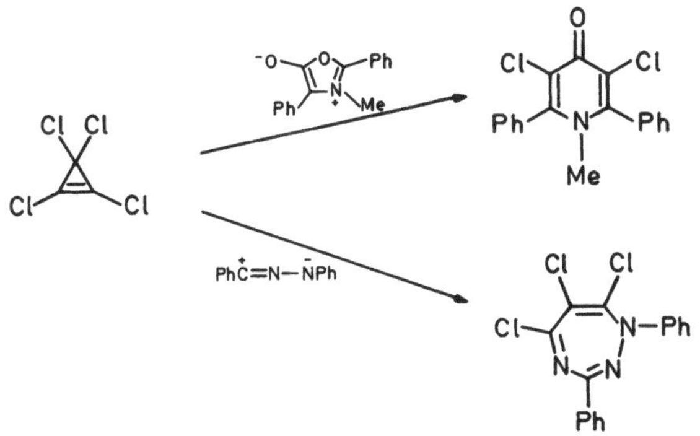

Ref. 254c

Ref. 254d

9 Oxidation

Peracid oxidation of alkylcyclopropenes leads to ring cleavage with the formation of an enone:

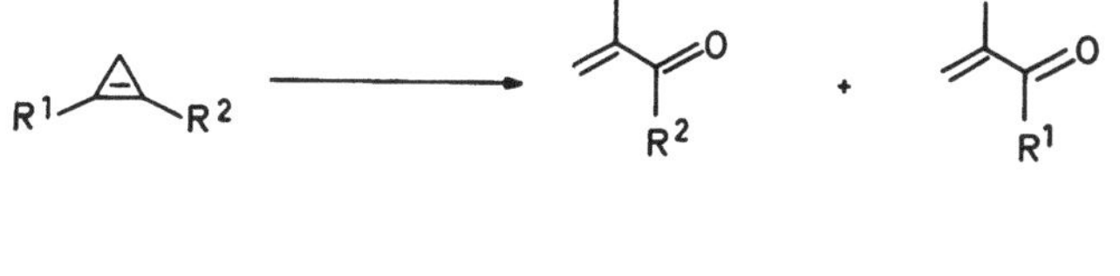

When, eg., R^1 = H and R^2 = Me the regioselectivity is low, and the reaction is explained in terms of formation and rearrangement of a 2-oxabicyclo[1.1.0]butane derivative [255]. The rate of reaction for cyclopropenes is very similar to corresponding cyclobutenes and cyclopentenes, in support of the formation of a highly strained intermediate [256], though the effect of 3-substituents can be interpreted in terms of σ- or π-attack [257]. In the case of the sterically hindered cyclopropenes (324) a different process occurs and two products apparently derived by decomposition of (325) are isolated. Apparently the initial reaction in this case involves abstraction of H-3 of the ring, the resulting cyclopropenium ion being trapped to produce (326). Replacing the 1-t-butyl group by phenyl leads to a dual pathway via hydrogen abstraction and oxidation to the oxabicyclo[1.1.0]butane [258].

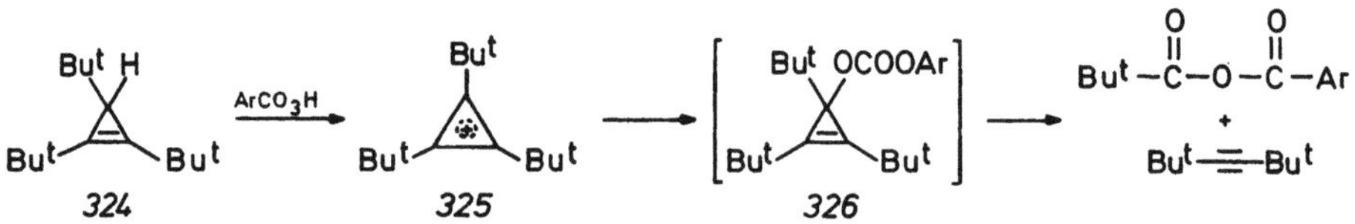

324 325 326

It is interesting to note that tri-t-butylcyclopropene shows an unusually high reactivity towards chromium(VI), and is again converted to the cyclopropenium ion in a process which shows a high deuterium isotope effect [259].

When the cyclopropene has a single carbomethoxy or aryl substitunt at the 3-position, the predominant stereoisomer of the enone obtained has the substituent cis-to the ketone [260, 261]. One possible explanation is the formation of a bicyclobutane by peracid attack from the side away from this substituent, followed by a stereocontrolled rearrangement. If the 3-substituent is hydroxymethyl, the reverse stereochemistry is observed in the enone, in agreement with a peracid attack directed by this group to the same face of the cyclopropene, followed again by rearrangement [262].

Analysis of the second-order rate constants for various methyl and phenyl substituted cyclopropenes shows effects similar to those observed with alkenes undergoing epoxidation, again in agreement with the intermediacy of the oxabicyclo[1.1.0]butane in cyclopropene oxidation [263].

Peracid oxidation of the 1-trimethylsilylcyclopropene (327) proceeds with good regiocontrol to produce the enone (328) in good yield; this is further oxidized by a second equivalent of reagent to (329):

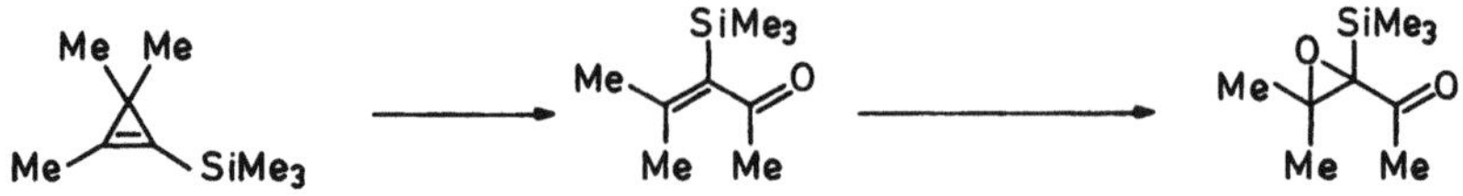

327 328 329

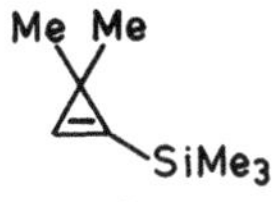

330

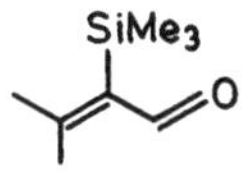

331

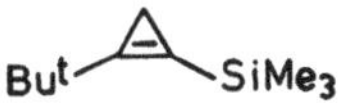

332

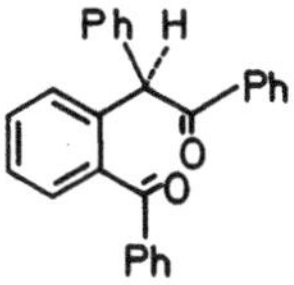

333

In the case of the dimethyl-substituted cyclopropene (330), the initial product is (331) but further oxidation occurs more easily. The presence of large substituents at C-2 reduces the regioselectivity of the initial oxidation, (332) producing a 4:1 mixture of isomeric enones [264].

Although 1-methylcyclopropene is inert to singlet oxygen [265], irradiation of 1,2,3-triphenylcyclopropene in methylene chloride with oxygen in the presence of methylene blue leads mainly to 1,2,3-triphenylpropandione, though in other solvents complex mixtures result. Similar treatment of tetraphenylcyclopropene leads principally to (333), although in this case there is no marked solvent effect [266]. The oxidation has been shown to proceed by a Type I rather than a Type II process on the basis of several observations. It is not affected by the singlet oxygen quencher DABCO, but the rate is reduced by radical inhibitors; moreover the cyclopropene does not react with singlet oxygen generated thermally from triphenylphosphite-ozonide. The photochemical process is accompanied by bleaching of the sensitizer dye, as is characteristic of radical processes, and the effects of solvents and temperature also support the intervention of radicals [267].

10 Solvolytic Ring Opening

Cycloprop-1-en-1-yl-substituted alcohols often undergo ring-opening on treatment with aqueous perchloric acid. In the case of (334) the initial product is apparently the allene (335), which is further converted to (336) [74, 268]:

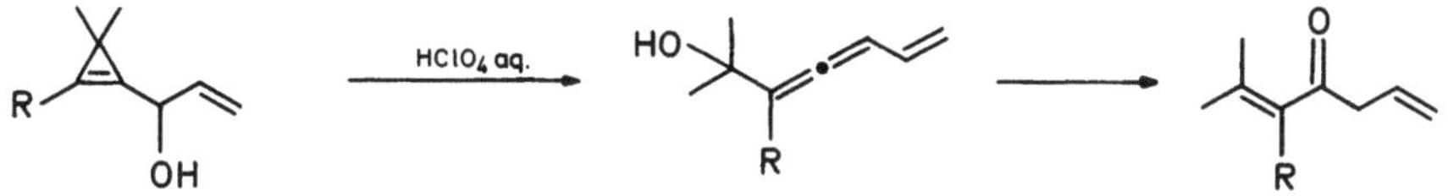

334 335 336

Reaction of 2-halomethylenecyclopropanes with silver ion in methanol can follow a similar course [269], although (337, X = Cl) rather unexpectedly leads to unrearranged product, (337, X = OMe) [270]:

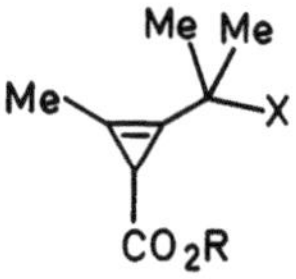

337

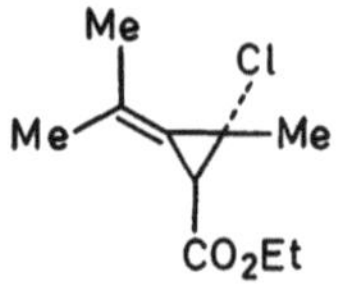

338

Indeed, the chloride (338) reacts under the same conditions with allylic rearrangement to produce the same cyclopropene (R = Et)[270], although there are several other related cases which lead to ring opening[269].

It is interesting to note that reaction of (339) with thiolate ion apparently proceeds by an S_N2' type of process, while the related dichloride (340) leads to products which can be explained in terms of an initial allylic rearrangement[182].

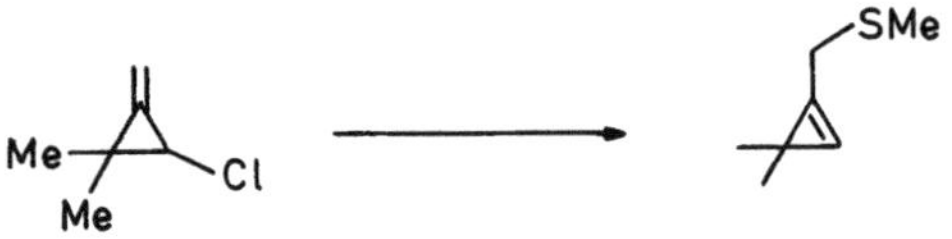

339

340

The reaction of cycloprop-1-enylpropan-2-ols with thionyl chloride is also highly dependent on substitution, presumably reflecting changes in the rate of cyclopropyl-allyl ring opening[165]:

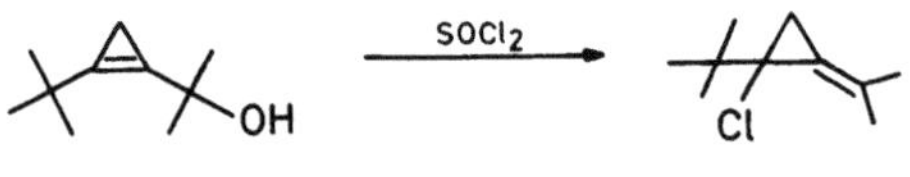

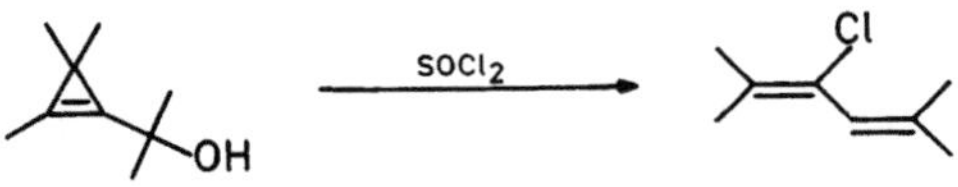

Cycloprop-2-en-1-ylmethanol derivatives (341) ring expand under acidic conditions to cyclobutenols in a reaction thought to proceed through a homoaromatic cyclobutenyl cation[271]:

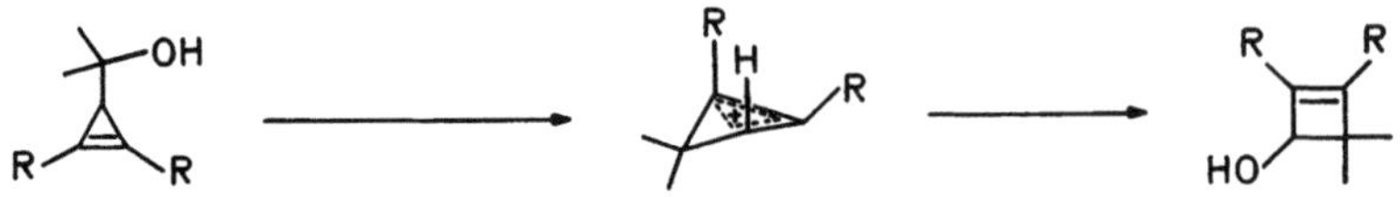

341

However, closely related alcohols are ring opened by reaction with perchloric acid in THF, suggesting a very strong substituent effect in this reaction[74]. In a similar process, dithianes such as (342, R = H), readily formed by trapping of cyclopropenium salts with the lithiodithiane, are efficiently converted to endiones, eg. (343), by reaction with calcium carbonate and methyl iodide [272]:

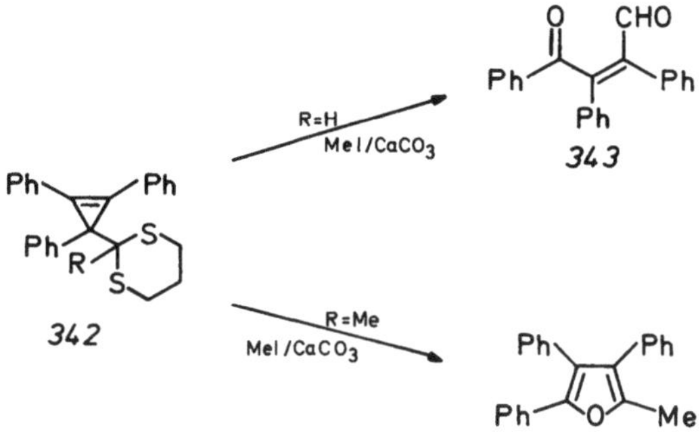

342 343

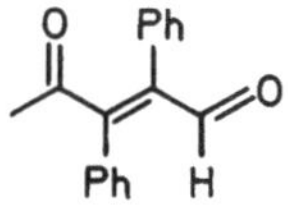

344

The corresponding methyl-derivative (342, R = Me) leads to a furan under these conditions but deprotection by mercuric oxide and boron trifluoride produces an E/Z mixture of diketones corresponding to (343). Replacing one of the phenyl substituents in (342, R = H) with methyl leads instead to the (E)-ketone (344). Various mechanisms are possible for these reactions, but an attractive one involves methylation on sulphur and ring expansion to (345), followed by trapping by water and ring opening [272]:

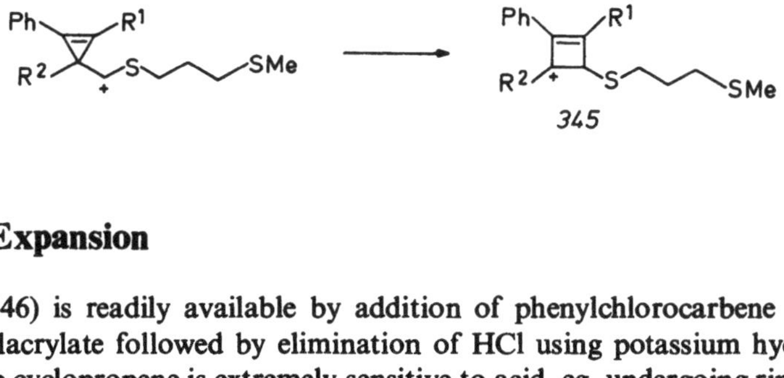

345

11 Ring Expansion

The acid (346) is readily available by addition of phenylchlorocarbene to methyl 3,3-dimethylacrylate followed by elimination of HCl using potassium hydroxide in toluene. The cyclopropene is extremely sensitive to acid, eg. undergoing ring opening on reaction with p-toluenesulphonic acid in toluene; more interesting is the ring expansion on treatment with thionyl chloride [273]:

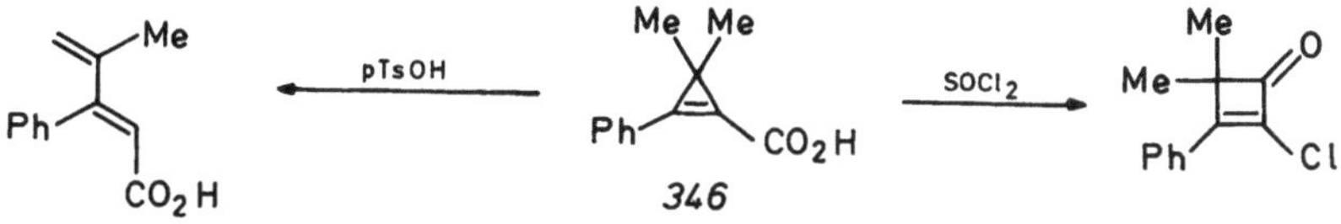

346

The dichloride (347) is converted to (348) on heating, apparently again through an intermediate cyclobutenyl cation. Hydrolysis of either compound with water leads to the ketone (349), which is reconverted to the cyclobutene by treatment with phosphorus pentachloride [274].

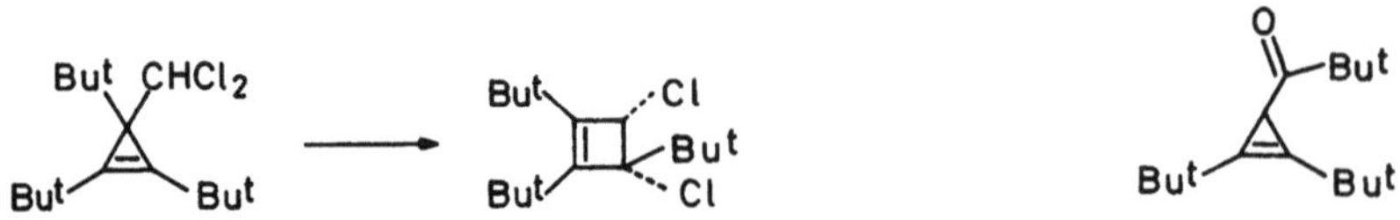

347 348 349

The hydroxy-cyclopropene (350) undergoes a ready base-induced ring expansion to (351) [275]:

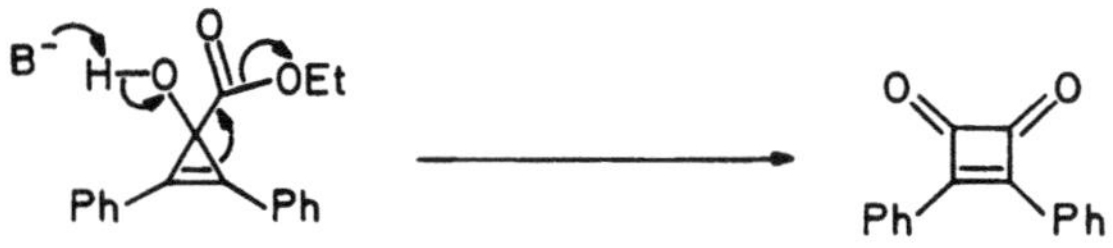

350 351

Reaction of cyclopropenes with cyclopropenium ions leads to a three carbon ring expansion to aromatic derivatives, presumably through ring opening of an initially formed cyclopropenyl-cyclopropylcation [276]:

In the case of trapping of cyclopropenium ions by azirines, the regiochemistry is interpeted in terms of an initial one-carbon ring-expansion [277]:

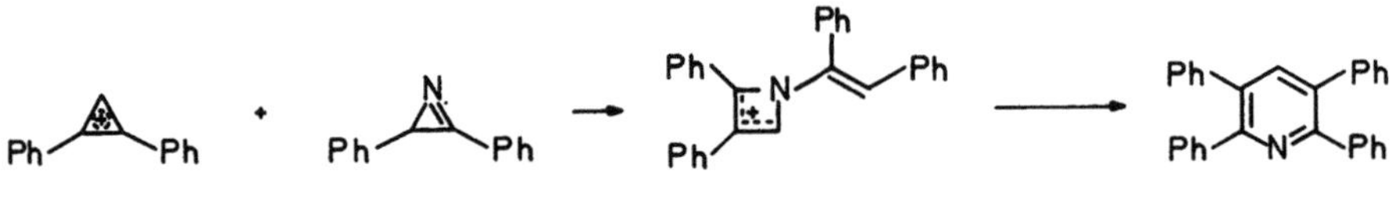

Similar ring expansions are observed when 3,3′-bicyclopropenyls are treated with silver ion (see section 9). Reaction of the propanoic acid derivative (352) with oxalyl chloride also leads to a three carbon ring expansion, although the detailed mechanism of the reaction is not clear [212]:

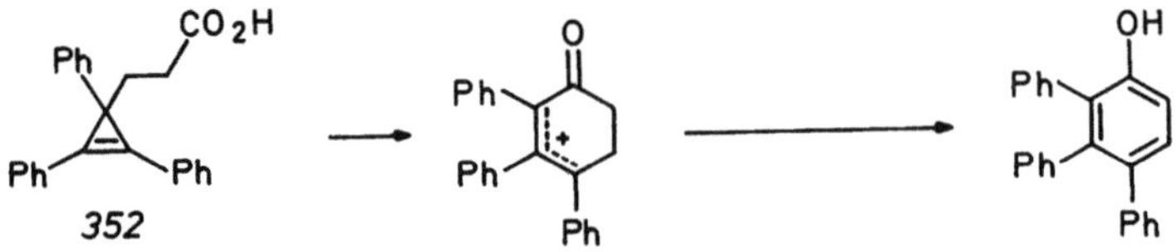

352

Three atom ring-expansion of cyclopropenyl azides and diazo-compounds provides a useful route to triazines and pyridazines respectively [278].

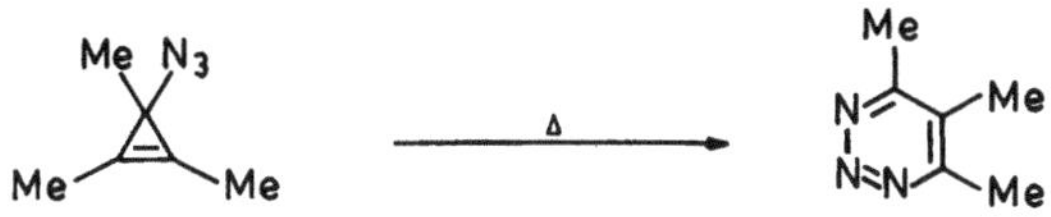

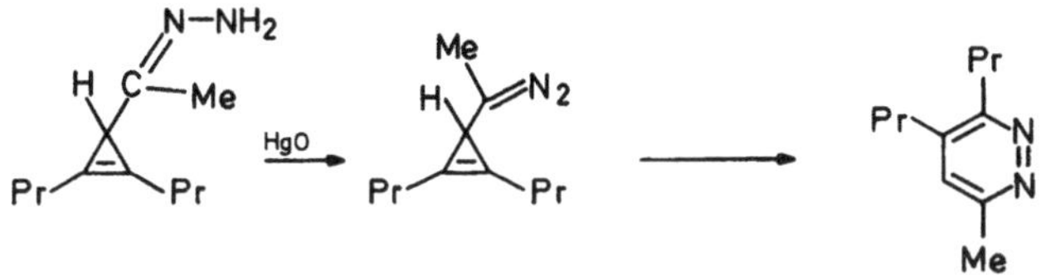

12 Rearrangement of 3,3′-Bicyclopropenyls

The isomerisation of 3,3′-bicyclopropenyls to benzene derivatives is highly exothermic and can be brought about under thermal, photochemical or metal catalysed conditions. The bicyclopropenyl (353, R = H) rearranges in the presence of silver ion to produce one regioisomer of the Dewar benzene (354) [279]:

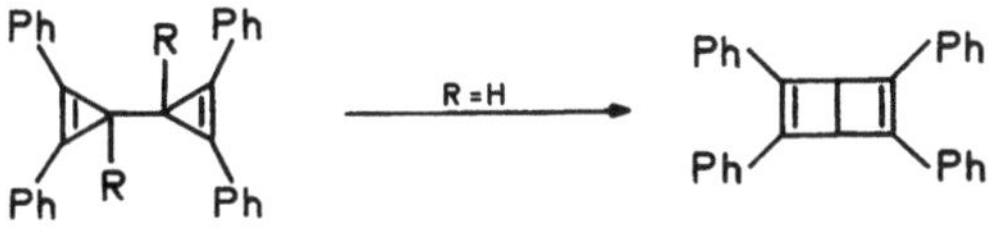

353 354

Product analysis shows that the thermal and metal induced reactions follow similar courses. The former reaction may proceed by opening of one ring to produce a formal vinyl carbene, followed by ring expansion; the latter may be rationalised in terms of electrophilic attack by silver ion at one of the cyclopropenes to produce a carbenium ion (355) — that is a silver-coordinated equivalent of the carbene — followed by ring expansion of the second cyclopropene.

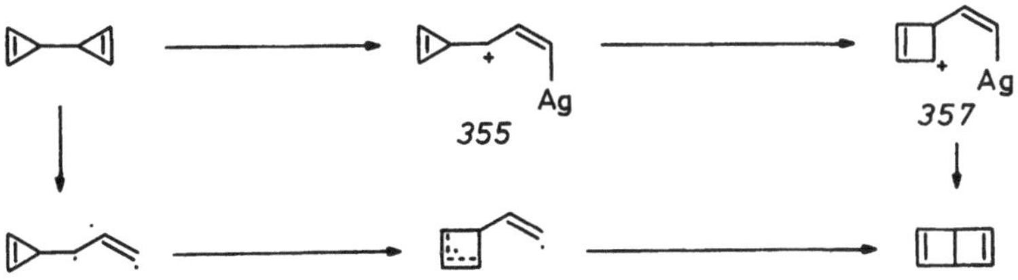

356

In agreement with this (353, R = CN) does not rearrange under either thermal or silver catalysed conditions, the substituent apparently stopping either step depending on the ring which is initially attacked [280]. The thermal reactions are complicated by the fact that, e.g., 3,3′-dimethyl-3,3′-bicyclopropenyl is converted to two diastereoisomers of the 1,1′-dimethylisomer on heating, and there has been considerable discussion of the mechanisms of the various processes involved [281]. Analysis of the thermal rearrangement of each isomer under more vigorous conditions to produce mixtures of isomeric xylenes shows that the diastereoisomers interconvert in competition with xylene formation through a path which does not involve direct bonding between the two rings. A process involving formation of a diradical (356) is proposed [282]. There is also evidence for discrete intermediates (355) and (357) in the silver ion reactions [283, 284]:

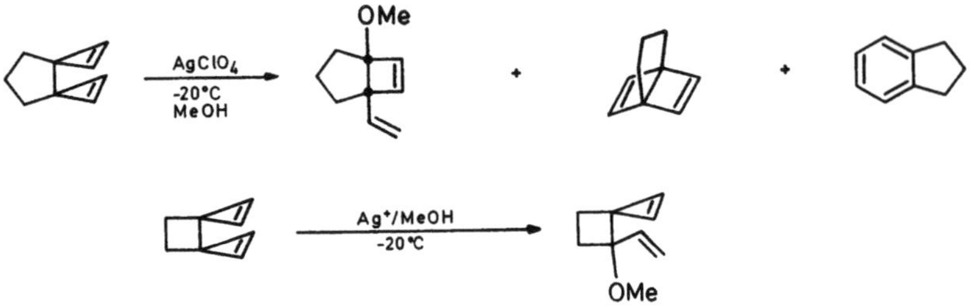

When the latter reaction is carried out in MeOD, the distribution of the label in the products is consistent with attack of silver ion at the π-bond and disrotatory opening of an incipient cyclopropyl cation towards the centre of developing positive charge [284]:

Although the mechanisms of these reactions may be complex, they do allow access to a range of highly strained intermediates, eg. [285]:

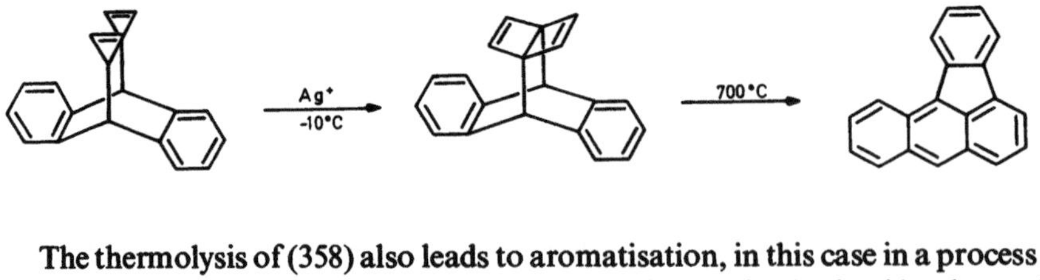

The thermolysis of (358) also leads to aromatisation, in this case in a process believed to involve an intermediate nitrile ylid. Evidence for this is obtained by thermolysis of a series of cyclopropenyl-substituted oxazolinones such as (359) for which cycloreversion with elimination of CO_2 is known to lead to a nitrile ylid. In some cases the ylid could be trapped by addition to methyl propiolate. Substituent effects suggest that the nitrile ylids undergo stepwise addition to produce a bicyclobutyl zwitterion which can either collapse to an azabenzvalene or rearrange to a cyclobutenyl cation [286].

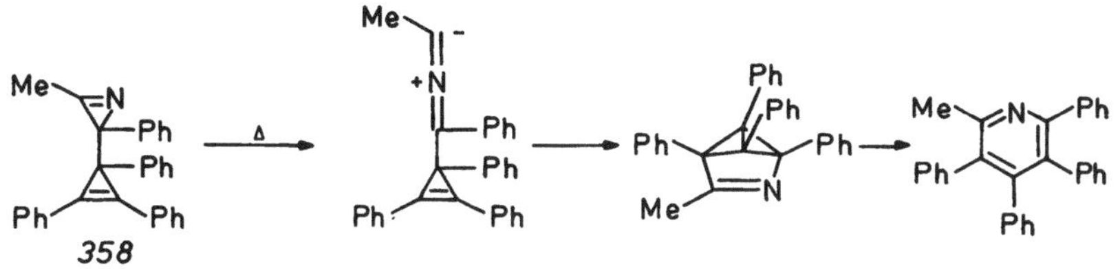

358

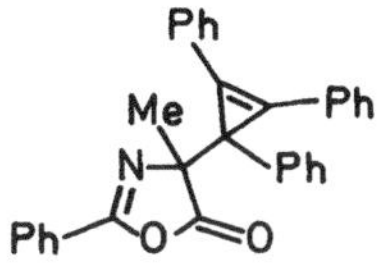

359

13 Radical Addition

Reaction of t-butyl hypochlorite with 1-methylcyclopropenes can lead to efficient formation of 2-chloro-1-methylenecyclopropanes through trapping of an intermediate allylic radical [287, 288]:

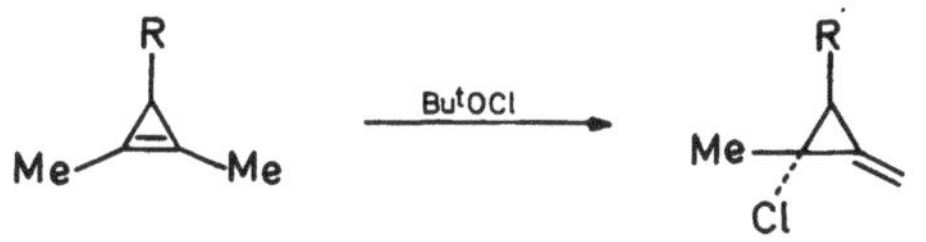

Radical reduction of 1-(halomethyl)cyclopropenes can however lead to complex products [288].

14 Sigmatropic Shifts

Thermolysis of 3-acyl- [289], or 3-trimethylsilylcyclopropenes [290], can lead to a formal 1,3-shift, eg.:

In the latter case this has been shown to involve an intramolecular silyl-shift. In the case of 3-azido-cyclopropenes, a similar migration apparently occurs through ionic intermediates [278].

15 Conclusion

Cyclopropenes are now readily available from a number of synthetic procedures and substituents may readily be introduced through 1-metallated species. The application of cyclopropenes in synthesis has to date provided routes to a very wide range of carbocyclic and heterocyclic systems derived by inter- or intra-molecular addition, and of extremely mild routes to unsaturated carbenes. 1,2-Addition to the alkene provides flexible approaches to a wide variety of *cis*- or *trans*-disubstituted cyclopropanes. The use of the cyclopropene ring as a template in enantiocontrolled synthesis has been largely neglected, but this may well prove to be an area of considerable future interest.

16 References

1. Closs, G. L.: Cyclopropenes, in "Advances in Alicyclic Chemistry" *1*, 53 (1966)
2. Deem, M. L.: Synthesis 675, 1972; ibid. 701, 1982; Billups, W. E.: Acc. Chem. Res. *11*, 245 (1978); Halton, B.: Chem. Rev. *73*, 113 (1973)
3. Padwa, A.: Acc. Chem. Res. 310, 1979; Padwa, A., Blacklock, T. J., Rieker, W. F.: Isr. J. Chem. *21*, 157 (1981)
4. Binger, P., Buch, H. M.: Topics in Current Chemistry 135 (1987)
5. Closs, G. L., Krantz, K. D.: J. Org. Chem. *31*, 638 (1966)
6. Koster, R., Arora, S., Binger, P.: Angew. Chem. *81*, 186 (1969); Angew. Chem. Int. Ed. Engl. *8*, 205 (1969)
7. Arora, S., Binger, P., Koster, R.: Synthesis 146, 1973
8. Koster, R., Arora, S., Binger, P.: Liebigs Ann. Chem. 1219, 1973
9. Koster, R., Arora, S., Binger, P.: Angew. Chem. *82*, 839 (1970); Angew. Chem. Int. Ed. Engl. *9*, 810 (1970)
10. Magid, R. M., Clarke, T. C., Duncan, C. D.: J. Org. Chem. *36*, 1320 (1971)
11. Berg, A. S.: Acta Chem. Scand. *B34*, 241 (1980)
12. Stoll, A. T., Negishi, E.: Tetrahedron Lett. 5671 1985
13. Negishi, E., Boardman, L. D., Tour, J. M., Sawada, H., Rand, C. L.: J. Am. Chem. Soc. *105*, 6344 (1983)
14. Breslow, R., Pecoraro, J., Sugimoto, T.: Organic Syntheses *57*, 41 (1977)
15. Kamyshova, A. A., Ryzhkova, T. A., Chukovskaya, E. T., Freidlina, R. K.: Doklady 1370, 1982; Kamyshova, A. A., Chukovskaya, E. T., Freidlina, R. K.: Izv. Akad. Nauk. SSSR, Ser. Khim. 2839, 1980; Chem. Abstr. *94*, 174401n (1981)
16. McDonald, R. N., Reitz, R. R.: J. Chem. Soc., Chem. Commun. 90, 1971
17. Al-Jallo, H. N., Al-Biaty, I. A., Al-Azawi, F. N.: J. Heterocycl. Chem. 1347, 1977
18. (a) York, E. J., Dittmar, W., Stevenson, J. R., Bergman, R. G.: J. Am. Chem. Soc. *94*, 2882 (1972); (b) ibid. *95*, 5680 (1973)
19. Binger, P.: Synthesis 190, 1974
20. Bovin, N. V., Surmina, L. S., Yakushkina, N. I., Bolesov, I. G.: Zh. Org. Khim. *13*, 1888 (1977); Chem. Abstr., *88*, 37290 (1978); D'yachenko, A. I., Agre, S. A., Rudashevskaya, T. Y., Shafran R. N., Nefedov, O. M.: Izv. Akad. Nauk. SSSR, Ser. Khim., 2820, 1984; Latypova, M. M., Plemenkov, V. V., Kalinina, V. N., Bolesov, I. G.: Zh. Org. Khim. *20*, 542 (1984); Bertrand, M., Monti, H.: C. R. Acad. Sci. Paris *264*, 998 (1967)
21. de Wolf, W. H., Stol, W., Landheer, I. J., Bickelhaupt, F.: Rec. Trav. Chim. Pays-Bas *90*, 405 (1971). See also de Wolf, W. H., Bickelhaupt, F.: ibid. 150
22. Sander, V., Weyerstahl, P.: Angew. Chem. *88*, 259 (1976); Angew. Chem. Int. Ed. Engl., *15*, 244 (1976); Hulskamper, L., Weyerstahl, P.: Chem. Ber. *118*, 3497 (1984); Hulskamper, L., Weyerstahl, P.: Chem. Ber. *117*, 3497 (1983)
23. Breslow, R., Cortes, D. A., Jaun, B., Mitchell, R. D.: Tetrahedron Lett. 795, 1982
24. Denis, J. M., Niamayoua, R., Vata, M., Lablache-Combier, A.: Tetrahedron Lett. 515, 1980; see also Raulet, C.: C.R. Acad. Sci. Paris *287*, 337 (1978)
25. Billups, W. E., Lin, L.-J., Casserly, E. W.: J. Am. Chem. Soc. *106*, 3698 (1984); Staley S. W., Norden, T. D.: ibid. 3699
26. Billups, W. E., Lin, L.-J.: Tetrahedron *42*, 1575 (1986)
27. Henseling, K.-O., Weyerstahl, P.: Chem. Ber. *108*, 2803 (1974)
28. Breslow, R., Eicher, T., Krebs, A., Peterson, R. A., Posner, J.: J. Am. Chem. Soc. *87*, 1320 (1965)
29. See also Jonczyk, A., Radwan-Pytlewski, T.: J. Org. Chem. *48*, 910 (1983)

30. Shields, T. C., Loving, B. A., Gardner, P. D.: Chem. Soc., Chem. Commun. 556, 1967. See also Shields, T. C., Billups, W. E.: Chem. and Ind., London 1999, 1967; Seyferth, D., Jula, T. F.: J. Organometal. Chem. *14*, 109 (1968)
31. Billups, W. E., Blakeney, A. J.: J. Am. Chem. Soc. *98*, 7817 (1976)
32. Crombie, L., Griffiths, P. J., Walker, B. J.: J. Chem. Soc., Chem. Commun. 1206, 1969
33. Baird, M. S., Buxton, S. R., Hussain, H. H.: J. Chem. Res. 310S, 1986
34. Jamal, S., Ahmad, I., Iqbal, J., Ahmad, M.: J. Chem. Res. 301S, 1983
35. Muller, P., Nguyen-Thi, H-C.: Helv. Chem. Acta *67*, 467 (1984)
36. See eg., Camaggi, G., Gozzo, F.: J. Chem. Soc. 178, 1970
37. Wiberg, K. B., Bonneville, G.: Tetrahedron Lett. 5385, 1982
38. Baird, M. S., Nethercott, W.: Tetrahedron Lett. 605, 1983
39. Baird, M. S., Hussain, H. H., Nethercott, W.: J. Chem. Soc., Perkin Trans. I. 1845, 1986
40. Chan, T. H., Massuda, D.: Tetrahedron Lett. 3383, 1975
41. Billups, W. E., Lin, L.-J., Arney, B. E., Rodin, W. A., Casserly, E. W.: Tetrahedron Lett. 3935, 1984
42. Billups, W. E., Casserly, E. W., Arney, B. E.: J. Am. Chem. Soc. *106*, 440 (1984)
43. Billups, W. E., Arney, B. E., Lin, L-J.: J. Org. Chem. *49*, 3436 (1984)
44. Dent, B. R., Halton, B., Smith, A. M. F.: Aust. J. Chem. *39*, 1621 (1986)
45. D'yachenko, A. I., Abramova, N. M., Rudashevskaya, T. Y., Nesmeyanova, O. A., Nefedov, O. M.: Izv. Akad. Nauk SSSR, Ser. Khim. 1193, 1982; Chem. Abstr. *97*, 109579z. (1982)
46. Dent, B. R., Halton, B.: Tetrahedron Lett. 4279, 1984
47. Weber, A., Neuenschwander, M.: Angew. Chem. *93*, 788 (1981); Angew. Chem. Int. Ed. Engl. *20*, 774 (1981)
48. Franck-Neumann, M., Buchecker, C.: Tetrahedron Lett. 15, 1969; Dietrich-Buchecker, C., Franck-Neumann, M.: Tetrahedron *33*, 745 (1977)
49. Franck-Neumann, M., Dietrich-Buchecker, C.: Tetrahedron Lett. 671, 1980
50. Schrader, L.: Chem. Ber. *104*, 941 (1971); Pincock, J. A., Morchat, R., Arnold, D. R.: J. Am. Chem. Soc. *95*, 7536 (1973)
51. Guiborel, C., Danion-Bougot, R., Danion, D., Carrie, R.: Tetrahedron Lett. 441, 1981
52. Pincock, J. A., Murray, K. P.: Can. J. Chem. *57*, 1403 (1979)
53. Heydt, H., Regitz, M.: J. Chem. Res. 326S, 1978; Welter, W., Hartmann, A., Regitz, M.: Chem. Ber. *111*, 3068 (1978)
54. Leach, C. L., Wilson, J. W.: J. Org. Chem. *43*, 4880 (1978); Komendantov, M. I., Bekmukhametov, R. R.: Chem. Abstr. *82*, 111337
55. Hendrick, M. E.: J. Am. Chem. Soc. *93*, 6337 (1971)
56. Pawlowski, N. E., Lee, D. J., Sinnhuber, R. O.: J. Org. Chem. *37*, 3245 (1972)
57. Shapiro, E. A., Dolgii, I. E., Nefedov, O. M.: Izv. Akad. Nauk SSSR, Ser. Khim. 2096, 1980
58. Dolgii, I. E., Okonnishnikova, G. P., Nefedov, O. M.: Izv. Akad. Nauk. SSSR, Ser. Khim. 822, 1979; Chem. Abstr. *91*, 38959m; Shapiro, E. A., Lun'kova, G. V., Nefedov, A. O., Dolgii, I. E., Nefedov, O. M.: ibid. 2535, 1981; Chem. Abstr. *96*, 142932m; USSR Patent 566830, Chem. Abstr. *88*, 50368; Maier, G., Hoppe, M., Reisenauer, H. P., Kruger, C.: Angew. Chem. *94*, 445 (1982); Angew. Chem. Int. Ed. Engl. *21*, 437 (1982)
59. Petiniot, N., Anciaux, A. J., Noels, A. F., Hubert, A. J., Teyssie, P.: Tetrahedron Lett. 1239, 1978; Domnin, I. N., Zhuravleva, E. F., Pronina, N. V.: Zh. Org. Khim. *14*, 2323 (1978); Chem. Abstr. *90*, 137423t
60. Zahra, J. P., Waegell, B.: Tetrahedron Lett. 2537, 1970
61. Pelletier, O., Jankowski, K.: Can. J. Chem. *60*, 2383 (1982)
62. Stang, P. J., Mangum, M. G.: J. Am. Chem. Soc. *97*, 3854 (1975)
63. (a) Longone, D. T., Strehouwer, D. M.: Tetrahedron Lett. 1017, 1970; (b) Perkins, W. C., Wadsworth, D. H.: J. Org. Chem. *37*, 800 (1971); Synthesis 205, 1972
64. Fiato, R. A., Williams, J. B., Battiste, M. A.: Synthesis 273, 1974
65. Domnin, I. N., Zhuravleva, E. F., Komendantov, M. I., Ritari, A. E.: Zh. Org. Khim. *13*, 1789 (1977); Chem. Abstr. *87*, 167673e
66. Schubert, H. H., Stang, P. J.: J. Org. Chem. *49*, 5087 (1984)
67. Yoshida, H., Kato, M., Ogata, T., Matsumoto, K.: J. Org. Chem. *50*, 1145 (1985)
68. Maier, G., Reuter, K. A., Franz, L., Reisenauer, P.: Tetrahedron Lett. 1845, 1985
69. Himbert, G., Giesa, R.: Liebigs Ann. Chem. 292, 1986
70. Donaldson, W. A., Hughes, R. P.: Synth. Commun. 999, 1981

71. Formanovskii, A. A., Leonora, L. I., Yakushkina, N. I., Bakhbukh, M., Grishin, Y. K., Bolesov, I. G.: Zh. Org. Khim. *13*, 1883 (1977); Chem. Abstr. *88*, 37289u
72. Levina, R. Y., Avezov, I. B., Surmina, L. S., Bolesov, I. G.: Zh. Org. Khim. *8*, 1105 (1972); Avezov, I. B., Bolesov, I. G., Levina, R. Y.: Zh. Org. Khim. *10*, 2114 (1974)
73. Vincens, M., Dumont, C., Vidal, M.: C.R. Acad. Sci. Paris *286*, 717 (1978)
74. Ceskis, B., Moiseenkov, A. M., Rudashevskaya, T. Y., Nesmeyanova, O. A., Samenovskii, A. V.: Izv. Akad. Nauk. SSSR, Ser. Khim. 1084, 1982; Chem. Abstr. *97*, 91692u
75. Moiseenkov, A. M., Ceskis, B., Rudashevskaya, T. Y., Nesmeyanova, O. A., Semenovskii, A. V.: Izv. Akad. Nauk SSSR, Ser. Khim. 1088, 1982; Chem. Abstr. *97*, 127827h
76. Dumont, C., Vidal, M.: Bull. Soc. Chim. Fr. 2301, 1973
77. Schipperijn, A. J.: Rec. Trav. Chim. Pays-Bas *90*, 1110 (1971)
78. Schipperijn, A. J., Smael, P.: Rec. Trav. Chim. Pays-Bas *92*, 1121, 1159 (1973)
79. Schipperijn, A. J., Smael, P.: Rec. Trav. Chim. Pays-Bas *92*, 1298 (1973)
80. Sorokin, V. I., Drozd, V. N., Akimova, N. P., Grandberg, I. I.: Zh. Org. Khim. *13*, 737 (1977)
81. Yakushkina, N. I., Zhurina, G. R., Surmina, L. S., Grishin, Y. K., Bazhenov, D. V., Plemenkov, V. V., Bolesov, I. G.: Zh. Obshch. Khim. *52*, 1604 (1982); Chem. Abstr. *97*, 182567
82. Kirms, M. A., Primke, H., Stohlmeier, M., de Meijere, A.: Rec. Trav. Chim. Pays-Bas *105*, 462 (1986)
83. Baird, M. S., Buxton, S. R., Whitley, J. S.: Tetrahedron Lett. 1509, 1984
84. Baird, M. S.: Tetrahedron Lett. 4829, 1984
85. Suda, M.: Tetrahedron Lett. 4355, 1980
86. Walborsky, H. M., Powers, E. J.: Isr. J. Chem. *21*, 210 (1981)
87. Padwa, A., Wannamaker, M. W.: Tetrahedron Lett. 5817, 1986
88. Dowd, P., Gold, A.: Tetrahedron Lett. 85, 1969
89. Weigert, F. J., Baird, R. L., Shapley, J. R.: J. Am. Chem. Soc. *92*, 6630 (1972); for related photochemical reactions see DeBoer, C., Breslow, R.: Tetrahedron Lett. 1033, 1967; Durr, H.: ibid. 1649, 1967
90. Aue, D. H., Helwig, G. S.: J. Chem. Soc., Chem. Commun. 604, 1975
91. Aue, D. H., Shellhamer, D. F., Helwig, G. S.: J. Chem. Soc., Chem. Commun. 603, 1975
92. Raasch, M. S.: J. Org. Chem. *37*, 1347 (1972)
93. Zotova, S. V., Bogdanov, V. S., Nesmeyanova, O. A.: Izv. Akad. Nauk. SSSR, Ser. Khim. 2706, 1979; Nesmeyanov, O. A., Zotova, S. V., Vostokova, E. I.: Izv. Akad. Nauk. SSSR, Ser. Khim. 2639, 1976
94. Razin, V. V., Gupalo, V. I.: Zh. Org. Khim. *10*, 2342 (1974); Chem. Abstr. *82*, 111632v
95. Birchall, J. M., Burger, K., Haszeldine, R. N., Nona, S. N.: J. Chem. Soc., Perkin Trans. I 2080, 1981
96. Durr, H.: Chem. Ber. *103*, 369 (1970)
97. Komatsu, K., Niwa, T., Akari, H., Okamoto, K.: J. Chem. Res. M2847, 1985
98. Baird, M. S., Hussain, H. H., Clegg, W.: unpublished results
99. Padwa, A., Rieker, W. F.: J. Am. Chem. Soc. *103*, 1859 (1981)
100. Padwa, A., Rieker, W. F., Rosenthal, R. J.: J. Org. Chem. *49*, 1353 (1984)
101. (a) Nefedov, O. M., Dolgii, I. E., Bulushheva, E. V., Shteinshneider, A. Y.: Izv. Akad. Nauk. SSSR, Ser. Khim. 1535,1979; 1901, 1976; Komendantov, M. I., Domnin, I. N.: Zh. Org. Khim. 939, 1973; Chem. Abstr. *80*, 26540c.; (b) Butler, G. B., Herring, K. H., Lewis, P. L., Sharpe V. V., Veazey, R. L.: J. Org. Chem. *42*, 679 (1977)
102. Schipperijn, A. J., Lukas, J.: Tetrahedron Lett. 231, 1972
103. Stechl, H.-H.: Chem. Ber. *97*, 2681 (1964); Tomilov, Y. V., Bordakov, V. G., Tsvetkova, N. M., Shteinshneider, A. Y., Dolgii, I. E., Nefedov, O. M.: Izv. Akad. Nauk. SSSR, Ser. Khim. 336, 1983
104. Doyle, M. J., McMeeking, J., Binger, P.: J. Chem. Soc., Chem. Commun. 376, 1976
105. Baird, M. S., Hussain, H. H., Clegg, W.: J. Chem. Soc., Perkin Trans. I, 1609, 2109 (1987)
106. Dolgii, I. E., Tomilov, Y. V., Tsvetkova, N. M., Bordakov, V. G., Nefedov, O. M.: Izv. Akad. Nauk. SSSR, Ser. Khim. 958, 1983
107. (a) Padwa, A., Kennedy, G. D., Wannamaker, M. W.: J. Org. Chem. *50*, 5334 (1985); (b) Padwa, A., Blacklock, T. J.: J. Am. Chem. Soc. *101*, 3390 (1979); Padwa, A., Rieker, W. F., Rosenthal, R. J. *105*, 4446 (1983); (c) Weiss, R., Schlierf, C., Kolbl, H.: Tetrahedron Lett. 4827, 1973
108. Pincock, J. A., Mathur, N. C.: J. Org. Chem. *47*, 3699 (1982). Calculations at the MINDO/3

level show that a 1-vinyl group on a cyclopropene lowers slightly the barrier to 1,3-bond cleavage, but raises the barrier to 2,3-cleavage (Pincock, J. A., Boyd, R. J.: Can. J. Chem. *55*, 2482 (1977)
109. Srinivasan, R.: J. Chem. Soc., Chem. Commun. 1041, 1971
110. Streeper, R. D., Gardner, P. D.: Tetrahedron Lett. 767, 1973
111. Hopf, H., Wachholz, G., Walsh, R.: Chem. Ber. *118*, 3579 (1985)
112. Padwa, A., Blacklock, T. J., Getman, D., Hatanaka, N., Loza, R.: J. Org. Chem. *43*, 1481 (1978). See also Chiacchio, U., Compagnini, A., Grimaldi, R., Purrello, G., Padwa, A.: J. Chem. Soc. Perkin Trans. I, 915, 1983
113. Padwa, A., Blacklock, T. J., Getman, D., Hatanaka, N.: J. Am. Chem. Soc. *99*, 2344 (1977)
114. Zimmerman, H. E., Aasen, S. M.: J. Amer. Chem. Soc. *99*, 2342 (1977); J. Org. Chem. *43*, 1493 (1978)
115. Pincock, J. A., Moutsokapas, A. A.: Can. J. Chem. *55*, 979 (1977)
116. Morchat, R. M., Arnold, D. R.: J. Chem. Soc., Chem. Commun. 743, 1978
117. Schrader, L., Hartmann, W.: Tetrahedron Lett. 3995, 1973
118. Razin, V. V., Rud, E. M.: Zh. Org. Khim. *12*, 689 (1976); Komendantov, M. I., Bekmukhametov, R. R., Domnin, I. N.: Zh. Org. Khim. *14*, 759 (1978); Chem. Abstr. *89*, 23503t
119. Leach, C. L., Wilson, J. W.: J. Org. Chem. *43*, 4880 (1978)
120. See footnote 26 in ref. 18b
121. Domnin, I. N., Kostikov, R. R., de Meijere, A.: Zh. Org. Khim. 2206, 1983
122. Liese, T., Splettstosser, G., de Meijere A.: Tetrahedron Lett. 3341, 1982; Kostikov, R., de Meijere, A.: J. Chem. Soc., Chem. Commun. 1528, 1984; Weber, W., de Meijere, A.: Chem. Ber. *118*, 2450 (1985); Liese, T., de Meijere, A.: Chem. Ber. *119*, 2995 (1986)
123. See de Meijere, A.: later article in this series
124. Franck-Neumann, M., Lohmann, J.-J.: Angew. Chem. *89*, 331 (1977); Angew. Chem. Int. Ed. Engl. *16*, 323 (1977)
125. Franck-Neumann, M., Lohmann, J. J.: Tetrahedron Lett. 2397, 1979
126. Franck-Neumann, M., Lohmann, J. J.: Tetrahedron Lett. 3729, 1978
127. Franck-Neumann, M., Lohmann, J. J.: Tetrahedron Lett. 2075, 1979
128. Franck-Neumann, M., Buchecker, C.: Tetrahedron Lett. 2875, 1973
129. Franck-Neumann, Miesch, M. M.: Tetrahedron Lett. 2909, 1984
130. Baird, M. S., Al Dulayymi, J. R., unpublished results
131. Padwa, A., Kennedy, G. D., Newkome, G. R., Fronczek, F. R.: J. Am. Chem. Soc. *105*, 137 (1983)
132. Boger, D. L., Brotherton, C. E.: J. Am. Chem. Soc. *106*, 805 (1984)
133. Boger, D. L., Brotherton, C. E.: Tetrahedron Lett. 5611, 1984
134. Boger, D. L., Brotherton, C. E., Georg, G. I.: Tetrahedron Lett. 5615, 1984
135. Binger, P., McMeeking, J.: Angew. Chem. *86*, 518 (1974); Angew. Chem. Int. Ed. Engl. *13*, 466 (1974)
136. Vidal, M., Vincens, M., Arnaud, P.: Bull. Soc. Chim. Fr. 657, 1972
137. Tomilov, Y. V., Shapiro, E. A., Protopopova, M. N., Ioffe, A. I., Dolgii, I. E., Nefedov, O. M.: Izv. Akad. Nauk. SSSR, Ser. Khim. 631, 1985; ibid. 700, 1983
138. Shapiro, E. A., Protopopova, M. N., Nefedov, O. M.: Izv. Akad. Nauk. SSSR, Ser. Khim., 2153, 1984: Shapiro, E. A., Lun'kova, G. V., Protopopova, M. N., Dolgii, I. E., Nefedov, O. M.: ibid. 2412, 1983
139. (a) Leftin. J. H., Gil-Av, E.: Tetrahedron Lett. 3367, 1972; (b) Padwa, A., Blacklock, T. J.: J. Am. Chem. Soc. *99*, 2346 (1971)
140. Billups, W. E., Lin, L. P., Chow, W. Y.: J. Am. Chem. Soc. *96*, 4026 (1974). See also Coburn, T. T., Jones, W. M.: J. Am. Chem. Soc. *96*, 5218 (1974)
141. Billups, W. E., Reed, L. E., Casserly, E. W., Lin, L. P.: J. Org. Chem. *46*, 1326 (1981)
142. Halton, B., Officer, D. L.: Aust. J. Chem. 36, 1167 (1983); Tetrahedron Lett. 3687, 1981
143. Billups, W. E., Reed, L. E.: Tetrahedron Lett. 2239, 1977
144. Vincens, M., Dumont, C., Vidal, M.: Bull Soc. Chim. Fr. 2811, 1974; Can. J. Chem. *57*, 2314 (1979); Vidal, M., Vincens, M., Arnaud, P.: Bull. Soc. Chim. Fr. 665, 1972
145. Thomas, E. W.: Tetrahedron Lett. 1467, 1983
146. Sternberg, E., Binger, P.: Tetrahedron Lett. 301, 1985
147. Billups, W. E., Shields, T. C., Chow, W. Y., Deno, N. C.: J. Org. Chem. *37*, 3676 (1972); Shields, T. C., Billups, W. E., Lepley, A. R.: J. Am. Chem. Soc. *90*, 4749 (1968); Shields, T. C.,

Billups, W. E.: Chem. and Ind., London 619, 1969; Billups, W. E., Chow, W. Y., Leavell, K. H., Lewis, E. S.: J. Org. Chem. *39*, 274 (1974)
148. Billups, W. E., Leavell, K. H., Chow, W. Y., Lewis, E. S.: J. Am. Chem. Soc. *94*, 1770 (1972); Tarakanova, A. V., Grishin, Y. K., Vashakidze, A. G., Mil'vitskaya, E. M., Plate, A. E.: Zh. Org. Khim. *8*, 1619 (1972)
149. Billups, W. E., Baker, B. A., Chow, W. Y., Leavell, K. H., Lewis, E. S.: J. Org. Chem. *40*, 1702 (1975)
150. Ransom, C. J., Reese, C. B.: J. Chem. Soc., Chem. Commun. 970, 1975
151. Arct, J., Migaj, B.: Tetrahedron *37*, 953 (1981)
152. Parker, R. H., Jones, W. M.: Tetrahedron Lett. 1245, 1984
153. Billups, W. E., Blakeney, A. J., Chow, W. Y.: J. Chem. Soc., Chem. Commun. 1461, 1971
154. Prestien, J., Gunther, H.: Angew. Chem. *86*, 278 (1974); Angew. Chem. Int. Ed. Engl. *13*, 276 (1974)
155. Browne, A. R., Halton, B., Spangler, C. W.: Tetrahedron *30*, 3289 (1974)
156. Browne, A. R., Halton, B.: Tetrahedron *33*, 345 (1977)
157. Kumar, A., Tayal, S. R., Devaprabhakara, D.: Tetrahedron Lett. 863, 1976
158. Billups, W. E., Chow, W. Y.: J. Am. Chem. Soc. *95*, 4099 (1973)
159. Davalian, D., Garratt, P. J.: Tetrahedron Lett. 2815, 1976
160. Grigorova, T. N., Komendantov, M. I.: Zh. Org. Khim. *17*, 317 (1981)
161. Komendantov, M. I., Domnin, I. N., Kenbaeva, R. M., Grigorova, T. N.: Zh. Org. Khim. *9*, 1420 (1973); Chem. Abstr. *79*, 91728p
162. Surmina, L. S., Novoselov, V. A., Bolesov, I. G.: Zh. Org. Khim. 1594, 1976; Chem. Abstr. *85*, 142697m
163. Kartashov, V. R., Skorobogatova, E. V., Akimkina, N. F., Zefirov, N. S.: Zh. Org. Khim. *18*, 38 (1982); Skorobogatova, E. V., Akimkina, N. F., Kartashov, V. P.: Zh. Org. Khim. *15*, 753 (1979); Chem. Abstr. *91*, 56403e
164. Kartashov, V. R., Akimkina, N. F., Skorobogatova, E. V.: Zh. Org. Khim. *16*, 889 (1980)
165. Baird, M. S., Hussain, H. H.: unpublished results
166. Grigorova, T. N., Ogloblin, K. A., Komendantov, M. I.: Zh. Org. Khim. *9*, 711 (1973); Chem. Abstr. *79*, 18618y; ibid. *8*, 2197 (1972)
167. Kartashov, V. R., Gal'yanova, N. V., Skorobogatova, E. V., Chernov, A. N., Zefirov, N. S.: Zh. Org. Khim. 2623, 1984
168. Shirafuji, T., Yamamoto, Y., Nozaki, H.: Tetrahedron Lett. 4713, 1971; Dombrovskii, V. S., Yakushkina, N. I., Bolesov, I. G.: Zh. Org. Khim. *15*, 1325 (1979)
169. Koster, R., Arora, S., Binger, P.: Angew. Chem. *81*, 185 (1969); Angew. Chem. Int. Ed. Engl. *8*, 205 (1969)
170. Bubnov, Y. N., Nesmeyanova, O. A., Rudashevskaya, T. Y., Mikhailov, B. M., Kazanskii, B. A.: Zh. Obshch. Khim. *43*, 135 (1973); Chem. Abstr. *78*, 124656y
171. Binger, P., Schafer, H.: Tetrahedron Lett. 4673, 1975
172. Richey, H. G., Kubala, B., Smith, M. A.: Tetrahedron Lett. 3471, 1981
173. Padwa, A., Blacklock, T. J., Loza, R.: J. Org. Chem. *47*, 3712 (1982)
174. Padwa, A., Blacklock, T. J.: J. Am. Chem. Soc. *99*, 2345 (1977)
175. Franck-Neumann, M., Miesch, M.: Tetrahedron Lett. 1409, 1982
176. Vidal, M., Vincens, M.: Bull. Soc. Chim. Fr. 675, 1972; Vincens, M., Dumont, C., Vidal, M., Domnin, I. N.: Tetrahedron *39*, 4281 (1983); Domnin, I. N., Dumont, C., Vincens, M., Vidal, M.: Izv. Akad. Nauk. SSSR, Ser. Khim. 1593 1985
177. Domnin, I. N., Dumont, C., Vincens, M., Vidal, M.: Izv. Akad. Nauk. SSSR, Ser. Khim. 1598, 1985
178. (a) Sander, V., Weyerstahl, P.: Chem. Ber. *111*, 3879 (1978); Gritsenko, E. I., Plemenkov, V. V., Butenko, G. G., Bolesov, I. G.: Zh. Org. Khim. *21*, 459 (1985); (b) Hauck, G., Durr, H.: J. Chem. Res. *180S*, 2227M (1981)
179. Shields, T. C., Shoulders, B. A., Krause, J. F., Osborn, C. L., Gardner, P. D.: J. Am. Chem. Soc. *87*, 3026 (1965)
180. Arct, J. Migaj, B., Zych, J.: Bull. Acad. Pol. Sci., Ser. Sci. Chim. *25*, 697 (1977)
181. Billups, W. E., Lin, L.-J.: Tetrahedron Lett. 1683, 1983
182. Billups, W. E., Blakeney, A. J., Rao, N. A., Buynak, J. D.: Tetrahedron *37*, 3215 (1981)

183. Frenkling, G., Hulskamper, L., Weyerstahl, P.: Chem. Ber. *115*, 2826 (1982); Hulskamper, L., Weyerstahl, P.: ibid. *114*, 746 (1981)
184. Hashem, M. A., Weyerstahl, P.: Tetrahedron *40*, 2003 (1984)
185. Padwa, A., Wannamaker, M. W.: Tetrahedron Lett. 2555, 1986
186. Grayston, M. W., Lemal, D. M.: J. Am. Chem. Soc. *98*, 1278 (1976)
187. (a) Baucom, K. B., Butler, G. B.: J. Org. Chem. *37*, 1730 (1972); Albert, R. M., Butler, G. B.: ibid. *42*, 674 (1977); (b) Tsuchiya, T., Arai, H., Igeta, H.: J. Chem. Soc. Chem. Commun. *550*, 1059 (1972)
188. Smith, K. A., Waterman, K. C., Streitwieser, A.: J. Org. Chem. *50*, 3360 (1985); Smith, K. A., Streitwieser, A.: J. Org. Chem. *48*, 2629 (1983)
189. Waterman, K. C., Streitwieser, A.: J. Am. Chem. Soc. *106*, 3874 (1984)
190. Svara, J., Fluck, E.: Chem. Ztg. *109*, 11 (1985)
191. Nefedov, O. M., Dolgii, I. E., Shvedova, I. B., Baidzhigitova, E. A.: Izv. Akad. Nauk. SSSR, Ser. Khim. 1339, 1978; Chem. Abstr. *89*, 108235; Zefirov, N. S., Averina, N. V., Boganov, A. M.: Zh. Org. Khim. *14*, 869 (1978)
192. Richey, H. G., Wilkins, C. W., Bension, R. M.: J. Org. Chem. *45*, 5042 (1980)
193. Welch, J. G., Magid, R. M.: J. Am. Chem. Soc. *89*, 5300 (1967)
194. West R., Goyert, W.: Tetrahedron Lett. 4067, 1970
195. Lukina, M. Y., Rudashevskaya, T. Y., Nesmeyanova, O. A.: Dokl. Akad. Nauk SSSR 1101, 1970; Chem. Abstr. *67*, 108102g; Nesmeyanova, O. A., Rudashevskaya, T. Y., Kazanskii, B. A.: ibid. *207*, 1362 (1972); Chem. Abstr. *78*, 110725h; Nesmeyanova, O. A., Rudashevskaya, T. Y.: Izv. Akad. Nauk. SSSR, Ser. Khim. 1562, 1978; Nesmeyanova, O. A., Rudashevskaya, T. Y., Grinberg, V. I.: Izv. Akad. Nauk. SSSR, Ser. Khim. 2590, 1977; Rudashevskaya, T. Y., Nesmeyanova, O. A.: ibid. 669, 1979; Bull. Acad. USSR Chem. *27*, 1364 (1979)
196. Lehmkuhl, H., Mehler, K.: Liebigs Ann. Chem. 1841, 1978; Rudashevskaya, T. Y., Nesmeyanova, O. A.: Izv. Akad. Nauk. SSSR, Ser. Khim. 1821, 1983
197. Lehmkuhl, H., Mehler, K.: Liebigs Ann. Chem. 2244, 1982
198. Richey, H. G., Bension, R. M.: J. Org. Chem. *45*, 5036 (1980)
199. Richey, H. G., Watkins, E. K.: J. Chem. Soc., Chem. Commun. 772, 1984
200. Nesmeyanova, O. A., Kudryavtseva, G. A., Chizhov, O. S.: Dokl. Akad. Nauk SSSR *210*, 862 (1973)
201. Ege, G., Gilbert, K.: Angew. Chem. *91*, 62 (1979); Angew. Chem. Int. Ed. Engl. *18*, 67 (1979)
202. Arct, J., Migaj, B., Leonczynski, A.: Tetrahedron *37*, 3689 (1981)
203. Dehmlow, E. V.: Liebigs Ann. Chem. *758*, 148 (1972); Chem. Ber., *107*, 2760 (1974)
204. Baird, M. S., Gerrard, M. E.: unpublished results
205. See, e.g., Sepiol, J., Soulen, R. L.: J. Org. Chem. *40*, 3791 (1975)
206. Padwa, A., Chou, C. S.: Tetrahedron *37*, 3269 (1981)
207. Fisk, T. E., Stang, P. J.: unpublished results in Stang, P. J.: Isr. J. Chem. 119, 1981
208. D'yakonov, I. A., Razin, V. V., Komendantov, M. I.: Zh. Org. Khim. *8*, 54 (1972)
209. Trost, B. M., Atkins, R. C.: J. Chem. Soc., Chem. Commun. 1254, 1971
210. (a) Greibrokk, T.: Acta Chem. Scand. *27*, 3207 (1973); (b) Dehmlow, E. V.: Tetrahedron Lett. 203, 1975
211. Lahti, P. M., Berson, J. A.: J. Am. Chem. Soc. *103*, 7011 (1981)
212. Monahan, A., Lewis, D.: J. Chem. Soc., Perkin Trans. I, 60, 1977
213. (a) Monahan, A. S.: J. Org. Chem. *33*, 1441 (1968); Nakatsuka, N., Masamume, S.: Org. Photochem. Synth. *2*, 57 (1976); Irngartinger, H., Goldmann, A., Schappert, R., Garner, P., Dowd, P.: J. Chem. Soc. Chem. Commun. 455, 1981; Dowd, P., Garner, P., Schappert, R., Irngartinger, H., Goldman, A.: J. Org. Chem. *47*, 4240 (1982); White, E. H., Winter, R. E. K., Graeve, R., Zirngibl, U., Friend, E. W., Maskill, H., Mende, U., Kreiling, G., Reisenauer, H. P., Maier, G.: Chem. Ber. *114*, 3906 (1981); Maier, G., Hoppe, M., Reisenauer, H. P.: Angew. Chem. *95*, 1009 (1983); Angew. Chem. Int. Ed. Engl. *22*, 990 (1983); (b) Dowd, P., Schappert, R., Garner, P., Go, C. L.: J. Org. Chem. *50*, 44 (1985)
214. Aue, D. H., Lorens, R. B., Helwig, G. S.: Tetrahedron Lett. 4795, 1973
215. Labows, J. N., Swern, D.: Tetrahedron Lett. 4523, 1971
216. La Rochelle, R. W., Trost, B. M.: J. Chem. Soc., Chem. Commun. 1353, 1970. See also Rendall, W. A., Torres, M., Strausz, O. P.: J. Org. Chem. *50*, 3034 (1985); Coxon, J. M., de Bruijn, M., Lau, C. K.: Tetrahedron Lett. 337, 1975
217. Battiste, M. A., Fiato, R. A.: J. Org. Chem. *43*, 1282 (1978)

218. Plemenkov, V. V., Breus, V. A.: Zh. Org. Khim. *10*, 1656 (1974); Chem. Abstr. *81*, 135517s
219. Wenzinger, G. A., Ors, J. A.: J. Org. Chem. *39*, 2060 (1974)
220. Reinhoudt, D. N., Smael, P., van Tilborg, W. J. M., Visser, J. P.: Tetrahedron Lett. 3755, 1973; van Tilborg, W. J. M., Smael, P., Visser, J. P., Kouwenhoven, C. G., Reinhoudt, D. N.: Rec. Trav. Chim. Pays-Bas, *94*, 85 (1975)
221. Reid, W., Herrmann, H. J.: Liebigs Ann. Chem. 1239, 1974; Plemenkov, V. V., Breus, V. A., Grechkin, A. N., Novikova, L. K.: Zh. Org. Khim. *12*, 787 (1976); Chem. Abstr. *85*, 5085 w
222. Hoffmann, R. W., Frickel, F.: Synthesis 444, 1975; Battiste, M. A., McRitchie, D. D., Gassman, P. G., Reus, W. F., Chasman, J. N., Haywood-Farmer, J.: Tetrahedron Lett. 2097 1979
223. Steigel, A., Sauer, J., Kleier, D. A., Binsch, G.: J. Am. Chem. Soc. *94*, 2770 (1972); Gockel, U., Hartmannsgruber, U., Steigel, A., Sauer, J.: Tetrahedron Lett. 595, 1980; Neunhoeffer, H., Schaberger, F.-D.: Liebigs Ann. Chem. 1845, 1983; Beynon, G., Figeys, H. P., Lloyd, D., Mackie, R. K.: Bull. Soc. Chim. Belges *88*, 905 (1979)
224. Fuhlhuber, H. D., Gousetis, C., Troll, T., Sauer, J.: Tetrahedron Lett. 3903, 1978; Fuhlhuber, H. D., Gousetis, C., Sauer, J., Lindner, H. J.: ibid. 1299, 1979
225. Schuster, H., Sichert, H., Sauer, J.: Tetrahedron Lett. 1485, 1983
226. Christl, M., Lanzendorfer, U., Hegmann, J., Peters, K., Peters, E.-M., von Schnering, H. G.: Chem. Ber. *118*, 2940 (1985); Christl, M., Lanzendorfer, U., Peters, K., Peters, E.-M., von Schnering, H. G.: Tetrahedron Lett. 353, 1983
227. Fujise, Y., Sakaino, M., Ito, S.: Tetrahedron Lett. 2663, 1977
228. Ito, S., Itoh, I., Saito, I., Mori, A.: Tetrahedron Lett. 3887, 1974
229. Dietrich-Buchecker, C., Martina, D., Franck-Neumann, M.: J. Chem. Res. 78, 1978
230. Galloway, N., Dent, B. R., Halton, B.: Aust. J. Chem. *36*, 593 (1983)
231. Dent, B. R., Halton, B., Smith, A. M. F.: Aust. J. Chem. *39*, 1621 (1986)
232. Sargeant, P. B.: J. Am. Chem. Soc. *91*, 3061 (1969)
233. Magid, R. M., Wilson, S. E.: J. Org. Chem. *36*, 1775 (1971)
234. Jefford, C. W., Acar, M., Delay, A., Mareda, J., Burger, U.: Tetrahedron Lett. 1913, 1979
235. See e.g. Browne, A. R., Halton, B.: J. Chem. Soc., Chem. Commun. 1341, 1972; J. Chem. Soc., Perkin Trans. I, 1177, 1977; Halton, B., Milsom, P. J., Woolhouse, A. D.: J. Chem. Soc., Perkin Trans. I 731, 1977; Halton, B., Officer, D. L.: Aust. J. Chem. *36*, 1291 (1983)
236. Muller, P., Nguyen-Thi, H-C.: Tetrahedron Lett. 2145, 1980; Isr. J. Chem. *21*, 135 (1981); Muller, P., Rodriguez, D.: Helv. Chim. Acta *66*, 2541 (1983); *69*, 1546 (1986)
237. Kobayashi, Y., Yoshida, T., Hanzawa, Y., Kumadaki, I.: Tetrahedron Lett. 4601, 1980; Kobayashi, Y., Yoshida, T., Nakajima, M., Ando, A., Hanzawa, Y., Kumadaki, I.: Chem. Pharm. Bull. *33*, 3608 (1985)
238. (a) Wiberg, K. B., Bartley, W. J.: J. Am. Chem. Soc. *82*, 6375 (1960); Breslow, R., Eicher, T., Krebs, A., Peterson, R. A., Posner, J.: J. Am. Chem. Soc. *87*, 1320 (1965); Cohen, H. M.: J. Heterocycl. Chem. *4*, 130 (1967); Prinzbach, H., Fischer, U.: Helv. Chim. Acta *50*, 1692 (1967); (b) Dehmlow, E. V., Naser-ud-din: J. Chem. Res. 40S, 1978
239. Eaton, D. F., Bergman, R. G., Hammond, G. J.: J. Am. Chem. Soc. *94*, 1351 (1972)
240. Gassman, P. G., Greenlee, W. J.: Synth. Commun. *2*, 395 (1972)
241. Komendantov, M. I., Bekmykhametov, R. R., Novinskii, V. G.: Zh. Org. Chim. *12*, 801 (1976); Chem. Abstr. *85*, 94297g
242. Aue, D. H., Lorens, R. B., Helwig, G. S.: J. Org. Chem. *44*, 1202 (1979); Aue, D. H., Helwig, G. S.: Tetrahedron Lett. 721, 1974
243. Schneider, M., Csacsko, B.: Angew. Chem. *89*, 905 (1977); Angew. Chem. Int. Ed. Engl. *16*, 867 (1977)
244. Regitz, M., Welter, W., Hartmann, A.: Chem. Ber. *112*, 2509 (1979)
245. Heydt, H., Busch, K.-H., Regitz, M.: Liebigs Ann. Chem. 590, 1980
246. Franck-Neumann, M., Buchecker, C.: Tetrahedron Lett. 2659, 1969; Franck-Neumann, M., Buchecker, C.: Angew. Chem. *85*, 259 (1973); Angew. Chem. Int. Ed. Engl. *12*, 240 (1973)
247. Baird, M. S., Hussain, H. H.: Tetrahedron *43*, 215 (1987); see also Avezov, I. B., Bolesov, I. G.: Vest. Mosk. Univ. Khim. *13*, 616 (1972); Chem. Abstr. *78*, 29666t
248. Zaitseva, L. G., Avezov, I. B., Subbotin, O. A., Bolesov, I. G.: Zh. Org. Khim. 1415, 1975
249. Maier, G., Wolf, B.: Synthesis 871, 1985
250. Pyron, R. S., Jones, W. M.: J. Org. Chem. *32*, 4048 (1967); Eicher, T., von Angerer, E.: Chem. Ber., *103*, 339 (1970)
251. Visser, J. P., Smael, P.: Tetrahedron Lett. 1139, 1973

252. Matsumoto, K., Uchida, T.: J. Chem. Soc., Perkin Trans. I 73, 1981
253. Ohsawa, A., Wada, I., Igeta, H., Akimoto, T., Tsuji, A.: Tetrahedron Lett. 4121, 1978
254. (a) Akmanova, N. A., Sagitdinova, K. F., Balenkova, E. S.: Khim. Get. Soedin 1192, 1982; Zaitseva, L. G., Berkovich, L. A., Bolesov, I. G.: Zh. Org. Khim. *10*, 1669 (1974): Chem. Abstr. *81*, 136026t; (b) Drozd, V. N., Bogomolova, G. S.: Zh. Org. Khim. *13*, 2012 (1977); (c) Deem, M. L.: Org. Prep. Proc. Int. *13*, 414 (1981); (d) Deem, M. L.: Synthesis 322, 1981
255. Ciabattoni, J., Kocienski, P.: J. Am. Chem. Soc. *91*, 6534 (1971); Crandall, J. K., Conover, W. W.: J. Org. Chem. *43*, 1323 (1978)
256. Friedrich, L. A., Fiato, R. A.: J. Org. Chem. *39*, 416 (1974)
257. Friedrich, L. E., Fiato, R. A.: J. Org. Chem. *39*, 2267 (1974)
258. Ciabattoni, J., Kocienski, P.: J. Am. Chem. Soc. *93*, 4902 (1971); J. Org. Chem. *39*, 388 (1974)
259. Muller, P., Khoi, T. T.: Tetrahedron Lett. 1939, 1977
260. Friedrich, L. E., Cormier, R. A.: J. Org. Chem. *35*, 450 (1970)
261. Prinzbach, H., Fischer, U.: Helv. Chim. Acta *50*, 1669 (1967)
262. Friedrich, L. E., Cormier, R. A.: Tetrahedron Lett. 4761 1971
263. Friedrich, L. E., Fiato, R. A.: J. Am. Chem. Soc. *96*, 5783 (1974)
264. Baird, M. S., Hussain, H. H.: Tetrahedron Lett. 5143, 1986
265. Jefford, C. W., Rimbault, C. G.: Tetrahedron Lett. 91, 1981
266. Politzer, I. R., Griffin, G. W.: Tetrahedron Lett. 4775, 1973
267. Frimer, A. A.: Isr. J. Chem. *21*, 194 (1981)
268. Moiseenkov, A. M., Ceskis, B. A., Rudashevskaya, T. A., Nesmeyanova, O. A., Semenovsky, A. V.: Tetrahedron Lett. 151, 1981
269. Leandri, G., Monti, H., Bertrand, M.: Tetrahedron *30*, 289 (1974)
270. Vidal, M., Dussauge, A., Vincens, M.: Tetrahedron Lett. 313, 1977; Arnaud, R., Dussauge, A., Faucher, H., Subra, R., Vidal, M., Vincens, M.: Tetrahedron *40*, 315, 1984
271. Vincens, M., Dumont, C., Vidal, M.: Bull. Soc. Chim. Fr. II-59, 1984
272. Padwa, A., Akiba, M., Chou, C. S., Cohen, L.: J. Org. Chem. *47*, 1183 (1982)
273. Schmitz, E., Sonnenschein, H., Kuban, R. J.: Tetrahedron Lett. 4911, 1985
274. Feiring, A. E., Ciabattoni, J.: J. Am. Chem. Soc. *95*, 5266 (1973); Ciabattoni, J., Feiring, A. E.: ibid. *94*, 5113 (1972)
275. DeBoer, C. H.: J. Chem. Soc., Chem. Commun. 377, 1972
276. Longone, D. T., Strehouwer, D. M.: Tetrahedron Lett. 1017, 1970; Denzel, T., Bestman, H.: ibid. 3817, 1969
277. Moerek, R. E., Battiste, M. A.: Tetrahedron Lett. 4421, 1973
278. Closs, G. L., Harrison, A. M.: J. Org. Chem. *37*, 1051 (1972); Feiring, A. E., Ciabattoni, J.: J. Org. Chem. *37*, 3784 (1972)
279. Weiss, R., Schlierf, C.: Angew. Chem. *83*, 887 (1971); Angew. Chem. Int. Ed. Engl. *10*, 811 (1971)
280. Weiss, R., Andrae, S.: Angew. Chem. *85*, 145, 147 (1973): Angew. Chem. Int. Ed. Engl. *12*, 150, 152 (1973)
281. Davis, J. H., Shea, K. J., Bergman, R. G.: Angew. Chem. *88*, 254 (1976); Angew. Chem. Int. Ed. Engl. *15*, 232 (1976); J. Am. Chem. Soc. *99*, 1499 (1977)
282. de Wolf, W. H., van Straten, J. W., Bickelhaupt, F.: Tetrahedron Lett. 3509, 1972; Landheer, I. J., de Wolf, W. H., Bickelhaupt, F.: ibid. 2813, 1974; de Wolf, W. H., Landheer, I. J., Bickelhaupt, F.: ibid. 179, 1975
283. Landheer, I. J., de Wolf, W. H., Bickelhaupt, F.: Tetrahedron Lett. 349, 1975
284. Peelen, F. C., Rietveld, G. G. A., Landheer, I. J., de Wolf, W. H., Bickelhaupt, F.: Tetrahedron Lett. 4187, 1975
285. Johnson, G. C., Bergman, R. G.: Tetrahedron Lett. 2093, 1979
286. Padwa, A., Cohen, L. A., Gingrich, H. L.: J. Am. Chem. Soc. *106*, 1065 (1984)
287. Vincens, M., Choubani, S., Vidal, M.: Tetrahedron Lett. 4695, 1981
288. Arnaud, R., Choubani, S., Subra, R., Vidal, M., Vincens, M., Barone, V.: Can. J. Chem. *63*, 2512 (1985)
289. Monti, H., Bertrand, M.: Tetrahedron Lett. 1235, 1969
290. Coleman, B., Conrad, N. D., Baum, M. W., Jones, M.: J. Am. Chem. Soc. *101*, 7743 (1979)

Author Index Volumes 101–144

Contents of Vols. 50–100 see Vol. 100
Author and Subject Index Vols. 26–50 see Vol. 50

The volume numbers are printed in italics

Bei Fragen zur Produktsicherheit wenden Sie sich bitte an:
If you have any questions regarding product safety,
please contact:

Walter de Gruyter GmbH
Genthiner Straße 13
10785 Berlin
productsafety@degruyterbrill.com